Carcinogenicity of Inorganic Substances
Risks from Occupational Exposure

This text is based on the International Seminar on Assessment of Carcinogenic Risk from Occupational Exposure to Inorganic Substances Held in Luxembourg at the European Centre (Kirchberg) from 17 to 20 October 1995

Compiled by F.M. Craig, Craig Publication Services, Scotland, UK on behalf of the Editorial and Scientific Board.

LEGAL NOTICE

None of the organizations responsible for organizing the International Seminar on Assessment of Carcinogenic Risks from Occupational Exposure to Inorganic Substances, of which this text reports the proceedings, nor any person acting on behalf of these organizations is responsible for any use which might be made of the information published here. The views expressed in these pages are those of the authors. The supporting organizations do not necessarily subscribe to or endorse these views.

Carcinogenicity of Inorganic Substances
Risks from Occupational Exposure

Chief Editor
J.H. Duffus
Edinburgh Centre for Toxicology, Heriot-Watt University, Edinburgh, UK

Editorial and Scientific Board
G.A. Aresini, E. Chantelot, M.H. Draper, W. Forth, A.M. Langer, L.G. Morgan, R.P. Nolan, K. Schumann and B. Swennen

THE ROYAL SOCIETY OF CHEMISTRY
Information Services

ISBN 0-85404-429-9

A catalogue record for this book in available from the British Library

© The Royal Society of Chemistry 1997

All rights reserved.

Apart from any fair dealing for the purpose of research or private study, or criticism or review as permitted under the terms of the UK Copyright, Designs and Patents Act, 1988, this publication may not be reproduced, stored or transmitted, in any form or by any means, without the prior permission in writing of The Royal Society of Chemistry, or in the case of reprographic reproduction only in accordance with the terms of the licences issued by the Copyright Licensing Agency in the UK, or in accordance with the terms of the licences issued by the appropriate Reproduction Rights Organization outside the UK. Enquiries concerning reproduction outside the terms stated here should be sent to The Royal Society of Chemistry at the address printed on this page.

Published by The Royal Society of Chemistry,
Thomas Graham House, Science Park, Milton Road,
Cambridge CB4 4WF, UK

Typeset by Land & Unwin (Data Sciences) Ltd, Bugbrooke, Northants NN7 3PA
Printed by Bookcraft (Bath) Ltd

Preface

The seminar was hosted by the European Commission, Directorate-General VF, Public Health and Safety at Work Directorate, at the Jean Monnet Building, European Centre, Luxembourg. The organizers are deeply indebted to Dr W.J. Hunter the Director of DG VF, for making this possible. It was attended by 110 participants from 21 countries, representing academia, regulatory and international advisory bodies, and industry.

Background

The assessment of carcinogenicity of chemical substances in general is a societal necessity, particularly in the context of workers' health. This topic is also of major importance for industry since the classification of a substance as a carcinogen at international or national level has serious consequences for both producers and consumers.

With regard to inorganic substances, it is vital to carry out this assessment on the basis of suitable criteria that are applicable to these substances with a proper understanding of their specific characteristics. In this context, three distinct areas require examination:

(1) There has been reliance on results of epidemiological studies which determine associations without confirming etiology. Relevant exposure data are often missing; confounders and possible synergisms are frequently not addressed adequately.
(2) There has been reliance on assays using heavily dosed rodents, involving inappropriate administration of the test substance with regard to human exposure; these factors compromise the usefulness of the data. Moreover, insufficient attention has been given to the differences in mechanistic and metabolic processes between humans and rodents, particularly at the cellular level in the lung.
(3) Many studies have failed to recognize the fundamentally different nature of bioinorganic chemistry from inorganic chemistry. This is reflected particularly in the absence of efforts and/or methodology to identify the nature of materials present in the exposure, especially the chemical species and physicochemical properties such as particle size.

Scientists close to the problem are aware that these shortcomings exist in many, frequently cited, studies. As these reports and their qualified conclusions work their way up the legislative chain, however, they receive less and less critical attention, and controversial conclusions become embedded in the system as received truths.

It thus emerged that an international forum raising the issues outlined above could have a decisive influence on thinking and practice in this area. Such a forum could recommend up-to-date scientific guidelines on how to cope with the perceived shortcomings and, in addition, formulate recommendations to assist regulatory agencies so that their decisions accurately reflect current scientific understanding. This international seminar aimed to be such a forum.

Contents

Setting the Scene: The Scientific Context   1
   J.H. Duffus, Edinburgh Centre for Toxicology, Edinburgh, UK

Scientific Presentations

General Overview of the Scientific Presentations   6
   M.H. Draper & J.H. Duffus, Edinburgh Centre for Toxicology, Edinburgh, UK

Bio-inorganic Chemistry and Cancer   19
   R.J.P. Williams, University of Oxford, Oxford, UK

Physico-chemical Properties of Inorganic Particles Controlling Biological Activity   39
   R.P. Nolan & A.M. Langer, Brooklyn College, New York, USA

Mineralogical Factors and the Relationship of Fibres and Dusts in Humans to Disease   58
   A. Churg, University of British Columbia, Vancouver, Canada

In Vitro Studies of Genotoxicity and their Significance   75
   M.-C. Jaurand, INSERM, Créteil, France

Sequence of Events in Lung Carcinogenesis   86
   C.H. Kennedy & J.F. Lechner, Inhalation Toxicology Research Institute, Albuquerque, USA

The Significance of the Toxicokinetics of Solid Particles in the Rat Lung   103
   H. Muhle, Fraunhofer Institut für Toxikologie und Aerosolforschung, Hannover, Germany

Mechanisms and Significance of Particle Overload   117
   P.E. Morrow, University of Rochester, New York, USA

Evaluating the Carcinogenicity of Crystalline Silica and Other Mineral Particles: Human, Animal, Cellular and Molecular Studies   134
   U. Saffiotti, National Cancer Institute, Maryland, USA

Overview of Epidemiological Studies on the Carcinogenicity of Metals   146
   A. Bernard, Catholic University of Louvain, Brussels, Belgium

Problems Encountered in Determining Metal Carcinogenesis through Epidemiological Studies	160
J.M. Harrington, Institute of Occupational Health, Birmingham, UK	
A Re-assessment of Respiratory Cancers at the Clydach Nickel Refinery: New Evidence of Causation	181
M.H. Draper, Edinburgh Centre for Toxicology, Edinburgh, UK	

Assessment of Carcinogenic Risk: International Activities, Regulations and the Industrial Situation

The Role of the International Agency for Research on Cancer (IARC)	211
P. Boffetta, International Agency for Research on Cancer, Lyon, France	
IPCS Activities on Risk Assessment of Inorganic Compounds with Reference to Potency of Carcinogenicity	213
B.H. Chen & E. Smith, IPCS, World Health Organization, Geneva, Switzerland	
The Role of the US Occupational Safety and Health Administration	220
W. Perry, OSHA, Washington DC, USA	
The Role of the European Commission, Directorate General V	222
G. Aresini, CEU Directorate General V, Luxembourg	
Chemical Substances – Classification and Labelling, Risk Assessment and Protection of Workers	224
J.M. Costa-David, CEU Directorate General XI, Brussels, Belgium	
Industry Views from a North American Perspective	229
I.M. Arnold, Noranda Inc., Toronto, Canada	
Industry Views from a European Perspective	235
L.G. Morgan, Consultant in Occupational Health, Swansea, UK	
Summary of the Panel and Plenary Discussion	243
M.H. Draper, Edinburgh Centre for Toxicology, Edinburgh, UK	

Working Party Reports

Working Party 1: Physico-chemical Characterization of Exposures	246
Working Party 2: Interpretation of *In Vitro* and *In Vivo* Experimental Studies	250
Working Party 3: Conduct and Intepretation of Epidemiological Studies	254

Conclusion

Looking to the Future	260
J.H. Duffus, Edinburgh Centre for Toxicology, Edinburgh, UK	
List of Seminar Participants	265
Subject Index	276

Organizers, Sponsors, Members of the Scientific and Editorial Board and Steering Organizing Committee

This international seminar was organized by **EDINTOX**, the Edinburgh Centre for Toxicology, Heriot-Watt University, with the support of the **EUROPEAN COMMISSION** DG VF – Public Health and Safety at Work Directorate, and the collaboration of **EUROMETAUX**, the European Association of Metals.

This seminar was held under the auspices of IUPAC, the International Union for Pure and Applied Chemistry, through the Commission on Toxicology.

It had co-sponsorship by NiDI (the Nickel Development Institute), NiPERA (the Nickel Producers Environmental Research Association), ICME (the International Council on Metals and the Environment), ILZRO (the International Lead Zinc Research Organization) and CDI (the Cobalt Development Institute).

Scientific Committee

John H. DUFFUS (*Chairman*)	Edintox, Heriot-Watt University, United Kingdom
Giorgio A. ARESINI	European Commission, DG V, Luxembourg
Morrell H. DRAPER	Edintox, Heriot-Watt University, United Kingdom
Wolfgang FORTH	Walther-Straub-Institut für Pharmakologie und Toxikologie, Germany
Arthur M. LANGER	Brooklyn College, City University of New York, United States
Lindsay G. MORGAN	Consultant in Occupational Health, United Kingdom
Robert P. NOLAN	Brooklyn College, City University of New York, United States
Klaus SCHUMANN	Walther-Straub-Institut für Pharmakologie und Toxikologie, Germany

Steering Organizing Committee

Giorgio A. ARESINI	European Commission, DG V, Luxembourg
Emmanuel CHANTELOT	Eurométaux, Belgium
John H. DUFFUS	Edintox, Heriot-Watt University, United Kingdom
Bert SWENNEN	Union Minière, Belgium

Setting the Scene: The Scientific Context

J.H. DUFFUS

EDINBURGH CENTRE FOR TOXICOLOGY, HERIOT-WATT UNIVERSITY, EDINBURGH, SCOTLAND, UK

Assessment of carcinogenic risk of inorganic substances still involves unproven assumptions based on an assumed analogy with the carcinogenicity of organic substances. The first aim of this seminar was to consider our scientific assumptions critically on the basis of the current knowledge presented by leading authorities in the relevant fields. Beyond this, the seminar considered the consequences for health and safety in the workplace. Can testing and assessment be improved? Are our regulatory and management practices appropriate? Can monitoring and control of exposures at work be made more effective?

Current approaches to the assessment of carcinogenic risk must be seen against the broader background of current thinking in toxicology. Toxicology as we know it today is essentially a development of the last 30 years or so. It is difficult to date the start precisely but possibly we could start with 1961 and the first report of teratogenic properties of thalidomide. This was followed in 1962 by Rachel Carson's 'Silent Spring', which publicized the dangers posed by indiscriminate use of pesticides. In 1972, and of special relevance to this seminar, came the first WHO IARC monograph on the evaluation of carcinogenicity of chemicals to humans. In 1973 Bruce Ames *et al.* published 'All Mutagens are Carcinogens'. This was followed in 1975 by another paper from the same group, 'Detection of Carcinogens as Mutagens – the Salmonella/Microsome Test. Assay of 300 Chemicals'. The concept of carcinogens as chemical mutagens was quickly accepted, especially in the light of papers such as that by Creech and Johnson (1974), 'Angiosarcoma of the Liver in the Manufacture of Polyvinyl Chloride'.

The scientific revolution in toxicology was followed by legislative developments and in 1977 we had the US Toxic Substances Act administered by USEPA. This was closely followed in 1979 by the Sixth Amendment of the EC Council Directive of 27 June 1967 on the approximation of laws, regulations and administrative provisions relating to the classification, packaging and labelling of dangerous substances (67/548/EEC), which set out the base set of data required for premarketing notification of a new substance. In this context, a new substance is defined as one placed on the market for the first time after 18 September 1981, as contrasted with existing substances that were already on the market before this date and that are included in the European Inventory of Existing Commercial Substances (EINECS). In 1981 came the promulgation of protocols for 'Good Laboratory Practice' and 'Quality Assurance'. Essentially, the legislation was concerned initially with hazard assessment, but subsequently it was realised that hazard assessment must be followed by risk assessment if potentially hazardous substances were to be handled safely. Risk in this context is an estimate of the probability of exposure to a hazard and, hence, an estimate of the likelihood of harm resulting in the circumstances of use. However, if there is no hazard, there is clearly no risk and so some consideration must be given to the problem of hazard or, in this case, carcinogen, identification.

Identification of carcinogens, the hazard we are concerned with here, is based on mutagenicity tests, rodent bio-assays and relevant epidemiological data. However, the lack of good epidemiological data, as well as making both hazard and risk assessment difficult or impossible, places undue emphasis on results of the mutagenicity and rodent tests and, in particular, the assumption that the molecular event starting carcinogenesis is mutation. Allowance is rarely made for the repair processes operating in human cells or for the role of regulatory receptors in controlling proliferation of human cells.

The key documents in identifying human carcinogens and the starting point for both hazard and risk assessment are the 'IARC Monographs on the Evaluation of Carcinogenic Risks to Humans'. These monographs define the known carcinogenicity of individual substances, groups of substances and industrial processes. Published in 1979, Summary Supplement no. 1 gave the following classifications. A total of 18 entities was identified for Group I carcinogens, and included (as groups) arsenic and arsenic compounds, chromium and certain chromium compounds, asbestos, and soots, tars and mixed oils and four industrial processes, including nickel refining and underground hematite mining. A further 18 entities were identified as Group II, including in Group IIA cadmium and certain cadmium compounds and nickel and certain nickel compounds. In Group IIB were beryllium and certain beryllium compounds. A caveat was added in relation to classification of groups of substances and processes – 'the specific compound(s) which may be responsible for a carcinogenic effect in humans cannot be specified precisely'.

By 1987, IARC Supplement 7, summarizing Volumes 1 to 42 (1972–1987), identified 50 entities as Group I carcinogens including (as groups), arsenic and arsenic compounds, chromium compounds (hexavalent), asbestos, nickel and nickel compounds (moved from Group IIA) and soots, tars and mixed oils, and 11 industrial processes, no longer including nickel refining but including three inorganic processes – aluminium production, iron and steel founding and underground hematite mining. Some 196 entities were identified as Group II including in Group IIA cadmium and certain cadmium compounds and beryllium and certain beryllium compounds (transferred from Group IIB). Lead and inorganic lead compounds now appeared in Group IIB.

It is of interest to consider the specific case of nickel as an example of the problems that can arise in identifying inorganic carcinogens and assessing hazard and risk. Historically, a very high incidence of nasal and lung cancers was associated with the Clydach refinery in Wales. Experimental studies showed that induction of sarcomas of various kinds followed subcutaneous, intramuscular and intraperitoneal injection of nickel compounds in rodents. In 1976, IARC evaluated the data as follows: 'epidemiological studies conclusively demonstrate an excess risk of cancer of the nasal cavity and lung in workers in nickel refineries. It is likely that nickel in some form(s) is carcinogenic to man'. By 1987, the evaluation in IARC Supplement 7 was somewhat different. Nickel and nickel compounds were placed in Group IA, indicating that the evidence for carcinogenicity to humans was sufficient, but this classification was accompanied by the caveat – 'This evaluation applies to the group of chemicals as a whole and not necessarily to all individual chemicals within the group'. It was commented that 'It is still not possible to state with certainty which specific compounds are human carcinogens and which are not. A large amount of evidence has accrued that nickel refining carries a carcinogenetic risk to workers. The risk is particularly high in those exposed during certain processes, mainly entailing exposure to subsulfides and oxides. The lung and nasal sinuses are the most clearly established target organs'. This risk assessment was based on the available epidemiological information.

In 1989, an IARC Working Group reconsidered nickel and nickel compounds with input from the first draft of 'Report of the International Committee on Nickel Carcinogenesis in Man', chaired by Sir Richard Doll. This group was still unable to establish the exact nature of the worker exposures and used concentration values derived from many assumptions that were essentially qualitative although numerical values are quoted in the final report. The final evaluations were as follows. 'Nickel compounds are carcinogenic to humans (Group I). Metallic nickel is possibly carcinogenic to humans (Group IIB)'. More specifically, 'There is sufficient evidence in humans for the carcinogenicity of nickel sulfate and of combinations of nickel sulfides and oxides encountered in the nickel refining industry. There is inadequate evidence in humans for the carcinogenicity of metallic nickel and nickel alloys'. An added comment stated the underlying concept that nickel compounds can generate nickel ions at critical sites in the target cells. This statement clearly implies that nickel ions are the carcinogenic species. However, it should be noted that the US National Toxicology Program results reported in 1995 have shown that nickel sulfate does not induce respiratory cancers in rats exposed by inhalation. Thus, externally applied nickel ions do not cause tumours, and any risk assessment based on the 1989 study must be re-examined.

The problems experienced in identifying nickel and nickel compounds as carcinogens suggest that it may be worthwhile looking carefully at the assumptions that underlie current methods for identifying and characterising carcinogens. They are as follows:

- positive results in short-term mutagenicity tests are directly relevant to effects on human cells *in vivo*;
- bio-assays in small mammals, especially rats and mice, are adequate models

for predicting cancer induction in humans;
- current epidemiological approaches provide adequate information to establish causal relationships and a measure of hazard and risk to humans.

These assumptions may not be wholly tenable. Short-term mutagenicity tests make little allowance for the ability of human cells to repair genetic damage and to cope with its consequences. Certainly these factors are currently ignored in all published approaches to hazard and risk assessment.

The rodent bio-assay dates from 1976 when the guidelines were established by the US National Cancer Institute. It has never been properly validated and was originally concerned only with oral administration. It is not clear what is the significance for humans of a positive result in the bio-assay in rats or mice. The default assumption has been that a positive result means that the substance tested is likely to be carcinogenic to humans although this has never been systematically checked.

Epidemiology as currently practised is essentially about the establishment of statistical associations. Whether causality may be deduced from such associations tends to be a subjective decision. Relevant information on industrial processes, work practices, the population at risk and even the details of disease development has frequently been ignored. For example, metal smelting, refining and processing leads to mixed exposures, including substances generated during the processing and even organics such as polycyclic aromatics from fuel or electrodes. Exposure measurements are often inadequate. If available, measurements of exposure to inorganic substances usually give only elemental analyses without any information on chemical speciation. Little consideration is given to full characterisation of industrial processes to determine what exposures are really likely.

I have not referred to the concept that there is no safe threshold for exposure to carcinogens, nor to the mathematical approaches to risk assessment that follow from this unproven assumption. However, these matters clearly warrant consideration by the working groups. There is also the question of relative carcinogenic potency. Not all carcinogens are equally potent and conceivably there may be human carcinogens of such low potency that a threshold of exposure may be established even for those acting by mutagenesis. In the context of this seminar, we must ask not only whether the current generally applied principles of carcinogenicity risk assessment are appropriate in the context of current scientific knowledge, but also whether the same principles can be applied to both organic and inorganic substances. Life evolved to include or exclude most inorganics that we know and thus may be well able to cope with a fairly wide range of exposures.

In setting the scene, I have concentrated on the scientific basis for our deliberations. However, this was not a purely scientific meeting. The consequences of bad risk assessment based on bad science may be incalculable. Regulators and management have difficult decisions to make and the rapid development of science has not made these decisions any easier. This seminar offered a rare chance for scientists, international agencies, regulators, management and workers' representatives to meet together.

Although there were three separate working parties: on physico-chemical characterisation of exposures, on conduct and interpretation of *in vitro* and *in vivo* experimental studies, and on conduct and interpretation of epidemiological studies, each with its own remit, there were a number of underlying questions, referred to earlier, which all three were asked to consider.

- What research is needed to establish better criteria for deciding whether, and in what forms, metallic elements may be involved in carcinogenesis?
- Is current knowledge being used correctly and, if not, what changes should be made?
- Are current test and assessment methods as good as they could be and, if not, how can they be improved?
- Are current regulatory measures and safety management appropriate to current knowledge and, if not, what changes are necessary and possible?
- Is current exposure monitoring making the best use of resources and providing the right information to protect the workforce?

There are no simple answers. However, this seminar gave the participants a chance to reach conclusions and recommendations that at least point to where answers may be found.

Scientific Presentations

General Overview of the Scientific Presentations

M.H. DRAPER AND J.H. DUFFUS

EDINBURGH CENTRE FOR TOXICOLOGY ON BEHALF OF THE SCIENTIFIC COMMITTEE

There were twelve formal presentations given in the first part of the seminar. Their purpose was to present authoritative overviews in four key areas to set the stage for the seminar group discussions. The key areas selected were:

(1) bio-inorganic chemistry, its fundamental role in all living processes and the relevance of such considerations to cancer causation (R.J.P. Williams);
(2) physico-chemical characteristics of exposure (R.P. Nolan, A.M. Langer, A. Robertson and A. Churg);
(3) experimental studies (M.-C. Jaurand, C. Kennedy, J. Lechner, H. Muhle and P. Morrow);
(4) epidemiological studies (A. Bernard, J.M. Harrington and M.H. Draper).

The overview below is an attempt to review some of the most important points relating to the aims of the seminar identified by the presenters. It is not a summary of their presentations and, while the authors of this synopsis have tried to reflect the views of the authors correctly, the reader is referred to the original papers to confirm exactly what the presenters said. This is particularly important because each paper has much to say and it is impossible to cover all the major points or give more than an indication of the content in such a brief synopsis.

1 Bio-inorganic Chemistry

The opening presentation on bio-inorganic chemistry and cancer (R.J.P. Williams) established the scientific basis for the seminar. The subject matter of bio-inorganic chemistry is the understanding of the roles inorganic elements play in biological homeostasis. In discussions about inorganic chemistry it must be understood that the shorthand usage of the name of a particular element in a context such as, 'the entry of magnesium into a cell' implies the presence of a specific molecular entity or species even if this is not described. As in other disciplines, to avoid misleading generalizations, this shorthand usage should be avoided and the specific molecular identities should be referred to whenever possible.

The human body has a balanced requirement for 15 – 20 elements (mostly 'metals'). Each of these must be maintained in a balanced system, since collectively they control a wide variety of biological functions. Living organisms have developed homeostatic mechanisms to protect themselves against deficiency, overload and unwanted elements, such as near neighbours in the periodic table, with similar chemistry which could cause toxicity by competition for reaction sites. However, if the cell environment differs significantly from the environment that drove its evolution, serious malfunctioning can occur.

Understanding the complex involvement of inorganic elements, particularly metals, in biological homeostasis requires consideration of the properties of the essential elements, which are summarized in the periodic table, based on their atomic numbers and electronic configurations. The Group 1 alkali metals sodium and potassium are primarily concerned with the maintenance of osmotic pressure and electroneutrality. The essential elements of Group 2 are magnesium and calcium. These two are linked to almost every process in the body, either directly or indirectly. Of particular interest with reference to cancer is their importance in the most common set of genes related to homeostasis, namely the genes for the various kinases and phosphatases that are involved in the phosphorylation or dephosphorylation of transcription proteins and those of the metabolism of the cyclic nucleotides, cyclases and esterases.

Zinc (Zn^{II}) in Group 12 (with cadmium and mercury) has a relationship to magnesium and calcium in that all three have no biologically significant oxidation/reduction chemistry. Zinc occurs in a great variety of enzymes, both inside and outside cells. Apart from its importance in catalytic enzymes, zinc is involved in some 20% of all transcription factors through zinc fingers. Because zinc has such a central role in cellular homeostasis, those elements which can displace it must be kept out of the cytoplasm. The major competitors for zinc sites are nickel and copper. Hence, these two in particular, as well as cadmium and mercury (Group 12), are actively excluded from the cytoplasm by attachment to a metallothionein (copper and cadmium) or through specific pump activity (nickel) into intracellular vesicles.

The final major elements of crucial significance are the transition metals of Groups 5 to 12, namely iron and copper and, to a lesser extent, manganese and cobalt. These elements function in a different manner from the other metals discussed in that they are the key elements in biological oxidation/reduction catalysts. However, owing to the complex inter-relations between the functions and the homeostatic mechanisms of the elements, it is not surprising that there are interactions between these metals and others from different groups of the periodic table. Thus, the homeostatic regulation of iron may be modified by cadmium, lead, nickel and even magnesium.

Apart from the Group 1 elements, the metallic elements discussed contribute to the vast constellations of metallo-enzymes that are essential to life at all levels. In turn these synthesize, *inter alia*, the controlling polypeptide factors that govern all cellular and extracellular activity. A cancer cell is one where homeostasis has been altered because these controlling factors are no longer operating properly. Investigating this feature of cancer offers an important way to understand better the fundamental processes that lead to eventual malignant growth.

2   Physico-chemical Characterisation of Exposures

Three papers concerning exposure characterisation were presented. The first (R.P. Nolan and A.M. Langer) dealt with the exact physico-chemical characterisation of particulates and fibres in relation to their likely biological effects. The second (A. Robertson), not available for publication in this book, described current practice in occupational hygiene in characterising particles in complex aerosols. Finally, the third paper (A. Churg) considered the relationship of the fibres and particulates present in human lungs at autopsy to the presence or absence of disease.

2.1   Surface Chemistry of Inorganic Dust in Relation to Biological Potential

If the cause of a respiratory disease is suspected to be a dust of fine inorganic particles, modern instrumental methods can be used to determine the particle type(s) and concentration(s) in the air. To understand the relationship between exposure to such airborne particles and disease requires an understanding of the physico-chemical properties of the particles.

Historical exposure indices based on particle numbers (or mass of respirable dust) in a given volume of air have been of limited value in estimating the risk of disease associated with particulate exposures. The limitations are particularly apparent when the index is used to compare the exposure response relationship between two similar cohorts, one with an excess of particular disease from exposure to the 'same' agent. One explanation for this is that a range of different particle size distributions can occur within the respirable fraction, allowing these materials to form a variety of aerosols differing in biological activity. The early industrial hygiene methods were insensitive to these differences in aerosol properties.

In addition to the variation in size distribution of the particles in an aerosol, the surface properties of the particles are variable. Two studies of non-asbestos fibrous particles in experimental animals have provided examples demonstrating the importance of surface properties in fibre carcinogenesis. Early experimental studies, using artificial routes of administration and high doses, produced tumours with a wide variety of inorganic fibres. Such findings erroneously suggested that surface properties were not important, probably because such properties were not significant in the particular experimental conditions of these studies.

In experiments where the route of administration was by inhalation, fibres of similar morphology were found to have strikingly different carcinogenic potency. For example, two aluminium silicate fibres erionite and refractory ceramic fibres (RCF), have produced a higher incidence of mesothelioma in experimental animals than any asbestos mineral previously tested. The two asbestos minerals crocidolite and chrysotile, that were used as positive controls in these studies produced no mesotheliomas and few if any lung cancers even though the concentrations of asbestos fibres used were much higher.

It now seems clear that the improvement in understanding of the physico-chemical properties of inorganic particles will be an essential element for the identification of aetiological agents in industrial exposures. This will lead to the establishment of meaningful exposure – response relationships for risk estimates and to the design of more appropriate animal experimental models, with inhalation at industrially relevant exposures as the route of administration.

2.2 Current Practice in the Characterisation of Particles in Complex Aerosols

Complex aerosols can be characterised physically by properties such as their size or shape, or chemically by, for example, analysis of their composition. Aerosol particle size is commonly expressed in terms of aerodynamic diameter. There are now internationally accepted conventions defining respirable, thoracic and inhalable fractions of airborne dusts. For most applications, dust samplers are used to collect dust of the appropriate size for subsequent laboratory analysis. Respirable dust samplers have long been available, but recent modifications to the definition of respirability will require some changes to sampler design or usage if the newly agreed criteria are to be met. Until recently, 'total' dust rather than 'inhalable' dust was used as a marker of exposure to aerosols. Research in the 1970s and 1980s demonstrated that many total dust samplers undersampled coarse particles and this led to the development of inhalable dust samplers. These are now specified in standard methods and their application continues to increase in industrial hygiene. Personal thoracic dust samplers are not widely available. Research at the Institute of Occupational Medicine in Edinburgh has led to the development of a family of dust samplers that can measure respirable, thoracic and inhalable dusts. Porous foam plugs separate dust into the different size fractions. These devices are currently being tested and validated in field trials. All of these instruments are designed to be reliable, inexpensive and easy to use. They are ideal for assessing exposure to a given size fraction of dust in well-defined situations and for assessing exposure in relation to occupational exposure limits. They do, however, characterise aerosols in a very limited manner. Much more detailed information on aerosol size distributions can be obtained using aerosol spectrometers. These are, however, research tools rather than routine monitoring instruments. They tend to be difficult to use and are not suitable for the collection of large numbers of samples.

Chemical characterisation of inorganic carcinogens is a wide and complex subject. Standardised analytical methods for organic compounds are available, relatively simple and easy to use, validated, reproducible and often the only means of reliably comparing exposures with occupational exposure limits. Many are backed up by extensive quality assurance schemes. For inorganics, the chemical characterisation is often simplistic. Speciation is rare and issues such as surface or chemical properties are not routinely addressed.

Many major improvements in worker health have been made without detailed knowledge of the exact nature of aerosol exposures. However, good occupational hygiene practice requires proper descriptions of processes generating carcinogens or potential carcinogens. This, in itself, can often provide sufficient information for an aerosol to be chemically characterised in some detail. Alternatively, a wide range of complex analytical procedures can be applied to speciate the bulk material or individual particles. Chemical properties can be measured and surface activity assessed. In general, however, these methods are complex and expensive to use. They are not usually suitable for making large numbers of measurements. Care is also necessary to ensure that the correct method is selected. It is tempting to select a complex analytical technique because it is available rather than because it can provide information to help answer properly formulated research questions. Research analytical methods may be used to support routine monitoring programmes with appropriate planning.

2.3 Analysis of Mineral Particles in Human Autopsy Lungs

It is a surprising fact that extensive analyses of human lung and associated tissues have shown that everyone in the population carries a numerically substantial burden (hundreds of millions of inorganic particles per gram of dried tissue) of mineral particles derived from ambient air in their lungs. For the most part this burden appears to produce no obvious disease. However, the recently established relationship of the fine particulate fraction of atmospheric pollutants (PM_{10}) to increased morbidity and mortality raises the possibility that this pulmonary particle load is not really innocuous. There is also some evidence of an association of high local particle concentration in the bronchial mucosa and lung cancer in the general population.

There is a great quantity of data on asbestos related malignancies in man. Disease in exposed cohorts occurs at fibre levels considerably greater than those found in the general population. Analysis of such cohorts shows that there are major differences between chrysotile and commercial amphibole (amosite and crocidolite) in regard to mesothelioma induction. For amosite and crocidolite, pleural mesothelioma appears at lung fibre concentrations one or more orders of magnitude less than those required to produce asbestosis, whereas for chrysotile, mesothelioma only appears at lung fibre concentrations comparable to those associated with asbestosis. Analysis by several different approaches suggests that in fact the real agent of mesothelioma in those with chrysotile exposure is the tremolite component of the chrysotile ore. While concentration effects are clear, it has generally not been possible to confirm predictions from animal studies that long, thin fibres are really the important agents of mesothelioma, and this issue needs further study.

Analysis of lung cancer cases in those with asbestos exposure is confounded by extensive cigarette smoking, selection bias and the association of lung cancer and asbestosis. Thus far, analysis of human lung tissue has failed to show any systematic associations of lung cancer and asbestos burden in either chrysotile or amphibole exposed populations. In the absence of asbestosis, parenchymal asbestos fibre concentrations in subjects with lung cancer are not significantly different from those in subjects with only pleural plaques.

Attempts to show an effect of cigarette smoking on parenchymal asbestos retention have been unsuccessful. However, recent experiments have shown that smoking increases the retention of fibres in the airway mucosa. Thus, smoking could potentiate asbestos disease by increasing fibre retention and, consequently, the effective dose to the lung.

3 Experimental Studies

Four papers describing experimental studies were presented. The first (M.-C. Jaurand) described *in vitro* studies of genotoxocity and their significance. The second (C. Kennedy and J. Lechner) described current knowledge of the sequence of events in lung carcinogenesis. The third paper (H. Muhle) reviewed the significance of the toxicokinetics of solid particles in rat lung. Finally, P. Morrow discussed the problem of particle overload.

3.1 *In vitro* Studies of Genotoxicity and their Significance

In vitro cell systems have been used to investigate the carcinogenic potency of inorganic fibres. A study of these systems can be instructive for understanding mechanisms of toxicity, including those of relevance to carcinogenesis, as they can show the nature of the interaction of a chemical with a wide variety of specific cell types.

The validation of short term *in vitro* tests ideally should include comparison with the carcinogenic potential in humans. However, the lack of both reliable qualitative data and exact quantitative data about past exposures associated with disease and failure to consider possible confounding factors have made it difficult to establish any meaningful correlations. However, comparisons between animal and *in vitro* data have been useful to establish the advantages and limitations of the two approaches. *In vitro* assays are based on the detection of changes associated with the carcinogenic process. Neoplastic transformation is a multistep process and the aim is to study in cell assay systems different markers of the events leading to invasive malignancy. Such markers include determination of DNA damage, repair of damage, chromosomal abnormalities and cell transformation. It is hoped that these techniques can be refined to include the new developments in molecular biology that have shown that neoplastic transformation results from the activation of oncogenes and inactivation of tumour suppressor genes associated with a modification of the cell cycle regulation.

Although most of the results from *in vivo* system studies of carcinogenetic processes relate to fibres, particularly asbestos fibres, some important generalizations have emerged. With well-characterised samples of chrysotile, great differences in carcinogenicity in animals were observed between samples. Such differences were also observed in *in vitro* reactivity. Some samples exerted a high activity, while others showed little or no activity, despite a greater number of fibres per unit weight. Thus different samples of a given fibre type may have different reactivity. This is the actual situation of workers exposed to asbestos, where they may be exposed to asbestos from mines, manufacturing industries and building industries. Thus, based on the findings with animal and *in vitro* experiments, it is clearly necessary to consider the exact material to which workers are, or were, exposed if a link is to be established between an incidence of tumours and the exact material inhaled.

3.2 Sequence of Events in Lung Carcinogenesis

It is now recognised that intrinsic tumour susceptibility genes can play a major role in defining the risk of whether or not a cancer will develop after exposure to a carcinogen. These genes operate as polymorphic variants of 'normal' genes or genetic loci that either positively (oncogenes) or negatively (tumour suppressor genes) change the probability that cancer will develop in a carcinogen-exposed individual. For example, individuals heterozygous (only one functional copy) for a particular tumour suppressor gene are more likely than those homozygous (two functional copies) for that gene to develop cancer because of the increased probability that a single somatic mutation will result in the total loss of activity of the tumour suppressor gene.

In addition to genes directly involved in cancer development, there are also the genes responsible for enzyme polymorphisms that may affect the rate of metabolism and/or detoxification of chemical carcinogens. Perhaps the most studied example of such polymorphism is cytochrome P-450 CYP2D6. Individuals homozygous for the variant of this enzyme, which readily metabolizes the anti-hypertensive drug debrisoquine, are at two- to five-fold greater risk for developing lung cancer than those individuals who are homozygous for the poor debrisoquine metabolizer phenotype.

Another recent development has been the realisation from molecular, cellular and epidemiological models of carcinogenesis, that cells undergo multiple genetic, epigenetic and morphological changes prior to the appearance of an invasive cancer. Furthermore, it appears that the absolute number and specificity of gene changes leading to tumour formation is likely to be different for tumours that develop independently from the original primary growth. This suggests that alternative patterns of gene dysfunction can result in the development of cancer. Thus in the case of lung cancer, cells with genetic alterations should populate the lung epithelium long before frank cancer is evident. This situation has been found in studies on lungs from smokers. Here, premalignant lesions, ranging from hyperplasia and metaplasia to severe dysplasia and carcinoma *in situ*, have been found to be distributed diffusely throughout the bronchial mucosa. The number and severity of these changes were directly related to cigarette smoke exposure. Furthermore, smokers diagnosed with lung cancer had more independent areas containing carcinoma *in situ* and premalignant lesions than heavy smokers with no evidence of tumour development. This phenomenon of generalized premalignant and early malignant changes throughout the bronchial mucosa of lung cancer patients is known as field cancerization.

Parallel with these advances has been a tremendous growth in knowledge about specific genes and their products and the integration of the biochemical activity of these products to control homeostasis. At the molecular level, it is when genetic changes alter this homeostasis and permit a clone of cells with a relative growth advantage to occur that neoplasia can arise. These genetic changes involve two different categories of genes that are important in the control of cellular proliferation, namely oncogenes and tumour suppressor genes. Proto-oncogenes encode proteins that stimulate cell division by functioning as growth factors, signal transducers and positive regulation of nuclear gene expression. Tumour suppressor genes encode proteins that inhibit cell division by functioning as negative regulators of cell cycle progress. Over fifty peptide growth factors have so far been identified. Overproduction or underproduction of some of these growth factors has been identified in a variety of cancer cells, particularly from human lung tumours.

Recently, a new category of oncogene has been discovered that inhibits programmed cell death or apoptosis. Thus it can enhance the survival of cells with carcinogenic genetic lesions that would normally lead to cell death.

Molecular studies of premalignant lung epithelial cells are beginning to reveal clues as to the nature and timing of early genetic alterations in lung cancer. Thus far, several frequent chromosomal alterations that may be markers for the dysplasia that precedes clinical lung cancer have been identified. Molecular and immunohistochemistry studies on sputum samples and lung bronchoscopy samples, collected from at-risk individuals prior to clinical cancer, should ultimately provide the information necessary to understand fully the nature and timing of genetic alterations in the development the lung cancers that develop most frequently in industrially exposed populations. This could in turn lead to chemopreventative strategies and to strategies that exclude persons with genetic risks from exposed working places. The ethics of the latter approach need very careful consideration.

3.3 The Significance of the Toxicokinetics of Solid Particles in the Rat Lung

The reactions of lung tissues to poorly soluble particles are quite different from those to particles that are soluble. Most importantly, insoluble materials have the potential to accumulate in lungs after chronic exposure. Further, because it is the entire particle that interacts with cells, the most important reactivity is that of the surface, which may be quite different from the bulk chemistry of the particle.

Lung clearance of solid particles with low solubility is dependent on the location of the deposition in the respiratory tract. The mucociliary escalator removes particles from the tracheobronchial region in about one day. In humans a further slow phase has been reported, but the significance of this is not fully understood. In the alveolar region, the macrophage-mediated clearance is the predominant mechanism for the removal of particles. This process is in general much slower than the ciliated clearance, and thus any impairment of this defence mechanism must be of concern. An understanding of the dynamics of clearance of particles from the lung is clearly important because it is an essential first line defence mechanism. Failure of the system must lead to an accumulation of particles somewhere in the system.

The kinetics of loading of lungs with particulate matter has been extensively studied in rats. Earlier toxicity studies used relatively high concentrations of dusts compared with the exposures usually experienced in industrial situations. A justification for this was that significant effects could be detected with a minimum number of animals. However, it is now apparent that the mechanism of damage at high exposures may be very different from those observed at low levels. The time course of retention following exposure and cessation of exposure has been determined by gravimetric analysis at appropriate intervals of the lungs of exposed animals. The macrophage-mediated component of clearance was determined by following the elimination of surrogate material, labelled with a gamma ray emitter, administered in parallel with the dust being studied.

3.4 The Problem of Particle Overload

Studies using a variety of particles with no intrinsic toxicity have demonstrated that a severe retardation in pulmonary particle clearance occurs at high pulmonary dust loads, accompanied by an accumulation of dust-laden macrophages and pathological changes in the lung tissues. Quantitative studies showed that there was a progressive decrease in alveolar clearance rates once an excessive pulmonary dust burden was attained. This condition of dust overloading represents a serious confounding factor in toxicological assessment, one in which the intrinsic toxicity of the material can be either masked or modified by the nonspecific effects of dusts on macrophage transport. Such findings have implications for human risk assessment, in that the clearance retardation induced by a high burden of particles with low toxicity in the lung will also affect the clearance of any accompanying, more toxic particles, which could be inhaled at very low concentrations.

When the dust load becomes excessive, the phenomenon referred to as overload begins to appear and, with increasing concentrations of dust, a series of irreversible changes becomes established. Recovery to normal function does not occur after removal from such dust exposure. Experimental studies in rats have shown that the onset of overload begins when the alveolar macrophages have accumulated some 6% by volume of particles. At this point, the migration of macrophages away from the alveolar region begins to be impaired and, with a 60% loading, alveolar macrophage transport virtually ceases. Parallel with this diminution in migration is a movement of particles into the interstitial tissues around the alveoli, with a subsequent appearance of particles in the draining lymph nodes. The presence of particles in the interstitium is accompanied by sustained pulmonary inflammation and in the longer term by pathological changes such as pulmonary fibrosis and/or nongenotoxic tumorigenesis, depending on the degree of dust overload.

It is generally agreed that the condition of dust overload in the rat induces nonspecific dysfunctional and pathological states. Thus, toner (an organic copolymer), talc (a magnesium silicate), titanium dioxide (a benign, inert dust), carbon black (a chemically inert material), poly(vinyl chloride) (a stable chloroethylene polymer) and volcanic ash (a complex carbonaceous and vitreous material) all give rise to a similar progression of respiratory responses and effects as overload increases. Fine fibrous particles and ultrafine isometric particles can produce the same general picture, but rather more rapidly.

Dust overload in the lungs is an experimentally induced condition of considerable interest, but its relevance to human exposure conditions has yet to be established. It is certainly relevant to those concerned with toxicity testing of insoluble particles in laboratory animals. Unfortunately, even in the many studies on coal miners there are no criteria that exactly match what is known about lung function and dust burdens in the rat. Recent studies have suggested that alveolar macrophage functions, which are important in rats, may not be important in clearance of human lungs.

4 Epidemiological Studies

Three papers on epidemiology were presented. The first (A. Bernard) provided an overview of the epidemiological studies available on metals. The second (J.M. Harrington) identified the problems which have made epidemiological studies difficult to interpret. The third paper (M.H. Draper) introduced improved methodology for obtaining and assessing existing data as applied to the 'epidemic' of nasal and lung cancers that occurred at the Clydach Nickel Refinery in the period up to the 1920s.

4.1 Overview of Epidemiological Studies on the Carcinogenicity of Metals

The nine 'classical' metals, antimony, cadmium, chromium, cobalt, copper, lead, mercury, nickel and zinc, were systematically reviewed, considering in particular the conclusions drawn from the available epidemiological evidence and the presence of important confounding factors that were often not adequately addressed in the original papers. The discussion on cadmium was particularly instructive in this regard. Considering the overviews as a whole, it becomes increasingly clear that the problem of speciation must be addressed as a priority. Of particular interest was the fact that for seven of the nine metals considered, arsenic was a significant but neglected confounder.

The five metals already evaluated by IARC (beryllium, cadmium, chromium, lead, nickel) were further discussed in mechanistic terms. This aspect of metal carcinogenesis is particularly complex, as was discussed in earlier seminar presentations, and may not be at all analogous to carcinogenesis by organic chemicals. Little attention has been paid to this in the epidemiological literature. Considerations of such aspects as valency, physical characteristics, route of administration (when extrapolating from animal studies) and metal/metal interactions are rarely mentioned. If suspect metal compounds and their workplace concentrations can be characterised (and even this is rare), that is usually as far as the epidemiologist goes.

4.2 Problems Encountered in Determining Metal Carcinogenesis Through Epidemiological Studies

Some general conclusions about the current state of epidemiological knowledge with regard to the carcinogenesis of metals can be drawn from critical consideration of the five metals referred to above (beryllium, cadmium, chromium, lead, nickel). In essence, the basic difficulty is one of interpretation, and these problems can arise in a number of ways:

(1) limited data base (*e.g.* beryllium);
(2) poorly defined exposure data (all five metals);
(3) inadequately recorded job history data (*e.g.* cadmium);
(4) inability to distinguish individual compounds or processes (*e.g.* chromium, nickel);
(5) ill defined risks from a ubiquitous agent (*e.g.* lead);
(6) problems of confounding exposures – cigarettes, arsenic, other metal interactions (all five metals);
(7) the inherent complexity of analysis of epidemiological data.

In addition, there are the general frustrating problems posed by the undoubted relevance of the epidemiological studies to the assessment of human carcinogenicity and the inherent complexities of the methodology. By contrast, there is the relative clarity of the animal data, but the inherent uncertainty of extrapolation to predict effects on humans. The failure to resolve these issues leaves industry uncertain of how best to minimize exposure to potential carcinogens, to handle employees confused and worried by an ill defined threat to their health, and to deal with the legislators in a quandary on how to control badly characterised exposure hazards to limit an ill defined but potentially life threatening risk.

It is suggested that the way forward should involve improvements in epidemiological quality. Much more attention should be paid by all concerned (including editors of scientific journals) to the guidelines of Good Epidemiological Practice (GEP), which are now available. There should be more consideration of mechanistic and the biological plausibility of interpretations of data as they relate to humans. Epidemiologists should join with toxicologists and chemists to consider how they can solve difficult issues. Too often, specialists work in isolation from each other. New developments such as 'molecular epidemiology' should be actively promoted. A particularly relevant practical issue is the problem of the extrapolation from the epidemiological data obtained from high exposure situations to the assessment of current worker risk, where workplace exposures are relatively low. This should be a priority issue. Finally, it is extremely important to improve lay understanding of the shortcomings and caveats that are needed when interpreting the results of epidemiological studies in industrial situations.

4.3 A Re-examination of the Cancer 'Epidemic' at the Clydach Nickel Refinery – a Metademographic Approach

At the Mond (now INCO Europe) Nickel Refinery at Clydach in South Wales, UK, an extremely potent carcinogenic agent was present from 1903 to some time between 1924 and 1930. Otherwise, there is little, if any, evidence of carcinogenicity, especially of nasal cancer where nickel metal and other nickel products are and have been extensively used. Thus, attention has to focus on the processes at Clydach. Here, the more detailed the examination, the less likely the conventional explanations appear. The study of the job sequences in particular has provided convincing evidence as to the reason why the 'epidemic' was confined effectively to the first two decades of operation. The lack of any correlation between the enormous throughput of the Bessemer matte feed stock (a complex mixture of nickel and copper sulfides, together with some metallic nickel and copper) and cancer incidence not only casts doubts on the involvement of sulfides, but also lessens the case for the oxides, because oxidic feed stocks continued in use long after the carcinogenicity issue had disappeared. The job sequence studies show that the hazard spread through the central processing units, but was essentially confined to those units that utilized the various sequential concentrates or residues that remained after each extraction cycle. These successive concentrates, known as 'conc 1, conc 2, *etc.* were collected and preserved for further extraction because of their cobalt and precious metal content (silver, platinum, palladium and gold) as well as any remaining nickel and copper.

There were 85 deaths from nasal cancer, and all but four of the cases were recruited before 1925; the first death occurred in 1923. Because nasal cancer is a rare disease it is reasonable to accept that all the cases can be attributed to exposure at the Clydach refinery. The same cannot be said for the 280 cases of lung cancer attributed to working at Clydach because lung cancer is a relatively common disease, and after about 1920, cigarette smoking became a significant confounding factor. It is probable that 210 of the 280 cases are 'true' Clydach cases and 82% of these were recruited before 1930. Of those recruited before 1920, the SMR for lung cancer was 550 (95% CI 440–680) and for nasal cancer a remarkable 34 000 (95% CI 26 000–44 000). During this period, the number of process workers probably never exceeded 600. For those recruited in the period 1930 to 1939, the comparable figures are: lung cancer, SMR 154 (CI 97-230); nasal cancer, one case, SMR 1440 (CI 36–8000).

In the course of the present investigation, some important samples of process material from the 1920s were discovered. Among these was a sample of a conc 4 from 1920, and a conc 3 from 1929. These samples were subjected to a wide range of physical and chemical analyses. There were striking differences between the two in the arsenic content and, to a lesser extent, in iron content. The 1920 sample had 9.6% and the 1929 sample 1.0% arsenic, where the respective iron results were 4.4 and 0.75%. This high arsenic result derived from the use of heavily arsenic contaminated sulfuric acid, discontinued in the early 1920s. Possibly the most surprising result was that the mean particle size was 3.0 micrometres for the 1920 sample and 1.5 micrometres for the 1929 sample, remarkably small particles for an industrial refinery process material. Electron microscopy showed that the particles consisted of clusters of ultramicroscopic crystals. Studies with laser beam ionization microprobe analysis (LIMA) showed that the particles had a uniform composition and that arsenic was a prominent surface feature among the expected nickel, copper, cobalt, iron, sulfur and precious metals. However, the most significant results were obtained from powder X-ray spectroscopy, because this showed that a prominent feature of the 1920 conc, but not the 1929 conc, was the presence of a molecule with the crystalline structure of a complex nickel arsenide mineral known as orcelite. The composition of orcelite is believed to be of the form $(NiFeCu)_{4.2}(AsS)_2$ with the 'defect' elements present maximally at concentrations of (w/w) Fe, 2.4%; Cu^0, 3%; S, 1.4%. With such a composition, a particle from a conc 4 would consist of some 25% of orcelite-like material.

A detailed study of the nickel refining process and the month by month job histories of the cancer cases shows a correlation of the virtual disappearance of respiratory cancers by 1930 with the elimination of the nickel arsenide compound resembling the mineral orcelite. When this observation is coupled with the fact that heavy exposure to nickel metal, nickel sulfides, nickel oxides and nickel sulfate continued for decades without significant associated respiratory cancer, there is a strong case for believing that the extremely potent carcinogen present at Clydach was this orcelite-like molecule.

The study of the complex situation at Clydach required collaboration from many scientific disciplines. This co-ordinated, multidisciplinary approach calls for a more specific designation, and it is suggested that the term 'metademography' might be suitable.

Bio-inorganic Chemistry and Cancer*

R.J.P. WILLIAMS

UNIVERSITY OF OXFORD, INORGANIC CHEMISTRY
LABORATORY, OXFORD, UK

In order that the relevance of the presence of inorganic elements in biological systems to our understanding of what causes cancer can be described, we need to have a basic attitude to the nature of the cancer cell as opposed to the normal cell, and thus to have knowledge of the ways in which inorganic elements are involved in living systems and/or could cause problems for them.

If we assume that cancer arises from a switch of metabolic homeostasis in a cell due to mutations or breakage of DNA, allowing transposition or viral infection, then the interest in inorganic elements can be in two different studies. The first is in the cause of the switch where the inorganic element brings about mutations or breakage. The second is in the requirements for proper balances of inorganic elements to maintain homeostasis in the non-cancerous condition, when the disturbance of such balances may lead indirectly to, or be symptomatic of, a cancerous condition. All changes of homeostasis are not cancerous of course, since such changes are the basis of differentiation.

My simple understanding of the failure of a required homeostatic condition is that mutated proteins, or proteins produced in wrong ratios, fail to maintain this state (generated by the normal proteins) and that as a consequence a quite novel, different homeostatic and cancerous state of a cell can arise. The failure is known to be due to very small initial DNA defects, say at five places, and hence in only a few proteins. Clearly, most such defects in DNA cause harmless changes in proteins or changes that are not cancerous, so that we must be concerned in this article with defects in specific proteins. Moreover, since only, say, five mutations cause the whole homeostasis to change, the particular proteins involved in change

---

* In this article I have not been able to describe speciation in any detail. I stress that when I refer to elements inside or outside living systems it is essential to analyse the form the element is in and its chemical speciation. This point is taken for granted in organic chemistry but it is often lost in the appreciation of inorganic chemical reactions in living systems. One simple example of an essential speciation is that of cobalt in vitamin B12 as found in organisms. Another is the use of *cis*-PtCl$_2$(NH$_3$)$_2$ as an anti-cancer agent – the *trans*-compound has no value. It is not a useful exercise to classify properties of any kind under an element, *e.g.* carbon or nickel. What is required for consideration of carcinogenic risk assessment or for value in life is a knowledge of the precise chemical compound under examination.

must belong to a network of interactive metabolic systems. Similarly, DNA rearrangement or viral incorporation must affect those proteins peculiarly related to homeostasis if they are to cause cancer. Again we cannot be sure that all regions of DNA are equally at risk with respect to damage by inorganic elements, nor can we be sure that given reagents, though not directly carcinogenic (mutagenic) in themselves, do not increase particular risk to zones of the DNA. Thus the relationship of damage to a risk of cancer is never a simple one, see Table 1. We sense that the involvement of inorganic elements in causing cancer could be direct or indirect and may be obvious or quite subtle. As examples, metal ions such as Hg^{2+} and Pt^{2+} in complexes attack DNA directly and could cause cancer during repair, whereas a deficiency of selenium in the diet may increase risk indirectly from a wide variety of oxidants, which could damage DNA.

Table 1 *Types of risk*

Character of element	Risks
No known requirements, *e.g.* Pb, Au Effect due to novel input	(1) Attack on DNA (direct)[a] (2) Attack on DNA (indirect)[a] (3) Damage to homeostasis leading to increase of (1) and (2) (4) Fail to repair DNA
Required for life, *e.g.* K, Mg Effect due to excess	(2) Attack on DNA (indirect) (3) Damage to homeostasis
Effect due to deficiency	(2) Attack on DNA (indirect) due to decrease in protection (3) Damage to homeostasis

[a] See text

Before we can understand the involvement of inorganic elements with cancer risks, we clearly must understand the different roles that inorganic elements play in cells. This is the subject matter of bio-inorganic chemistry.

Let us not consider a developing organism, but one which is fully adult. We may then say that it is in a steady homeostatic state. It is this steady state that is at risk, since it is a system of balanced chemical reactions, which can become readjusted to a cancerous state. The steady state is one of constant flow of organic and inorganic materials from the diet, and of course the intake of any material is not free from risk even when it is part of the requirement to maintain the steady state. Thus, we must not believe that there are simply good and bad chemicals. There may well be some chemicals that are simply bad, but in general the risk associated with a chemical is dose dependent, see Figure 1. We need an understanding of risk within the steady state and this requires us, especially at this meeting, to appreciate the role of the nature and the concentration levels of the inorganic elements in living systems. How are they involved in maintaining steady states?

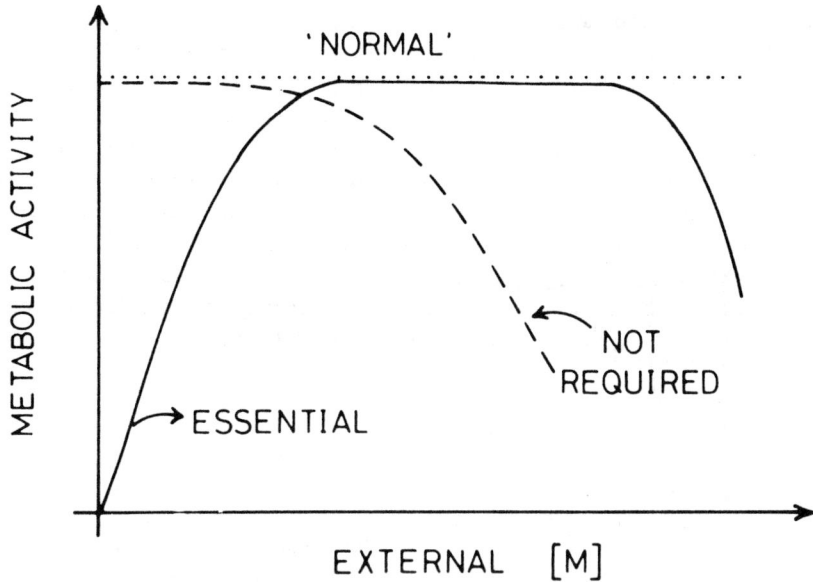

Figure 1 *The requirement for 'normal' homeostasis is met by the correct supply of an essential element M in the diet while risk or failure is increased by deficiency or excess. Essential elements are shown in Figure 2. All other elements not required by the living system become increasingly damaging above a low level.*

1 The Essential Elements

Figure 2 shows the essential elements for life, including human life. There are close on twenty. Each one must be maintained in a balanced system, since, collectively they control a variety of functions. Weakness in one function can lead to increased risk elsewhere, so that we need to see the functional roles of these elements both separately and co-operatively. I shall follow the Periodic Table in a description of their roles. I shall try to separate the roles as far as possible before indicating the co-operativity in a cellular system. While referring to any one element I shall refer to risks from other members of the same group and to related states of elements from other groups.

1.1 Alkali Metals (Group 1)

Their principal function is to maintain osmotic and electroneutrality, though they have a role in some structures of DNA, RNA and proteins. We all know that the concentrations of Na^+ and K^+ are tightly controlled in man. As an immediate example of risk, we know that too high an intake of common salt increases the risk of stroke for those with high blood pressure. One measure of a good homeostatic condition is blood pressure of course. The other elements in Group 1 are Li, Rb and Cs, but to some extent Tl(I) is similar. Both lithium and thallium interact with the nervous system and can be used to good or ill effect. Lithium carbonate is a common drug used to stabilise hyperactive patients. Since various cell

1	2	3	4	5	6	7	8	9	10	11	12	13	14	15	16	17	18	n
1 H																	2 He	1
3 Li	4 Be											5 B	6 C	7 N	8 O	9 F	10 Ne	2
11 Na	12 Mg											13 Al	14 Si	15 P	16 S	17 Cl	18 Ar	3
19 K	20 Ca	21 Sc	22 Ti	23 V	24 Cr	25 Mn	26 Fe	27 Co	28 Ni	29 Cu	30 Zn	31 Ga	32 Ge	33 As	34 Se	35 Br	36 Kr	4
37 Rb	38 Sr	39 Y	40 Zr	41 Nb	42 Mo	43 Tc	44 Ru	45 Rh	46 Pd	47 Ag	48 Cd	49 In	50 Sn	51 Sb	52 Te	53 I	54 Xe	5
55 Cs	56 Ba	57 71 La-Lu	72 Hf	73 Ta	74 W	75 Re	76 Os	77 Ir	78 Pt	79 Au	80 Hg	81 Tl	82 Pb	83 Bi	84 Po	85 At	86 Rn	6
87 Fr	88 Ra	89 103 Ac-Lr	104 Unq	105 Unp	106 Unh	107 Uns	108 Uno	109 Une	110 Uun	111 Uuu	112 Uub	113 Uut	114 Uuq	115 Uup	116 Uuh	117 Uus	118 Uuo	7

57 La	58 Ce	59 Pr	60 Nd	61 Pm	62 Sm	63 Eu	64 Gd	65 Tb	66 Dy	67 Ho	68 Er	69 Tm	70 Yb	71 Lu	6
89 Ac	90 Th	91 Pa	92 U	93 Np	94 Pu	95 Am	96 Cm	97 Bk	98 Cf	99 Es	100 Fm	101 Md	102 No	103 Lr	7

Figure 2 *The Periodic Table. The elements in heavy boxes are essential to man or in one or two cases to the ecosystem,* i.e. *V, (Cr), and B.*

mechanisms are exposed to risk when exposed to lithium or heavy alkali metals, we need to ask whether enzymes or membranes can be damaged by them, and can this damaged condition lead to problems at the DNA level and then go on to cause cancer? I return to this point later. In passing, note that lithium is teratogenic and obviously is therefore interactive with some major growth processes.

1.2 Group 2 Elements

The essential elements are Mg and Ca. Almost every process in the body is linked, directly or indirectly, to these two elements as we know from muscle, nerve, bone, skin, *etc.* activities and problems caused by faulty handling of these elements. Deficiency or excess of either or both can be extremely damaging. It pays us to give special attention to the elements themselves, but also to their link with central cellular processes. I cannot describe all of these processes, see Table 2, in a short article and so I shall choose a particular set of reactions, which involve both elements.

Table 2 *Parts of homeostatic enzyme action linked to magnesium and calcium*

Enzyme action	Involvement of metals
Nucleotide triphosphate dependent	MgNTP, Ca triggering
Cyclic nucleotide phosphate dependent	Mg produced, Ca triggering
External digestion, proteases, *etc.*	Ca utilizing
Many pumps	Ca and MgATP dependent

N.B. Calcium in particular is also linked to cell–cell adhesion, *e.g.* via lectins, and directly to signals, *e.g.* CaEGF

The most common set of genes related to homeostasis is that composed of various kinases and phosphatases, involving phosphorylation or dephosphorylation of transcription proteins and those of the metabolism of the cyclic nucleotides (which activate by binding other sets of proteins), cyclases and esterases, see Figure 3. Mutations in the kinases, or the protein transcription factors affected by them, are well-known causes of cancer. We need to see that, in fact, the required homeostasis of the action of these proteins is linked to Mg^{2+} and Ca^{2+} levels. We take first the fact that probably all, and certainly most, ATP reactions are Mg ATP reactions, including those that lead to cyclic nucleotides. A schematic diagram is given in Figure 4. The other nucleotide triphosphates are similarly involved with Mg^{2+} through binding.

The involvement of calcium in many of these reactions is in response to a stimulus. The calcium enters the cell on a given signal. Clearly, the level of

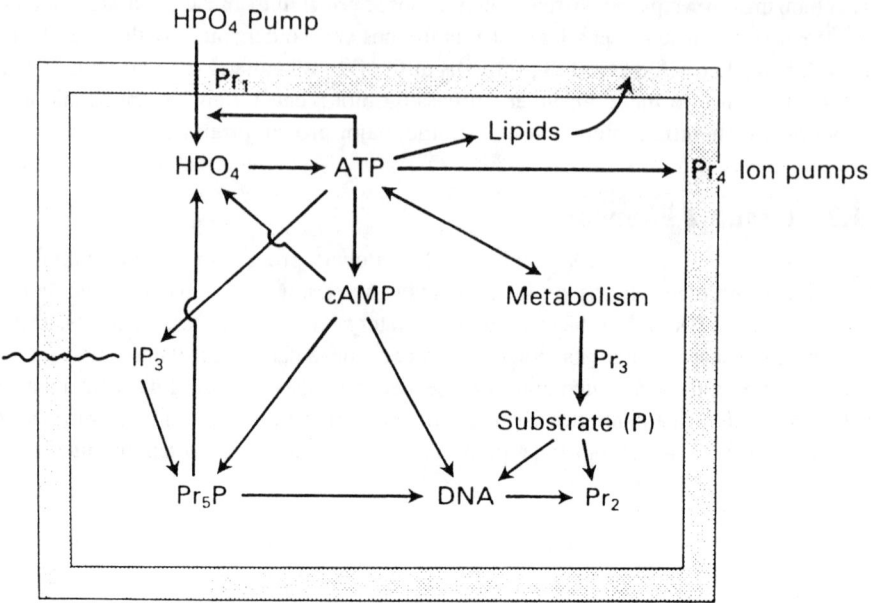

Figure 3 *A much simplified view of the involvement of phosphate in virtually all cellular activities. Pr_n indicates different proteins. All activities are in balance but not in the same balance in different differentiated cells or in a cancer cell.*

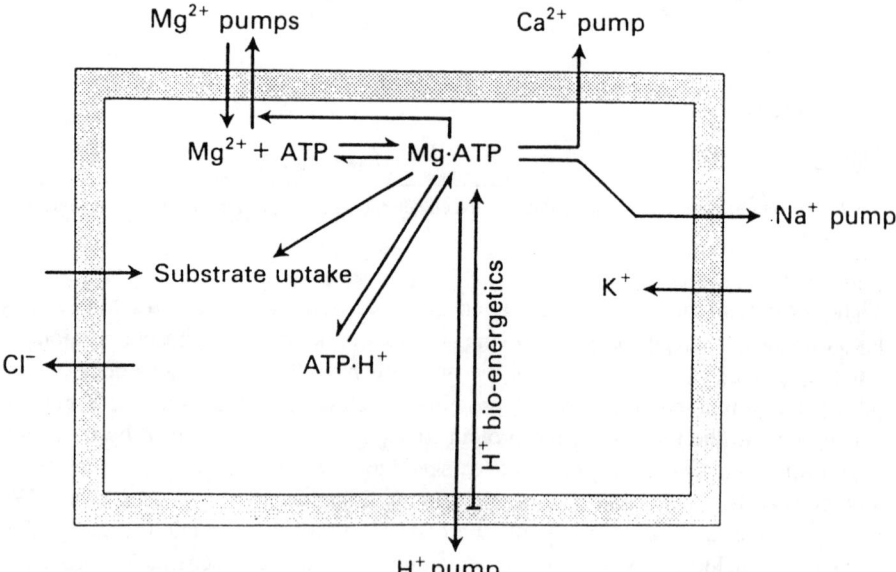

Figure 4 *A schematic figure to show the dependence of Na, K, Cl, H, Mg and Ca upon phosphate metabolism. All are in balance, which is different for each differentiated cell or a cancer cell. The balances depend on the levels of peptide hormones.*

calcium in the absence of a stimulus, and the level after stimulus, are important, but the nature of the stimulus and its repetitive nature must also affect the homeostatic state. It is of interest that repeated stimulation at given frequencies can change a fast to a slow muscle and the reverse. This is a change of homeostasis. The conclusion is that protein homeostasis is dependent indirectly on calcium through the reactions of bound phosphate. The calcium binding proteins do not bind directly to DNA but activate phosphorylation and dephosphorylation of proteins that do act on DNA. Thus, homeostasis depends on:

$$\text{Ca} + \text{Phosphate Reactions (Mg)} \rightarrow \text{Transcription Factors} \rightarrow \text{Proteins}$$

The phosphate reactions include many involving nucleotide tri- and di-phosphates and cyclic nucleotide phosphates. Hence, we must always be aware of the intracellular calcium, see Figure 5. The description just given is an example of a network of factors at work, and we suspect faults in networks cause cancer.

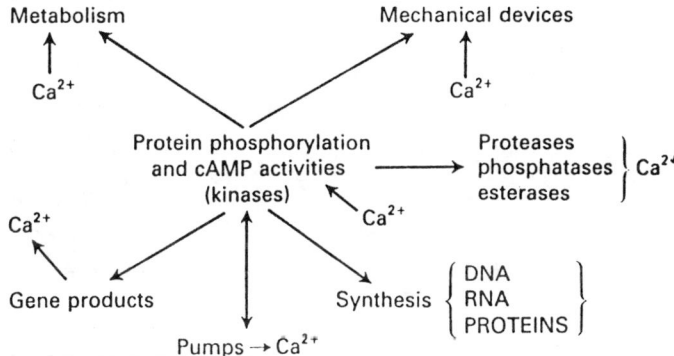

Figure 5 *A further illustration of the intimate integration of calcium and phosphate activities in higher cells. The entry 'Gene products' refers to such peptide hormones as calcitonin, which interacts with both calcium and cAMP systems.*

The outside of the cell has many calcium dependent features, and of special interest to us are the calcium dependent activities that affect extracellular matrices. These matrices hold cells in place and can prevent cell migration, and hence can be involved in metastatic activity. Again, the levels of calcium in the body fluids are related to those in the cell, and just as there is a connection of calcium to phosphate and its compounds in the cell, so calcium outside the cell is homeostatically controlled by bone, an insoluble form of calcium phosphate. Bone itself is controlled by a network of hormones such as vitamin D, which requires an iron enzyme for its synthesis and it interacts with a zinc transcription factor, see Figure 6.

Other elements in this group are Sr and Ba, which can interfere with calcium activity, but more noteworthy is the blocking of K^+ channels by Ba^{2+}. Finally in this group is Be, and it is known to be hazardous, especially again, owing to the interaction with phosphorylation reactions.

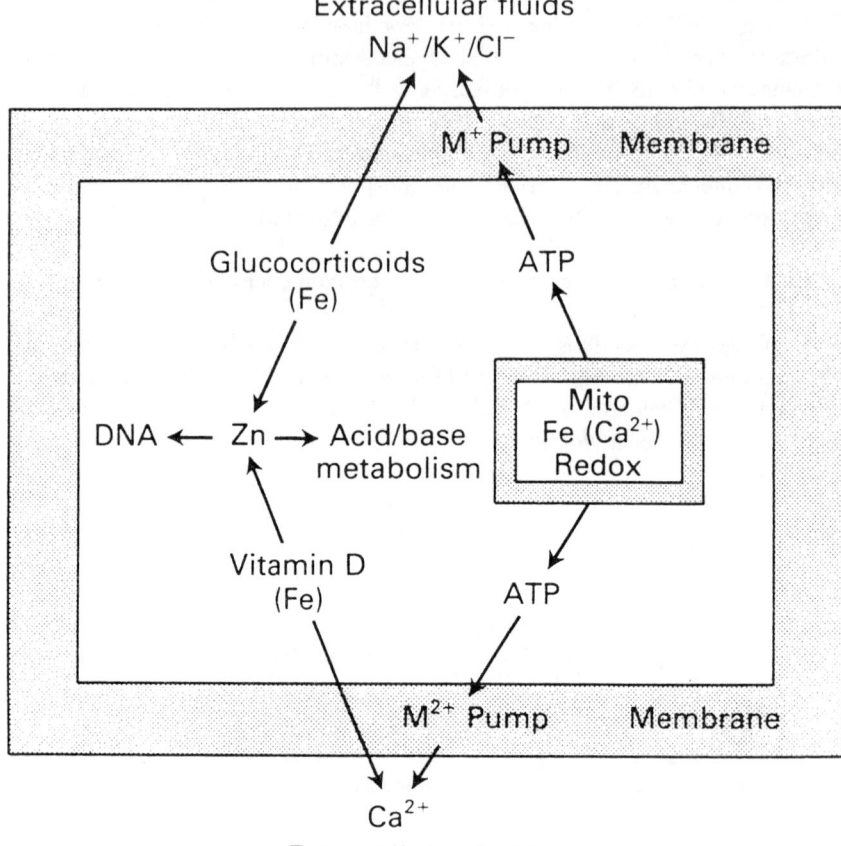

Figure 6 *The connection between zinc and a multitude of activities. Zinc is an internal hormone linking the homeostasis of the cell to many other hydrophobic hormones.*

1.2.1 Control, Regulation and Group 2 Elements

It follows from the discussion of the network of interactions between H^+, Ca^{2+}, Mg^{2+} and HPO_4^{2-}, including all the proteins with which they interact, that in the resting state of a cell, a very large number of continuous activities are connected. There are two different types of connection. The first, which we shall call control, is over metabolic pathways and part of the function of the four elements listed, and small compounds such as IP_3, MgATP, c-AMP and the proteins to which they bind, such as calmodulins and phosphorylated enzymes, is to act in feedback allosteric control in each and every pathway so that all the pathways, even those producing useful energy (in mitochondria), are in a connected, homeostatic, single system.

This system is not independent from the production of proteins themselves, but at first it is simpler to isolate the organized production of proteins under a separate heading, regulation. It is implied that mainly through phosphate in various forms, such as phosphoproteins and cyclic nucleotides (c-NMP), the expression of RNA and proteins is managed. The management is linked to the concentrations of transcription proteins, which interact directly with phosphate or c-NMP when it is their levels that regulate. Since the ability to make c-NMP or phosphoproteins is connected to the controls over Mg^{2+}, H^+, Ca^{2+} and small phosphates (ATP), there is a network for the control of, and the activity of, the pathways of metabolism to and from the regulation at the level of DNA. Given such a situation, it is clear enough that proteins (enzymes) critical in the network will be kinases, phosphatases, cyclases (to make c-NMP) and such items as pumps and channels for calcium. The type of cellular defect that then is brought about by damage to any of these will be to the homeostasis, and a small defect can utterly change a cell. Some may cause one disease, some another, but maybe many will cause cancers. There may be many defects that will allow a cell to change to a new homeostatic condition, allow a new growth rate and an ability to migrate. In this article I stress deliberately the role of inorganic ions in homeostasis. For example, malignant hypothermia is caused by a mutation in protein channels for handling calcium.

1.3 Zinc – Group 12 Elements

We switch to Group 12 since in chemistry Zn^{2+} has an obvious relationship to Mg^{2+} and Ca^{2+} in that all three have no oxidation/reduction chemistry.

The well-known activities of zinc are in a great variety of enzymes, since Zn^{2+} is a powerful acid. Thus, it is found both inside and outside cells (see Table 3), and in many digestive proteases, amylases and so on. However, it is involved directly in other important functions, such as RNA synthetase, phosphokinase C and in some reverse transcriptases. Thus, faults in zinc metabolism could relate both directly or indirectly to polynucleotide synthesis. Zinc is involved again in the production of many peptide hormones and also in their destruction. It is therefore central to the homeostasis of many higher plants and animals.

Table 3 *Zinc and copper enzymes of the external matrix*

Enzyme	Metal	Activity
		Hydrolysis of
Elastase	Zn	Elastins
Collagenase	Zn	Collagens
Stromelysin	Zn	Proteoglycans
Gelatinase	Zn	Gelatins
Protease 24.11	Zn	Peptide hormones
		Oxidation
Lysine oxidase	Cu	Cross-links collagen and elastin
Laccase	Cu	Cross-links via phenols
Phenol oxidase	Cu	Cross-links via phenols
Amine oxidase	Cu	Removes amine hormones

However, the important involvement of zinc in relation to cancer may not be with catalytic activity or control of metabolism, but with regulation. Zinc is involved in perhaps 20% of all transcription factors through zinc fingers. It has a major connection to hydrophobic hormones, unlike calcium and phosphorylation, which link mainly to hydrophilic hormones, including many of the peptides produced by zinc enzymes. The hydrophobic hormones, sterols, retinoic acids, thyroxines and so on bind directly to zinc finger transcription factors. We do not know for certain how this system is built into a feedback network, but it is possible that it is through free zinc, see Figure 6. These hormones are concerned with the expression of a range of proteins, some interacting with the calcium/phosphate system, *e.g.* calbindin and osteocalcin, which is again an indication of the fact that activities in a network are not independent. (They cannot be, since together they produce unique growth patterns.) A recent example of a zinc protein transcription factor (not a true zinc finger) is p-53. Here, mutations very close to the zinc are known to be contributors to the onset of cancer. Do the mutations adjust zinc binding and therefore expression of proteins in the network?

Notice that in the above there can be a careful separation within a cell of different roles of one inorganic element. A cell can separate a catalytic function from a regulatory function on the basis of binding constant or by removing certain enzymes into vesicles. It appears that the binding constant of zinc in an enzyme may well be orders of magnitude higher than that in a transcription protein, and, in fair part, this stability difference can be due just to the protein fold strength. Thus, zinc can exchange between transcription factors to maintain homeostasis, but not with enzymes. Again, zinc in a vesicle does not equilibrate with zinc in the cytoplasm, *e.g.* zinc in the insulin vesicles. This returns us to the organization of other elements so as to avoid the risk of confusion of functions, since all metal ions compete to some degree for all sites.

If zinc has such a central role, then those elements that can displace it must be kept out of the cytoplasm. We know that a quite general stability order for the binding of divalent metal ions to organic ligands is:

Mg < Mn < Fe < Co < Ni < Cu < Zn <

The major competitors for zinc sites are nickel and copper. Hence, nickel and copper have to be banished from the cytoplasm. It is known, with but rare exceptions, that copper proteins are extracellular or in vesicles, and the one known nickel enzyme, urease, found in plants, is in vesicles. Other elements that present a threat are the other elements of Group 12, Cd and Hg. They too must be removed if they appear in quantity. In fact, we know that living cells remove such unwanted elements by the two mechanisms described above:

(1) Through a tight-binding protein, *e.g.* metallothionein, a special heavy metal binding protein that removes Cu(I) and Cd(II)
(2) Through pumps that place Ni(II) in vesicles and Cu(I) outside the cytoplasm.

What happens when such systems fail? Is a new homeostasis possible? Since it involves zinc, is there an adjustment of extracellular tissue around a damaged cell such that the cell can migrate?

1.4 Groups 3, 4 and 13

I take the elements Sc, Y, La, Ti, Zr, Hf, Al, Ga, In, Tl together as they are the last elements which have virtually no oxidation/reduction activity. Also, none of them has a biological function. In fact, in higher concentrations, they poison cells and generally are excluded from organisms. It is very likely that this is because as trivalent ions they bind strongly to phosphate and carboxylate centres and therefore can damage homeostasis directly or indirectly by displacing Mg^{2+} and/or Ca^{2+} from their functional sites. (Al^{3+} is known to cause mental disease problems, dialysis dementia, but may not be related to Alzheimer's disease.) This is an illustration of direct poisoning and killing of cells, as opposed to causing a cell to change its nature. There is, however, no reason in chemistry to suppose that elements such as these cannot cause DNA damage. Whether or not cancer results is perhaps happen-stance.

1.5 Transition Metals Groups 5 to 11

These are the familiar heavier elements in three rows, see Figure 1, and I shall stay with the first row for much of my discussion, since the elements are often biologically essential. The elements are:

V, Cr, Mn, Fe, Co, Ni, Cu

We have already noted that, in man, Ni as Ni^{2+} is removed by separation into vesicles. We can add that Cr as Cr^{3+} is also avoided perhaps for the same reason as that proposed for avoiding all other M^{3+} ions. I return to higher oxidation states later. In man, vanadium has no known function, while the elements Mn, Fe, Co and Cu are all extremely important and are termed essential. In passing, note that molybdenum from the next lower row of transition elements is also required. The major source of the ions from the first row is that of the divalent ions. As mentioned these ions can displace zinc increasingly along the series:

Mn < Fe < Co (< Zn) < Cu

The elements are required, however, in quite different functions as oxidation/reduction catalysts. To deal with the problem of use while avoiding competition, cells have a variety of tricks, which have been described above under speciation, and which have two parts:

(i) special binding, and
(ii) isolation in space.

The special binding depends on control over the oxidation/reduction chemistry of the elements and the synthesis of special small chelating agents and chelating proteins. Thus, cobalt, a very minor element, is effectively held as Co^{3+} in vitamin B_{12}. Much of iron is held in porphyrins and in special Fe^{3+}/Fe^{2+} sulfide units in the cytoplasm. The best trick for separation, however, uses *relative binding strength* in the presence of slight *excess ligand production*, stimulated by the presence of the given element and switched off once the metal ion concentration falls.

Let us assume that a cell controls the input to the cytoplasm of two metals, M_1 and M_2, and the synthesis of ligands L_1 and L_2. There are four possible complexes, M_1L_1, M_1L_2, M_2L_1 and M_2L_2, but only two M_1L_1 and M_2L_2, are wanted. If M_1 always binds more strongly than M_2 and binds L_1 most strongly, then as long as L_1 and L_2 exceed M_1 in concentration and there is little excess L_1, M_1 will be found as M_1L_1 and M_2 will be forced to appear with L_2 as M_2L_2. Generally, there is a strong binding preference of the later metals in the series for N- or S-centres, while elements earlier in the series bind to O-centres. This leads to a neat separation along the lines:

O-binding	O/N-binding	S/N-binding
Ca, Mg	Mn, Fe, Co	Ni, Cu, Zn

For a fuller understanding, including extra selectivity factors, the reader must consult the references.

As far as a selective use of space is concerned, biological cells pump out very largely certain metals. The separation is roughly:

Vesicles and Extracellular	Cytoplasmic
Na, Ca, Ni, Cu, (Mn)	K, Mg, Fe, Zn

Once isolated in space or by chemical combination the elements are employed by incorporation into proteins to make enzymes. These metallo-enzymes are central to life at all levels. Those of you who are unfamiliar with inorganic chemistry must see that organic chemistry in cells cannot look after itself. In the first instance it is organic chemistry in water and is based on anions. It then needs a balance of inorganic cations Na^+, K^+, Mg^{2+}. (In evolution these ions plus Ca^{2+} became the major messengers for muscle, nerve and brain.) The second type of activity where organic chemistry is not valuable is in catalysis. Metal ions exceed, in both acid strength and redox change possibilities, the corresponding properties of organic groups. Thus, some 15–20 inorganic elements are essential but they introduce risk.

Now, the elements under discussion in this section are selectively used in oxidation/reduction catalysis. Many reactions in the cytoplasm are reductive electron transfer catalysts in such metabolic pathways as the citrate cycle, which generates energy. There is always a danger of linkage to dioxygen to produce peroxide and superoxide, which are a major cancer-inducing hazard. Further reaction in the cytoplasm is simple detoxification as in the case of catalase and superoxide dismutase. The enzymes are not in a network, but failure to produce them can lead to damage through peroxides or superoxide.

Equally importantly and now outside the cytoplasm, these metals, especially Cu and Fe, stabilise extracellular fluids and produce a wide variety of hormones (see Table 4) and hence become part of networks, see references.[1,2]

Table 4 *Some hormones produced by oxidative metabolism*

Hormone	Metal element involved
Sterols	Iron in cytochrome P-450
Adrenaline	Copper and iron in adrenal gland
Oxidized peptides	Copper in glycine oxidation
Thyroxine	Iron (iodine) in peroxidase
Prostaglandins	Iron in peroxidase

N.B. Hormone levels must be related to the proper balance of inorganic elements. There are corresponding hormones in plants and insects.

1.6 Summary of Single Element Use

In summary, some 15–20 elements are of known required functional significance in man. Almost every function is connected with some level of risk to DNA, but the real position and significance of the elements in cellular metabolism cannot be seen by taking them one at a time outside the context of the co-operative chemistry of cells. I turn then to the way in which elements co-operate in feedback networks to establish homeostasis. I take as an example the chemistry of extracellular tissue and messages using the particular case of copper and zinc, which will lead on to a discussion of the involvement of iron, calcium, phosphate and so on.

1.7 Co-operative Functional Value of Zinc and Copper Outside Cells

Outside cells, enzymes have been required for digestion from a period very early in evolution. Many digestive enzymes utilize the attacking strength of zinc. The function has been extended in advanced organisms to the production of hormonal peptides from pro-proteins and their destruction, and to the management of connective tissue, see Table 3. A major feature of metastasis is a requirement of digestive enzymes from a cancer cell to break down connective tissue, allowing the cell to migrate and allowing blood capillaries to grow to the cancer cell. The cancer producing defect must then be connected to the regulation of zinc protease production, and not I suggest through zinc itself, but quite possibly through a change in a transcription factor involving phosphorylation.

Copper must be placed outside the cell cytoplasm too, *i.e.* in vesicles or in extracellular space, and is used in parallel with zinc to produce (and destroy) hormones, and now to produce (not destroy) connective tissue with crosslinks, see Table 3. Thus, it is necessary for the copper/zinc activities to be harmonised to get appropriate cellular growth within the synthesis/hydrolysis of connective tissue.

Again, the enzymic use of copper (outside the cytoplasm) must be regulated by protein production to control inside the cell, and clearly must also be linked to the handling of zinc.

A possibility is the use of metallothionein, to measure both zinc and copper (and cadmium) levels and to inform the DNA of their presence. Metallothionein is a strong binding agent for both (and cadmium) but its fold is metal dependent. Thus, each metal is handled separately but linked through one protein. Metallothionein is a cytoplasmic buffer and binds more weakly than the metalloenzymes but similarly to zinc fingers. (Note the use of sulfur binding to these metals).

Could mutational defects in those proteins lead to cancer? The answer may be 'no' to a direct effect, but we must worry about the effects on cell migration. Defects known now in copper metabolism include two familiar copper dependent diseases, and both are due to mutations in the copper pump. One is reminded of the mutation in the calcium pump that leads to *malignant* hypothermia.

Now I have given two examples of networks of controls: (1) Mg, Ca, phosphate and (2) Cu, Zn. They come together in the extracellular matrix, for not only is collagen production linked to Cu/Zn, but it links to Ca, (Mg) and phosphate in bone, see Figure 7. The network is more extensive in that bone growth is dependent on vitamin D (which controls Ca^{2+} uptake) and this vitamin requires iron for its production while acting on a zinc finger transcription factor. Quite clearly, we can continue this analysis from one element to the next so that all 15–20 elements of man must be controlled together.

In higher oxidation states the elements Mn, Fe, Cu and Mo especially are used to oxidize organic material using agents as O_2 and H_2O_2. All this chemistry is full of risk to DNA. DNA is a primitive code designed for use in a reductive medium. The number of oxidation enzymes is very large, and once again the reader must refer to a standard text for details.

1.8 Groups 14–17 – Non-metals

The non-metals of concern in this article are Si, Ge and P on the one hand, since they have no oxidation/reduction functions, and (S), Se and Cl, Br, I on the other as they are redox active. The function of Si is as a simple biomineral in plants, and it appears to be harmless in man though always present. Of course, solid silica and silicates are not harmless and cause aggravation in the lungs especially. Phosphorus as phosphate is essential in many biological reaction pathways, see Figure 3. Its activities have been mentioned already in the context of homeostasis, to which I return again later.

The dangerous chemicals are those associated with oxidation. Much though selenium is used in detoxification, the intermediates of its reactions are always a risk. The only halogen used biologically in man is iodine in the thyroxine hormone. However, chloride is essential as an electrolyte in all cells and body tissues. There is always a chance that any one of the halides Cl^-, Br^- or I^- will be oxidized to aggressive inorganic or organic compounds, and all such compounds can attack DNA. We must be constantly aware of the risks from the halogenated organic molecules, much though they are useful in drugs and disinfectants.

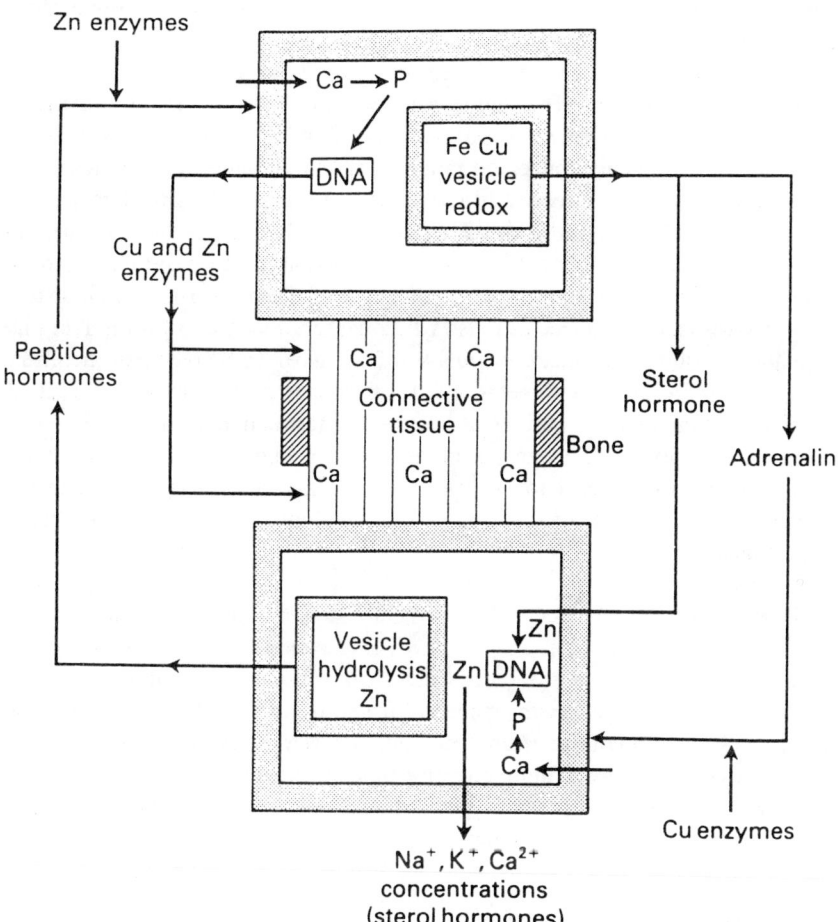

Figure 7 *An extension of Figure 6 to include two cells, which allows the role of iron and copper to be seen in relation to the role of zinc in the control of extracellular signals and tissues and shows also the connection to calcium and phosphate compounds from these controls. Note the further involvement of proteins, especially now in the precipitation reactions.*

I turn my attention to what can go astray in two parts: (1) attack on DNA, and (2) failure of the network to protect DNA and to allow normal functioning of the genes, *i.e.* failure of homeostasis.

2 Attack on DNA

The attack on DNA to cause mutation can be divided into (i) direct attack and (ii) attack through poisons from reactions. We have noted above that metal ions can be divided into three major groups, those of Groups 1 and 2 of the Periodic Table, those of Groups 11 to 15 and the transition metals. The first group is largely of very low attacking power and is used by cells generally. The only element that causes any major threat is beryllium, which binds powerfully to phosphates and can break phosphate ester bonds. This could also lead to disturbances in phosphate metabolism (see below).

The second group has much greater attacking power, largely through co-ordinating strongly to bases. This Lewis acid strength is also common to many of the elements in the third group, so we treat this aspect of their chemistry immediately. The elements concerned range from Mn all along the first long row of the Periodic Table, *i.e.* Group 7 to Group 15. All have some ability to bind to DNA bases through N as well as O atoms. The DNA is then more open to mutation. A typical attack is the well-known action of the anti-cancer drug *cis*-platin. The Pt^{2+} ion attacks the N(7) of guanine. There is an old saying that those agents which cure cancer can also cause cancer. It is based on the fact that curing cancer requires an attack on DNA (of a cancer cell), but cell chemistry being what it is means that there is a risk of mutagenesis or rearrangement of DNA of a normal cell as well as a chance of destruction of the DNA of a cancerous one. If you look at the books on The Toxicology of Chemicals Series One, Carcinogenicity, Vols. I to IV,[3] you will see that as well as beryllium the metal elements in the order of the groups of the periodic table already considered with accompanying hazards are,

Be, B, P, Cl, Ti - Cr - Mn - Co - Cu, Zn, Ga - As - Br - Sb - I - Tl - Pb

and undoubtedly many more will be added in subsequent volumes. I would suggest that at least all the *metal* elements from Group 3 to 15 have some risk associated with them due to their ability to attack DNA. Thus, it is the size of the risk from individual elements and their compounds that we must assess.

There are features of attack which should be separated, however. The distinction to be made is between elements attacking more at O-atoms than at N- or S-atoms of organic molecules as mentioned above. Factors such as high charge and small size favour attack on O-atoms, while high electron affinity (measured by the energy required to go from the metal atom to the ion, the ionization potential) and large size leads to attack on N- and S-centres. I would then say that Al^{3+}, Ti^{4+}, Ga^{3+} and similar ions are not unlike beryllium, Be^{2+}, in that they will attack at phosphate or aromatic oxygen centres. On the other hand, increasingly along the series,

$Mn^{2+} < Fe^{2+} < Co^{2+} < Ni^{2+} < Cu^{2+} < Zn^{2+}$

or down groups,

$Ni^{2+} < Pd^{2+} < Pt^{2+}$
$Zn^{2+} < Cd^{2+} < Hg^{2+}$

attack will be on N- or S-centres. The orders also give a good guide to the strength of the binding and hence the risk of reaction with DNA. Attack at N-sites appears most risky.

The third group of elements can also attack directly or indirectly as redox reagents. Thus we are all aware that redox metals have the possibility of generating free radicals in DNA, even directly and especially in higher oxidation states. Chromates, permanganates and ferrates are very aggressive oxidants. The oxidizing attack by side reactions of these same elements is through the initial reaction with small molecules such as O_2 and NO_3^-, but there are many more, giving rise to reagents such as $O_2\cdot{-}$, H_2O_2 and NO_2. The danger is obvious. Finally there is the possibility of attack by O-atom transfer from such units as CrO_4^-. This is called two-electron oxidation.

When we compare the carcinogenic risks due to metals, they are parallel in many ways to those introduced by non-metals. Attack on DNA by alkylating agents is closely paralleled by attack by metal ions in that the attack is really by R^+, compare CH_{3+} and Pt^{2+}. (The R^+ attack arises from the halogen organic compounds in particular.) Redox attack is through a one-electron reagent such as $O_2\cdot{-}$, compare Fe^{3+}. Two-electron attack can be from $>$SeO going to $>$Se, parallel with attack by FeO^{2+} going to Fe^{2+}.

I hope to have established that virtually all elements, except perhaps those in Groups 1 and 2 (but not Be), are associated with some risk of attack on DNA. There is also a degree to which some elements are protective. Thus Cu, Fe, Mn and Se act to remove O_2, $O_2\cdot{-}$, H_2O_2 and NO within the cell. There is then a balance between risk and use in protection. Many carcinogenic organic compounds are removed by cytochrome P-450, but not without introducing a lower level of risk from other chemicals, *e.g.*:

Aromatic + O_2 → Epoxide → Destruction or Removal

↓   P-450   ↓

Cancer   Cancer

This pathway is not different in basic chemistry from the prostaglandin series of reactions. In fact, since many hormones are oxidized organic molecules they must have a carcinogenic risk. Cancer is in part an enhanced self-induced risk built into the systems of the living cell once they began to use oxidation.

Clearly what is required is a control over risk. The body must handle the inorganic elements to optimize use and minimize risk. This is a part of bio-inorganic chemistry that relates to homeostasis.

3 Buffering and Homeostasis of Ions

The two simple modes of protection are (1) to keep the systems around the DNA as free from oxidants (major risk) as possible, and (2) to bind potential reagents, which could cause damage while making them useful. Under (1) the cell has two management strategies. It can keep the elements which pose a risk in a separate compartment or vesicle and/or it can keep the region around the DNA at a low redox potential. Under (2) the inorganic element can be bound by a buffering protein or other chelating agent. An example shows all three in operation. We take the case of the extremely dangerous (to DNA) Cu^{2+} ions. The scheme for reducing toxicity is:

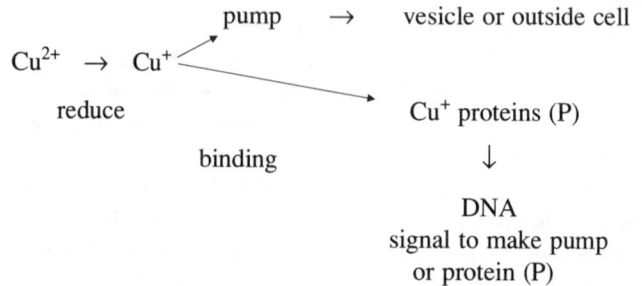

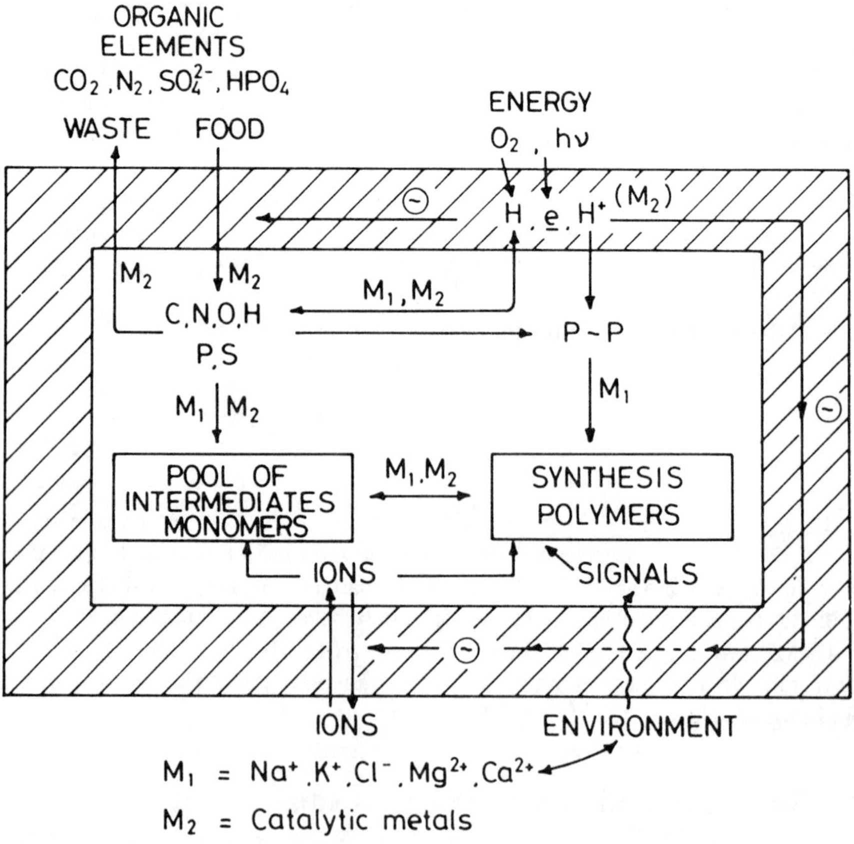

Figure 8 *A generalized description of a cell. The homeostatic requirement of 15–20 elements is shown. The major non-metals are C, N, O, H, P and S of course but we must remember Se and probably Si. The four major metals, M_1, are shown with chloride. These thirteen elements require at least Mn, Fe, Co, Ni, Cu, Zn and Mo (M_2) in balanced amounts as catalysts, making twenty in total. There may well be others.*

Inside the vesicle or outside the cell, copper as cuprous is perhaps the major oxidizing centre of multicellular organisms and can work there safely. The copper proteins (P) in cells bind Cu^+ very strongly ($K > 10^{15}$), and apart from acting as buffers (metallothioneins) or transorption factors to DNA (Ace-1) they can be used in detoxification, even in the cytoplasm, *e.g.* superoxide dismutase, so long as the metal ion is buried deeply in a protein with no chance of adventitious oxidation.

The fact that the human body has a balanced requirement for 15–20 elements (Figure 8) and that some of these elements have near neighbours in the Periodic Table, which are very similar in chemistry to them, means that there is always a risk if the internal network is not thoroughly protected by the internal systems. The risk lies in the natural internal processes as well as in the environmental dangers, see Table 5, since they can suffer from excess or deficiency. The stress on the network is increased by the environment if it differs from expectation based on Earth's history. The risk for an organism is greater the more complex the organism, leaving man quite exposed, but he has also a high degree of protection. Thus, we have to see that for every cell there is a risk opposite every chemical act. Put directly, there is a chance that any known cell that can replicate can become cancerous. Our problem with the inorganic elements and their compounds is to discover the risks (and of course this applies to organic compounds too). A cancer cell is in a homeostatic 'inorganic' growth condition, just as is a normal cell. We need to uncover the weak spots of its homeostasis to kill it. Here the arsenal of inorganic drugs has been poorly investigated. It must be that the inorganic status, the composition of the cancer homeostasis, is different from that of the normal cell. If we investigate this feature of cancer we may find new ways to tackle it.

Table 5 *Effects of excesses of non-biological metals on humans and animals*

Metal	Effect of excess	Observations
Al	Implicated in dyalysis encephalopathies and in Alzheimer's disease	Interaction with phosphate? Cross linking of proteins
Cd	Reduction of effective filtration capacity of glomerulus (renal toxicity)	Blocks sulfhydryl groups in enzymes and competes with zinc for its sites. Stimulates metallothionein synthesis and interferes with Cu^{2+}, Zn^{2+} metabolism
Hg	Damage to the central nervous system; neuropsychiatric disorders	CH_3Hg^+ compounds are lipid-soluble
Pb	Saturnism; injuries to the peripheral nervous system; disturbs haem synthesis and affects the kidneys	Pb^{2+} may replace Ca^{2+} with loss of functional (and structural) integrity. Replaces Zn^{2+} in δ-aminolaevulinic acid dehydratase; reacts with sulfhydryl groups
Tl^+	Poisonous to nervous systems; enters cells via K^+ channels	Tl^+ is similar to K^+ and binds more tightly, including to N and S ligands

4 Conclusion

There is an inevitable association of both DNA and inorganic elements with life and of life with the environment. There is a consequential risk of damage of all kinds since life is an energised process and all elements are to some degree catalytic. The risk was increased by the introduction of oxidative chemistry some 1–2 billion years ago. As complexity of organization increased, the damage done could be more or less permanent since advanced cells differentiate into permanent patterns and reversion is almost impossible. Cancer develops from a cell that is differentiated (irreversibly) by damage. Man's chemistry has introduced into the biosphere a vast new range of chemicals and risks. Each such chemical, inorganic or organic, has a risk of generating cancer. The assessment of risk against benefit is complex since it is age-dependent and depends on preventative measures. There is no way man can eliminate cancer, but he may be able to reduce risk and he will be able to improve his attack on the cancer cell (but all such attacks will also have risks) and to control his environment if he wishes. Cancer is not so much a disease as an inevitable consequence of multicellular differentiated life forms which we must accept as we accept death, dealing with both in an appropriate manner.

References

The references given are intended to lead the reader to a wide range of material and are just good starting points.

1. Frausto da Silva, J.R.R. and Williams, R.J.P. (1991) 'The Biological Chemistry of the Elements', Oxford University Press, Oxford.
2. Williams, R.J.P. and Frausto da Silva, J.R.R. (1996) 'The Natural Selection of the Chemical Elements', Oxford University Press, Oxford.
3. Aresini, G., Berlin, A., Draper, M.H., Duffus, J.H., Krug, E., Roi, R. and van der Venne, M.Th. (1989–1993) 'The Toxicology of Chemicals', Series One, Carcinogenicity, Vols. I–IV. Commission of the European Communities, Luxembourg.

Physico-chemical Properties of Inorganic Particles Controlling Biological Activity

R.P. NOLAN and A.M. LANGER

ENVIRONMENTAL SCIENCES LABORATORY, BROOKLYN COLLEGE, BROOKLYN, NEW YORK, USA

1 Introduction

Exposure to certain inorganic particles has been associated with the occurrence of human disease. The principal route of exposure identified has been inhalation, the particles dispersed in an aerosol. Most often the anatomical site where pathological changes occur is in the pulmonary architecture and pleura, although diseases of the nasopharynx, gastrointestinal tract and other sites have also been reported. Inorganic particles differ from many other toxicants in that they do not act as individual molecules, *i.e.* metabolic by-products, but rather as solids, made respirable by having at least one dimension less than 5 µm in size.[1] Host degradation of the inorganic particle may occur.

Both man-made and naturally occurring inorganic particles have been shown to possess biological activity.[2-4] These particles may be either amorphous or crystalline, but display two important common features: morphological appearance (form) and surface structure(s), which imparts a significant surface chemistry.

Particle dimension(s) has been shown to be an important determinant of biological activity in experimental studies, *e.g.* using *in vitro* assays and experimental animal studies, and in epidemiological studies of exposed populations.[5-7] Aerosol stability is important for both inhalation and site of deposition, while particle dimensions (which control surface area) and surface chemistry are important once the particle is at the target tissue.[1,8]

If the variation in particle size distribution of the same type of inorganic is sufficiently different between two populations, then a single substance can possess a range of biological activities.[9] In addition to the variation associated with size distribution, the particle surface can exist in a number of physico-chemical states producing different biological potentials. The variation in particle size distribution and surface structure is determined by a particular specimen's mechanical/thermal history, growth conditions and any chemical modification that may have occurred. A quantitative description of non-fibrous/fibrous particle morphology and selected examples of how particle size distribution and surface properties influence biological activity are the subject of this paper.

2 Quantitative Description of Particle Morphology

The contribution of particle morphology to the total surface area per unit mass and particle number per unit mass can be easily calculated if the density of the particle is known and a shape is assumed. As a practical matter the actual shapes of a population of particles are rarely known and determining their size distribution can be a rather tedious procedure. However, a quantitative description of idealized particles is useful for evaluating the range of biological activity, which may be associated with morphology.

2.1 Non-fibrous Particles

For a particle to be considered respirable at least one dimension must be less than 5 μm.[1] However, the results of human lung content analysis indicate inhaled particles are generally considerably smaller. For the purpose of analysis, we will assume the non-fibrous particles to be perfect cubes no larger than 3 μm. The size of the cube will be varied within the respirable size range from 0.05 to 3 μm and the surface area and the mass of an individual particle assuming three different densities are calculated (Tables 1 and 2). As the cube size increases from 0.05 to 3 μm, the surface area per particle and mass per particle increases 3,600- and 217,000-fold, respectively (at least for a particle of density 2.33 g cm^{-3}; Table 1). The number of cubic particles in millions contained in a single milligram of dust with a density of 2.33 g cm^{-3} decreased from 3,450,000 to 16 over the same size interval (Table 3).

The significant range in particle size distributions that can occur even within the respirable size range can be associated with a corresponding range in

Table 1 *For non-fibrous particles with cubic morphology of various sizes and densities, the surface area/particle and mass/particle are calculated*

Cube size (μm)	0.05	0.10	0.20	0.40	0.80	1.00	2.00	3.00
Surface area/ particle (μm²)	0.015	0.06	0.24	0.96	3.84	6.00	24.00	54.00
Mass/particle [a]								
Density (g cm^{-3})								
2.33	0.00029	0.0023	0.019	0.15	1.2	2.3	18.6	62.9
2.65	0.00033	0.0027	0.021	0.17	1.4	2.7	21.2	71.6
4.13	0.00052	0.0041	0.033	0.26	2.1	4.1	33.0	111.5

[a] Mass given in picogram (10^{-12} g) assuming 2.33, 2.65 and 4.13 g cm^{-3}, the approximate density of cristobalite, quartz and rutile, respectively.

Table 2 Density of the non-fibrous particles, fibrous minerals and man-made vitreous fibre

	Non-fibrous particles	Density
(a)	Silica polymorphs	(g cm^{-3})
	α-Quartz	2.65
	α-Tridymite	2.28
	α-Cristobalite	2.33
	Titanium dioxide polymorphs	
	Anatase	3.79
	Rutile	4.13
(b)	**Naturally occurring mineral fibres**	
	Asbestos minerals	
	Actinolite	2.9–3.2
	Amosite	3.4–3.5
	Anthrophyllite	2.8–3.1
	Chrysotile	2.53–2.55
	Crocidolite	3.3–3.4
	Erionite	2.02
	Halloysite	2.11
	Palygorskite	2.4
	Wollastonite	2.80–3.10
(c)	**Man-made vitreous fibre**	
	Alumina–silicate–zirconia	2.2
	Alumina	3.2
	Boron nitride	1.9
	Glass/mineral wool	2.5
	Refractory ceramic fibre	2.7
	Silicon nitride	3.2
	Xonotlite	2.7
	Zirconia	4.8

biological activity. This can be experimentally determined using erythrocyte hemolysis, one of the many *in vitro* assays used to characterise inorganic particles. The assay quantifies the ability of pathogenic minerals to alter the permeability of erythrocyte membranes. A single quartz specimen was size fractionated on the basis of Stokes's settling velocity into three respirable size fractions (Table 4). Each Stokes's diameter approximates a sphere of equivalent diameter falling through a fluid.[10] Although a useful separation technique, the actual size distribution of each fraction was determined by means of direct visual measurement on transmission electron photomicrographs.[11] Two equivalent masses of respirable quartz can have different biological activities due to differences in particle size distribution. The ratio of the membranolytic potency (HC50) of the quartz specimens as a function of Stokes's diameter is roughly equivalent to the ratio of the change in surface area.[11] Commonly used exposure methods, those

Table 3 Number of non-fibrous particles/mg when their morphology is cubes of various sizes

Cube size	0.05	0.10	0.20	0.40	0.80	1.00	2.00	3.00
Density ($g\ cm^{-3}$)			Millions of particles/mg					
2.33	3 450 000	435 000	52 600	6670	2000	435	54	16
2.65	3 030 000	370 000	47 600	5880	1750	370	47	14
4.13	1 920 000	244 000	30 300	3850	476	244	30	9

based on determining either particle number or mass, would be insensitive to the differences in the physico-chemical properties of these specimens (see Langer and Nolan, 1994 for a more detailed discussion).[12]

The practical consequences of these important physical characteristics are the limitations they place on the early industrial hygiene data used to evaluate exposures. To determine simply the number of particles in a given volume of air, e.g. millions of particles per cubic foot of air, is a poor description of exposure. Identical dust values, obtained on aerosols with significant variation of particle size within the respirable range, do not adequately reflect true risk. Additionally, these measurements were carried out using optical microscopy methods, which had insufficient resolution to resolve fine respirable particles. The method was also non-specific and could not be used to determine the identity of the particle counted.

Table 4 The $HC50s^a$ particle size distribution of a fractionated quartz sample

Stokes diameter (µm)	No. of particles sized	Percentage of particles with largest dimension in the range (µm)				HC50* ($mg\ ml^{-1}$)
		<1	1.1–5.0	5.1–10.0	10.1–15.0	
1	586	93.2	6.8	0.0	0.0	0.23 ± 0.09
2	417	3.6	92.6	3.8	0.0	3.14 ± 0.13
5	598	0.0	62.9	3.6	0.8	5.85 ± 0.35

[a]The HC50s are the quartz (sigma silica) concentrations required to lyse ~ 50% of the human erythrocytes in a standard suspension of 1.8×10^8 cells ml^{-1} (see Nolan et al., 1987 for details).

A different approach to quantifying exposure commonly used involves determining the mass of the particles in a given volume of air. This methodology has been used, for example, to quantify exposure to quartz dust. The respirable fraction of quartz in an aerosol is collected on a membrane filter, which is analysed by continuous scan X-ray diffraction. The intensities of selected X-ray peaks associated with quartz are used to quantify the mass of the mineral quartz on the filter, making the assay more specific than simply counting the particles.

However, the quartz X-ray peaks may be overlapped when other mineral phases (for example, mica) are present, indicating a greater concentration of respirable quartz than is actually present. The actual particle size distribution within the respirable fraction is not determined. Knowing the number of particles or the respirable mass of particles in a given volume of air is insufficient information for exposure–response studies.

2.2 Fibrous Particles

A population of fibres can also be described on the basis of morphology. The regulatory criterion for an object to be referred to as a fibre is that its length is 5 µm or greater and that its long axis exceeds its width by a ratio equal to or greater than 3:1.[13] The ratio of length to width is referred to as the aspect ratio. A fibre population will contain a distribution of fibres of various lengths and diameters. Often only these two dimensions are used to describe three dimensional fibres due to assuming a particular cross-sectional configuration. The cross-sectional configuration of the fibres can be variable and difficult to determine, therefore they are not often described.[14] The extent of the variation in surface area and particle number per unit mass of the dust will be calculated for various fibre width, densities and cross-sectional configurations of the idealized fibre.

For the purpose of analysis, an idealized fibre population with a uniform length of 9 µm is considered. The cross-sectional configuration is assumed to be squares varied across the range of respirable sizes (Table 5). Both the surface area and the mass of an individual fibre increase as the square becomes larger and the aspect ratio decreases. The mass of the fibre is dependent on the density (mass per unit volume) (Table 2). Populations of different types of fibres of identical fibre number and dimensions can have different masses, which can be calculated using density data (Table 6).

The surface area of this fibre is completely defined by the sum of all its dimensions. A fibre has at least two distinctly different surface areas. One is the

Table 5 *For a 9 µm fibre (with a square cross-sectional configuration) the surface area/fibre, mass/fibre and aspect ratio as a function of a square size and density are calculated*

Square size (µm)	0.05	0.10	0.20	0.40	0.80	1.00	2.00	3.00
Surface area/fibre (µm)	2.50	4.01	8.08	16.32	33.28	42.00	88.00	138.00
Mass/fibre[a] Density (g cm^{-3})								
2.00	0.045	0.180	0.72	2.80	11.5	18.0	72.0	162
2.53	0.057	0.229	0.91	3.64	14.6	22.8	91.1	205
3.40	0.077	0.306	1.22	4.90	9.6	30.6	12.2	275
Aspect ratio	180	90	45	22.5	11.3	9.0	4.5	3

[a]Mass given in picograms (10^{-12} g), assuming 2.00, 2.53 and 3.40 g cm^{-3} the approximate density of erionite, chrysotile and crocidolite, respectively.

Table 6 *Number of fibres/mg when the fibres have a constant length of 9 μm and various square cross-sectional sizes*

Cross-sectional square size (μm)	0.05	0.1	0.2	0.4	0.8	1.0	2.0	3.0
Density (g cm^{-3})			Millions of fibres/mg					
2.00	21 700	5560	1390	347	87	56	14	6
2.25	19 300	4940	1230	309	77	49	12	5
2.50	17 400	4440	1110	278	69	44	11	5
2.75	15 800	4040	1010	253	63	40	10	4
3.00	14 500	3700	926	231	58	37	9	4
3.25	13 400	3420	855	214	53	34	9	4
3.50	12 400	3170	794	198	50	32	8	4

sum of all surface areas expressed by planes along the length of the fibre axis, and the other is the sum of both cross-sectional areas on the fibre ends. The percentage of the total surface area of the fibre that is cross-sectional varies as a function of the end-square size. For small diameter fibres (small end areas) with high aspect ratios the surface area at the ends contributes very little to the total fibre surface area. Conversely, as the diameter increases, this contribution becomes more significant (Table 7). Therefore, the surface area of a long, narrow diameter, fibre is very dependent on the planes defining the fibre sides (length) and not the planes defining the ends (Table 8).

Table 7 *The percentage of the total surface area that is on the ends of a 9 μm fibre with a varying square cross-sectional configuration*

Length of square (μm)	% of total surface area which is end
0.05	0.3
0.10	0.6
0.20	1.1
0.40	2.2
0.60	3.2
0.80	4.3
1.00	5.5
2.00	10.0
3.00	14.3

As with non-fibrous particles it is useful to understand surface area and fibre number in terms of populations of fibres. As with non-fibrous particles a convenient method for selecting a population of fibres is by mass. When comparing two populations of fibres of identical dimensions both the fibre number per unit mass and the surface area per unit mass will vary with the density of the fibre. Fibres of high density require fewer particles to constitute the same mass as

Table 8 *Surface area per gram of fibres having a constant length of 9 µm and square cross-sections of various sizes*

Cross-sectional square size (µm)	0.05	0.1	0.2	0.4	0.8	1.0	2.0	3.0
Density (g cm^{-3})	\multicolumn{8}{c}{Surface area (m^2 g^{-1})}							
2.00	39.2	20.1	10.1	5.1	2.6	1.1	1.1	0.8
2.25	34.9	17.9	9.0	6.5	2.3	1.9	1.0	0.7
2.50	31.4	16.1	8.1	5.9	2.1	1.7	0.9	0.6
2.75	28.5	14.6	7.4	3.7	1.9	1.5	0.8	0.6
3.00	26.1	13.4	6.7	3.4	1.7	1.4	0.7	0.5
3.25	24.1	12.4	6.2	3.1	1.6	1.3	0.7	0.5
3.50	22.4	11.5	5.8	2.9	1.5	1.2	0.6	0.4

compared with fibres of low density. Therefore, it is useful to calculate these values for a range of densities as well as width sizes. Fibres with the smallest width size and the lowest density will yield a dust with the greatest surface area and the largest number of fibres per unit mass. As the density increases, each individual fibre increases in mass and the number of particles required to constitute a given quantity of dust decreases. The number of fibres and surface area will decrease by almost 43% for a 9 µm fibre of any given width size as the density increases from 2.00 to 3.50 g cm^{-3} (see Tables 6 and 8).

First, consider the number of fibrous particles per mg, again assuming the fibres are 9 µm in length with square cross-sectional configurations of various sizes. The number of particles is reduced by 99.97% (over a 3500-fold reduction) as the diameter of the fibre increases from 0.05 to 3.0 µm (Table 6). The contribution of fine diameter fibre can be very great. For example, airborne chrysotile can have a median width of 0.05 µm and the historical industrial hygiene methods would not count these fibres.[14] The number of fibres, within each square population, changes with density, although the percentage change in the number of fibres as a function of square size is the same for each density. Secondly, the surface area (m^2 g^{-1}) is reduced by a factor of 50-fold as the square size increases from 0.05 to 3 µm for each given density (Table 8). Impurities within the fibre, for example chrysotile fibre bundles intergrown with magnetite or nemalite, can significantly alter its density.[15]

The assumption that the fibres have square cross-sectional configurations is useful to simplify the problem for a calculational analysis. For a 9 µm long fibre with a cross-sectional configuration of a circle (diameter = 0.05 µm), square (0.05 by 0.05 µm) and rectangle (0.05 by 0.025 µm), the total surface area is 1.41 µm^2, 1.81 µm^2 and 1.35 µm^2, respectively. The error in the surface area calculation may be large if an incorrect cross-sectional configuration is assumed. Therefore, to assume a fibre cross-sectional configuration would require justification.

Real fibres differ from ideal fibres in many ways. Asbestos occurs with many individual fibrils closely packed to form fibre bundles. A property described as polyfilamentous, these fibres often display splayed ends and contoured surfaces.

Manipulation to disaggregate these fibre bundles will significantly alter the size distribution. Therefore, the size distribution of the asbestos minerals is strongly dependent on the extent of mechanical manipulation to which the fibres have been subjected.[16]

3 Carcinogenicity of Fibrous Particles

The hypothesis that morphology is the single most important determinant of the carcinogenic activity of fibrous particles is often referred to as the Stanton Hypothesis. Stanton and his co-workers concluded that the probability of pleural sarcoma induction (a tumour considered equivalent to a pleural mesothelioma) was highest in specimens containing the greatest number of fibres of 0.25 µm or less in diameter and more than 8 µm in length.[17] These narrow, fibrous particles of course have the greatest surface area and particle number per unit mass of dust.

The experimental data, which form the basis of the hypothesis, were obtained in a model where tumors were induced after the surgical implantation of a thin pledget of coarse fibrous glass (30mm x 30mm x 2 mm, weighing about 45 mg and coated on one side with the fibre to be evaluated) into the pleura of the animal. Generally, a 40 mg dose of the fibrous particles to be evaluated was allowed to harden, from a gelatin suspension, onto the pledget surface before being placed directly against the visceral pleura of the left lung. The purpose of the pledget was to increase the effective fibre dose by limiting the exposure to a 900 mm^2 area of the lung surface. The approach is very artificial and only one other research group has ever reported results using this model (see Nolan and Langer, 1993 for a further discussion).[8]

The doses used in the Stanton studies were very high and it is generally thought an intraperitoneal injection of 10 mg is an adequate dose.[18] Using the Stanton model, a significant number of tumours was generated with a wide array of fibrous particles. Stanton *et al.*, 1981 interpreted their data as indicating that the most significant requirement for a population of fibrous particles to produce a tumour was morphology. The results of these and other experimental animal studies (where the route of administration was intraperitoneal injection) led many investigators to conclude that the surface properties of the fibrous particles were unimportant in tumour induction. An additional variable commonly thought to be of importance in determining a fibre's biological activity is biodurability.[19]

3.1 Importance of Dose and Route of Administration in Biological Outcome

Fibre morphology, as the primary determinant of carcinogenic activity, has been clearly associated with tumour production when the route of administration was surgical implantation or intrapleural/intraperitoneal injection and the dose is high.[17,20,21] The intraperitoneal route of administration can produce an incidence of mesothelioma approaching 100% in animals. No human exposure to fibrous particles has ever lead to such a high mesothelioma incidence. The experimental methodology used in these studies led Stanton and Wrench[22] to conclude in 1972,

'Direct application of our results to the problems in man would be unwise because the method of application and the high doses used are remote from the usual exposure in man to fibre . . . '. Human exposure by the routes of administration used in these studies occur extremely rarely (for reviews see references 23–25).

The studies using these artificial routes and generally high doses indicate that many different fibre types which contain a proportion of fibres of a dimension > 8 μm in length and ≤ 0.25 μm in diameter are carcinogenic.[6] The results *do not* indicate that *all* durable fibres of these dimensions are equally *potent* carcinogens. Nor do they indicate that these properties determine the carcinogenic activity of *all* fibre types using inhalation as the route of administration.

3.2 Limited Correlation of Crocidolite Asbestos Carcinogenesis with the Stanton Hypothesis

If one selects from the wide array of fibrous samples evaluated by Stanton *et al*., 1981,[17] the thirteen experiments using crocidolite for evaluating the relationship between morphological and carcinogenic activity, the results are of particular interest. Crocidolite is a fibre type known to cause mesothelioma in man.[26,27] By comparing the different crocidolite specimens, the elemental composition and structure remain similar, while the morphology is varied. The thirteen specimens used were obtained from three sources and manipulated by various mechanical and physical methods. The percentage tumour probability varied from 94 ± 6% to 0% for the specimens (Table 9). Crocidolite 1, 2 and 3 had virtually the same tumour probability and an 8-fold range in the number of Stanton fibres. Crocidolite 5 produced 78 ± 10% tumour probability with 1950 Stanton fibres μg^{-1}, while crocidolite 12 with 5370 Stanton fibres μg^{-1} produced only 10 ± 7%. A tumour probability of less than 30% was reported not to differ significantly from background.[17] Crocidolite 8 with no Stanton fibres produced a 53 ± 13% tumour probability. For a single fibre type – crocidolite – the number of Stanton fibres was a poor indicator of tumour probability. Other experiments indicated a range of tumour probability between fibre types with a similar number of Stanton fibres. For example, two tremolite asbestos samples produced a 100% tumour probability, while a talc sample having a greater number of Stanton fibres produced no tumours (see Nolan and Langer, 1994 for a further discussion).[12]

The correlation between the number of Stanton fibres and the tumour probability is variable. Mesotheliomas have been induced in rats at a 2.5-fold higher incidence than the fibrous starting material using a crocidolite milled until approximately 99% of the fibres were 'isometric'.[28] Although equant fibres, produced by milling UICC crocidolite, have been correlated to plating efficiency of V79-4 cells *in vitro*, the *in vivo* activity of the different milled specimens remained very similar to the unmilled starting material.[29,30]

Table 9 *Selected samples from Stanton et al. 1981 grouped by mineral name*

Mineral	Tumour incident	Tumour probability % ± SD	No. of fibres µg^{-1} ≤ 0.25mm × >8mm	Total no. of fibres µg^{-1}	Percentage of Stanton fibres in total no. of fibres
Crocidolite					
1	18/27	94 ± 6	162,000	5.88 × 10^6	2.8
2	17/24	93 ± 7	19,900	1.68 × 10^6	1.2
3	15/23	93 ± 7	102,000	6.57 × 10^6	1.6
4	15/24	86 ± 9	134,000	3.06 × 10^5	43.8
5	14/29	78 ± 11	1,950	3.54 × 10^5	0.6
6	9/27	63 ± 14	339,800	9.69 × 10^5	4.1
7	11/26	56 ± 12	447	8.13 × 10^9	<<0.1
8	8/25	53 ± 13	0	1.09 × 10^6	-
9	8/27	33 ± 10	17,700	8.17 × 10^5	2.1
10	6/29	37 ± 14	1,230	6.21 × 10^5	0.2
11	4/29	19 ± 9	0	6.84 × 10^3	-
12	2/27	10 ± 0	5,370	9.28 × 10^4	5.8
13	0/29	0	0	2.74 × 10^6	-
Tremolite					
1	22/28	100	1,380	1.41 × 10^5	0.9
2	21/28	100	692	6.86 × 10^4	1.0
Talc					
6	0/30	0	1,995	3.76 × 10^5	0.5

3.3 Effect of Poly(2-vinylpyridine *N*-oxide) (2-PVPNO) on the Induction of Mesothelioma

The hydrogen bonding polymer poly(2-vinylpyridine-*N*-oxide) (2-PVPNO), is known to bind to the surface of asbestiform amphiboles.[31] Three hundred micrograms of asbestiform actinolite injected intraperitoneally into 29 rats produced 79.3% tumours with the first tumour occurring at 45 weeks. Injection of 2-PVPNO separately from the mineral reduced the tumour incidence slightly, while the weeks to first tumour and average latency remained fairly constant. Injection of the same dose of asbestiform actinolite with 2-PVPNO bound to the surface and additional 2-PVPNO injected separately reduced the tumour incidence to 48.3%, while the time to first tumour increased by 52 weeks.[20] Further experiments using the same model at a slightly lower dose, 250 µg, produced a 59% reduction in tumour incidence by binding 2-PVPNO to the mineral's surface without repeated injections of the polymer. The time to first death and average latency were greater for the rats injected with 2-PVPNO modified actinolite.[32] The bonding of a high molecular weight polymer to the surface of the mineral reduced the carcinogenic potency of the fibre while the morphology remained unchanged.

3.4 Fibrous Particles Producing Mesothelioma Incidence Beyond that Expected on the Basis of Morphology

The carcinogenic activity of the fibrous zeolite mineral erionite has been evaluated by inhalation in rats.[33] The positive control was the amphibole asbestos mineral crocidolite, which was the first asbestos fibre associated with human mesothelioma.[26] Epidemiological studies have shown that occupational exposure to crocidolite can lead to a mesothelioma mortality in excess of 10% (for a review the Langer and Nolan, 1989[27]). The dust clouds for both the erionite and crocidolite exposures containing 10mg/ m^{-3} of dust, which produce fibre levels of 354 fibre cm^{-3} and 1630 fibre cm^{-3} ≥ 5 µm in length, respectively. The erionite cloud contained fewer fibres and more equant particles. However, the erionite mineral has a lower density, and an equal mass of fibrous aerosol should contain more fibre than crocidolite. Of the 28 rats exposed to erionite by inhalation, 27 developed mesothelioma, while among the 28 rats with a ~4.5-fold higher exposure to crocidolite *none* developed mesothelioma. A non-fibrous zeolite specimen of identical elemental composition to erionite produced one mesothelioma among the 28 rats exposed.

Wagner *et al.*, 1985 reported, 'A similar range of fibre sizes was contained in the crocidolite and erionite cloud'.[33] Even with a smaller number of fibres in the aerosol, erionite produced mesothelioma in 98% of the rats. A high incidence of mesothelioma in rats from inhalation of erionite under similar conditions has also been reported by Johnson and Wagner, 1989.[34] The histopathology of the mesotheliomas induced by erionite resemble those produced by injection of asbestos.[35] The significantly higher mesothelioma incidence in the experimental animals exposed to erionite by inhalation cannot be explained on the basis of fibre dimensionality. Of the 648 rats exposed to the different commercial asbestos varieties using inhalation as the route of administration, Wagner *et al.*, 1974 reported only 11 mesotheliomas.[30] The erionite group had a 96% mesothelioma incidence, while the asbestos group had a 1.7% incidence.

Additional studies using intrapleural inoculation in rats have shown 'the oncogenic potential of erionite relative to crocidolite asbestos to be much greater (230 to 1)'. The relative increase in oncogenicity of the erionite fibres was not due to a greater proportion of the more carcinogenic long, thin fibres[17] being present in the specimen (the data for its fibre dimension relative to UICC crocidolite showed that there were *less* by weight present in the erionite than crocidolite).[36] Similar findings have also been reported after intraperitoneal injection of erionite in rats by Davis *et al.*, 1991.[18] Therefore, the carcinogenic potency of fibrous erionite cannot be explained on the basis of morphology.

The recognition of this extremely dangerous fibre was associated first with 40% incidence of mesothelioma after human exposure.[37,38] Using inhalation as the route of administration, the erionite fibres were clearly more dangerous than crocidolite. The early injections studies did not predict the carcinogenic potency of erionite. After the results of human exposure, and the inhalation study in experimental animals was completed, injection studies were designed to demonstrate the increased risk associated with erionite exposure.[18,36]

Recently, the carcinogenic activity of another non-asbestos fibre kaolin-based refractory ceramic fibres (RCF), a man-made vitreous fibre, has been evaluated by inhalation in hamsters.[39] The positive control was the serpentine asbestos mineral chrysotile, which, under similar exposure conditions in previous studies, the sample had produced 25–40% bronchoalveolar tumours in rats. The hamsters were exposed to a dust cloud of 30 mg m^{-3} containing 220 fibres cm^{-3} of refractory ceramic fibres.

The dust was lofted from a sample specifically prepared to generate an aerosol with a size distribution similar to that experienced by occupationally exposed workers. Therefore, the aerosol contained a significant proportion of fibres with ~25 μm lengths and ~1 μm diameter (for recent exposure data, see Rice et al., 1994[40]). The chrysotile aerosol contained 10 mg m^{-3} and ~102 000 fibres cm^{-3}. Here the length was 2.2 μm and the diameter 0.09 μm. Among the RCF exposed hamsters, no pulmonary tumours occurred, although 42 of 102 (41.2%) animals developed mesothelioma, while an additional 33 hamsters developed mesothelial hyperplasia.

It is noteworthy that no pulmonary tumours developed in the erionite or RCF-exposed animals, although a high incidence of mesothelioma occurred in both groups (see Glass et al., 1995, for a recent review of the earlier animal studies).[41] The fibrosis in the chrysotile exposed hamsters was more severe, without the occurrence of pulmonary or pleural tumours. Radiographic evidence indicates increased pleural changes and pleural plaques in workers with 20 years or more of exposure while manufacturing refractory ceramic fibres and products.[42]

3.5 Epidemiological Studies of Populations Exposed to Fibrous Particles

Even when epidemiological methods have established an association between exposure to an agent in the environment and an adverse health effect, such as an increased incidence of cancer or lung fibrosis, limitations in our knowledge still exist. Among these are the unambiguous identification of the aetiological agent(s) responsible for the disease and determining the extent of exposure associated with the increased incidence of a particular adverse health effect. The epidemiological studies to determine the lung cancer risk associated with chrysotile asbestos exposure and mesothelioma risk associated with exposure to the commercial varieties of amphibole asbestos provide examples of the practical consequences of these limitations.

Owing to chrysotile being a naturally occurring fibre, occupational exposure occurs both during mining and milling of the ore and asbestos product manufacture. The unique commercial properties of chrysotile led to its use in virtually thousands of products, although groups of workers large enough for epidemiological studies, predominantly exposed to chrysotile asbestos only, occurred principally in three manufacturing industries: cement, textile[43-45] and friction products.[46,47] Owing to the complex exposures in the cement industry, our discussion will be limited to the last two manufacturing areas and mining/milling of chrysotile ore.

A review of the epidemiological studies concerning chrysotile indicates that exposure to the dust formed during mining and milling or the manufacture of asbestos products can cause disease.[48] The most important results of the epidemiological studies of these three chrysotile-exposed groups are that the lung cancer risk varies considerably and between the groups and that the exposure

Table 10 *Historical exposures to chrysotile in textile manufacturing, mining/milling and friction products given in millions of particles per cubic foot and conversion to fibres $cm^{-3} \geq 5$ μm*

	No. of air samples	Time period	Exposures (equivalent to 1 mppcf)	
Textile	5952	1930–75	All operations except preparation	3.0 fibres cm^{-3}
			Preparation	8.0 fibres cm^{-3}
Mining and milling	4500	1949–66	15% Fibre in the dust	5.3 fibres cm^{-3a}
			5% Fibre in the dust	1.8 fibres cm^{-3}

[a] In the old mills, this fibre level was associated with 5 mppcf, while in the better mills 2 mppcf.

index alone is not a predictor of risk. Within each individual study, the lung cancer risk does increase with the index of exposure. All of these studies were carried out retrospectively and exposures occurred decades earlier. The limited industrial hygiene data available were most often collected with midget impingers and were expressed in millions of particles per cubic foot of air (mppcf). The limitations of these measurements discussed earlier are apparent and they are extremely limited as an index of lung cancer risk. The range of conversion factors from an exposure level of 1 mppcf was 1.8 to 8 fibres $cm^{-3} \geq 5$ μm (Table 10). An exposure level of 1 mppcf is equivalent to ~35 particles cm^{-3}, therefore only 5–23% of the particles present in the air are counted as chrysotile asbestos. The increase in relative risk of lung cancer per fibre cm^{-3} × year was ~25-fold greater for textile workers than among miners/millers when the same index of exposure was applied (Table 11).

The historical index of exposure (mppcf), even when converted to fibre $cm^{-3} \geq 5$ μm, did not explain the variation in lung cancer risk between the Québec miners/millers and the Charleston, SC chrysotile textile workers. Another index of exposure involves the determination of the actual fibre concentration found in lung tissues obtained at autopsy or during a surgical procedure. The concentration of chrysotile asbestos and other types of fibres present in the lung tissue of these workers occupationally exposed to chrysotile during mining/milling and textile manufacture was quantified using analytical transmission electron microscopy. The geometric mean concentrations of chrysotile and fibrous tremolite were 9.7-fold and 23-fold higher, respectively, in Québec, where the lung cancer risk was lower than Charleston.[49] The pattern of disease associated with chrysotile asbestos exposure is difficult to interpret, and the results of the experimental animal studies add further complexity to the situation.[12,16]

The amphibole asbestos minerals vary in their ability to produce human mesothelioma. Mesothelioma was first associated with asbestos exposure in the Republic of South Africa.[26,50] Reports from there indicate that in mining areas that were geographically separated, although the mining activity of both was very

Table 11 Exposure to chrysotile in textile manufacture, mining/milling and friction product manufacture, and risk of lung cancer

	No. of lung cancers obs/exp	SMR	Increase in relative risk of lung cancer per fibres $cm^{-3} \times$ years	
Textile	35/11	315	0.01	Dement et al. 1982b[44] McDonald et al., 1983[45]
Mining/milling	321/228	140	0.0004	McDonald et al., 1980[63] McDonald et al. 1993[64]
Friction	240(11)[a]/221.4 Men (10 yrs) 14(2)[a]/2.1 Women (10 yrs)	108 66	Effectively zero	Berry and Newhouse, 1983[46] Newhouse and Sullivan, 1989[47]
	97(9)[a]/82.3 Men (20 yrs) 5(1)[a]/4.5 Women (20 yrs)	118 110		

[a]( ) = Number of pleural mesotheliomas.

similar, very different incidences of mesothelioma have occurred. Crocidolite from Cape Province has been responsible for most of the mesotheliomas in South Africa (some 1800), while crocidolite and amosite from the Transvaal region has produced as few as five mesotheliomas, up to 1979.[7]

Factors such as medical services and extent of mining have been discounted, and the most likely explanation for this disparate pattern of disease appears to be differences in fibre diameter.[51] In the northwest Cape Province the average crocidolite diameter is 0.073 µm, while in the Transvaal, crocidolite and amosite show average diameters of 0.212 and 0.243 µm, respectively. Fibres from both the Cape Province and the Transvaal have induced mesotheliomas in rats by intrapleural inoculation. The Cape fibre was more effective than the Transvaal in producing mesotheliomas, 59% compared with 40% in one experiment, and 68% compared with 31% in another experiment.[52] Using the intrapleural model, the crocidolite/amosite from the Transvaal was less active.

A similar argument can be made for the lack of mesotheliomas among the Finnish anthophyllite workers.[53-55] The mine opened in 1918 and produced 120,000 tons of anthophyllite prior to closing in 1975. The mineral was widely used, particularly in construction material. Twenty years after the closing of the mine, no case of anthophyllite-only mesothelioma has been reported.[56] The Paakkila anthophyllite mine in Finland has only about 1% of the fibres with diameters of less than 0.1 µm.[7,57]

Other agents besides asbestos have been known to induce mesothelioma.[58-60] Recently, mesotheliomas have been induced in rats by intrapleural injection of ferric saccharate.[61] Apparently, no fibres at all were required to produce these tumours. The role of iron as a catalyst of free radical reactions was suggested by Okada et al., 1989[61] as an explanation of the carcinogenic effect. The iron present in the crocidolite fibres, both Fe^{2+} and Fe^{3+}, could be an important factor.

Kennedy et al., 1989, have reported that amosite, crocidolite and chrysotile function as catalysts to generate hydroxyl radicals from hydrogen peroxide and a reducing agent.[62]

4 Conclusions

Often, the clinical features and pathology of a particular disease provide little or no evidence of the agent responsible for its origin. If the aetiological agent is a fine inorganic particle present as an aerosol in the environment, modern instrumental methods can be used to determine the particle type(s) and concentration(s) in the air. To understand the relationship between exposure to these aerosols and disease, an understanding of the physico-chemical properties of the particles is required.

Historical exposure indices based on particle number (or mass of respirable dust) in a given volume of air have been of limited value in estimating the risk of disease associated with particular exposures.[63,64] The limitations are particularly apparent when the index is used to compare the exposure–response relationship between two cohorts with an excess of particular disease from exposure to the 'same' agent. One explanation for this is that a range of different particle size distributions can occur within the respirable fraction, allowing these materials to form a variety of aerosols differing in biological activity. The early industrial hygiene methods were simply insensitive to these differences in the aerosol.

In addition to the variation in the size distribution of the particles in the aerosol, the surface properties of the particles are variable. The results of the two studies of non-asbestos fibrous particles in experimental animals provide interesting examples for the importance of surface properties in fibre carcinogenesis. Owing to early experimental animal studies using artificial routes of administration and high doses, producing tumors with almost any type of inorganic fibre, the surface properties were not considered important.

Under the experimental conditions utilized in the injection studies, these properties may not be important. However, when the route of administration is inhalation (or injection at low doses), fibres of similar morphology can have strikingly different carcinogenic potency. Two aluminum silicate fibres erionite and refractory ceramic fibres (RCF), have produced a higher incidence of mesothelioma in experimental animals than any asbestos mineral previously tested.[33,39] Two asbestos minerals crocidolite and chrysotile, were used as positive controls in these studies and produced no mesotheliomas and few, if any, lung cancers. The concentration of asbestos fibres in each inhalation study was higher in the case of chrysotile, the exposure was 500-fold greater than the RFC.

Improving our understanding of the physico-chemical character of inorganic particles is key to identifying aetiological agents, determining meaningful exposure–response relationships for risk estimates and to design experimental animal studies, particularly when inhalation is the route of administration. This knowledge is crucial to the elucidation of mechanisms of action of inorganics.

References

1. Lippmann, M. (1994) Nature of Exposure to Chrysotile. *Ann. Occup. Hyg.,* **38** (No. 4), 459–467.
2. Liddell, D. and Miller, K. (1991) 'Mineral Fibres and Health', CRC Press, Boca Raton, FL, pp. 1–381.
3. Ross, M., Nolan, R.P., Langer, A.M. and Cooper, W.C. (1993) Health Effects of Mineral Dusts other than Asbestos. In, 'Health Effects of Mineral Dusts', Guthrie, G.D. and Mossman, B.T., eds., Reviews in Mineralogy Vol. 28. Mineralogical Society of America, Washington, DC, pp. 361–407.
4. Harber, P., Schenker, M.B. and Balmes, J.R., eds. (1996) 'Occupational and Environmental Respiratory Disease', Mosby, St. Louis, pp. 1–1038.
5. Brown, R.C., Hoskins, J.A. and Johnson, N.F. (1991) 'Mechanisms in Fibre Carcinogenesis', NATO ASI Series, Plenum, New York, pp. 1–589.
6. Johnson, N.F. (1993) The limitations of inhalation, intratracheal and intracoelomic routes of administration for identifying hazardous fibrous material. In, 'Fibre Toxicology', Warheit, D.B., eds. Academic Press, San Diego, pp. 43–72.
7. Harington, J.S. (1981) Fibre Carcinogenesis: Epidemiological Observations and the Stanton Hypothesis, *J. Natl. Cancer Inst.,* **67**, 977–989.
8. Nolan, R.P. and Langer, A.M. (1993) Limitations of the Stanton Hypothesis. In, 'Health Effects of Mineral Dusts', Guthrie, G.D., Mossman, B.T., eds., Reviews in Mineralogy Vol. 28. Mineralogical Society of America, Washington, DC, pp. 309–326.
9. Nolan, R.P. and Langer, A.M. (1996) Mineralogy. In, 'Occupational and Environmental Respirable Disease', Mosby, St. Louis, pp. 173–186.
10. Nolan, R.P. and Langer, A.M. (1983) Quartz and hemolysis: Physicochemical factors controlling membrane activity. In, 'Health Issues Related to Metal and Nonmetal Mining', Wagner, W.L., Rom, W.N., Mererchant, J.S., eds., Butterworth Publishers, Boston, pp. 63–81.
11. Nolan, R.P., Langer, A.M., Eskenazi, R.A. and Herson, G.B. (1987) Membranolytic Activities of Quartz Standards, *Toxic in vitro,* **1**, 239–245.
12. Nolan, R.P., Langer, A.M. and Addison, J. (1994) Lung content analysis of cases occupationally exposed to chrysotile asbestos, *Environ. Health Persp.,* **102** (Suppl. 5), 245–250.
13. Langer, A.M., Nolan, R.P. and Addison, J. (1991) Distinguishing between amphibole asbestos fibres and elongate cleavage fragments of their non-asbestos analogues. In, 'Mechanisms of Fibre Carcinogenesis', Brown, R.C., Hoskins, J.A., Johnson, N.F., eds., NATO ASI Series, Plenum, New York, pp. 253–265.
14. Veblen, D.R. and Wylie, A.G. (1993) Mineralogy and amphiboles and 1:1 layer silicates. In, 'Health Effects of Mineral Dusts', Guthrie, G.D., Mossman, B.T., eds., Reviews in Mineralogy Vol. 28 Mineralogical Society of America, Washington, DC, pp. 61–137.
15. Liebling, R. and Langer A.M. (1972) Optical Properties of Fibrous Brucite from Asbestos Québec, *Am. Mineral.,* **57**, 857–864.
16. Langer, A.M. and Nolan, R.P. (1994) Chrysotile: Its occurence and properties as variables controlling biological effects, *Ann. Occup. Hyg.,* **38** (No. 4), 427–451.
17. Stanton, M.F., Layard, M., Tegeris, A., Miller, E., May, M., Morgan, E. and Smith, A. (1981) Relationship of Particle Dimension to Carcinogenicity in Amphibole Asbestoses and Other Fibrous Minerals, *J. Natl. Cancer Inst.,* **67**, 965–975.
18. Davis, J.M.G., Addison, J., McIntosh, C., Miller, B.G. and Niven, K. (1991) Variation in the carcinogenicity of tremolite dust samples of differing morphology. In, 'The Proceedings of the Third Wave of Asbestos Disease: Exposure to Asbestos in Place',

Public Health Control, Landrigan, P.J., Kazemi, H., eds., *Annal. NYAS*, **643**, 473–490.
19. Bignon, J., Saracci, R. and Touray, J.-C., eds. (1994) Biopersistence of Respirable Synthetic Fibres and Minerals, *Environ. Health Perspect.*, **102** (Suppl. 5), 1–283.
20. Pott, F., Ziem, U., Reiffer, F.J., Huth, F., Ernst, H. and Mohr, U. (1987) Carcinogenicity studies on fibres, metal compounds and some other dusts in rats. *Exp. Pathol. (JENA)*, **32**, 129–152.
21. Coffin, D.L., Palekar, L.D., Cook, P.M. and Creason, J.P. (1989) Comparison of Mesothelioma Induction in Rats by Asbestos and Non-Asbestos Mineral Fibres. Possible Correlation with Human Exposure Data. In, 'Biological Interaction of Inhaled Mineral Fibres and Cigarette Smoke', Wehner, A.P., Felton, D.L., eds., Battelle Press, Columbus, OH, pp. 347–354.
22. Stanton, M.F. and Wrench, C. (1972) Mechanism of Mesothelioma Induction with Asbestos and Fibrous Glass, *J. Natl. Cancer Inst.*, **48**, 797–821.
23. Pott, F., Roller, M., Rippe, R.M., Grumann, P.G. and Bellman, B. (1991) Tumors by the intraperitoneal and intrapleural routes and their significance for the classification of mineral fibres. In, 'Mechanisms in Fibre Carcinogenesis', Brown, R.C., Hoskins, J.A., Johnson, N.F., eds., NATO ASI Series, Plenum, New York, pp. 547–565.
24. Rossiter, C.E. (1991) Fibre Carcinogensis: Intra-caritary studies cannot access risk to man. In, 'Mechanisms in Fibre Carcinogenesis', Brown, R.C., Hoskins, J.A., Johnson, N.F., eds., NATO ASI Series, Plenum, New York, pp. 567–578.
25. McClellan, R.O., Miller, F.J., Hesterberg, T.W. *et al.* (1992) Approaches to Evaluating the Toxicity of Carcinogencity of Man-Made Fibres: Summary of Workshop held 11–13 November, 1991, Durham, North Carolina, *Regul. Toxicol. Pharmacol.*, **16**, 321–364.
26. Wagner, J.C., Sleggs, C.A. and Marchand, P. (1960) Diffuse pleural mesothelioma and asbestos exposure in the northwest Cape Province, *Br. J. Ind. Med.*, **17**, 260–271.
27. Langer, A.M. and Nolan, R.P. (1989) Fibre Type and Mesothelioama Risk. In, 'Symposium on Health Aspects of Exposure to Asbestos in Buildings', Harvard University, Energy and Environmental Policy Center, pp. 91–140.
28. Kolev, K. (1982) Experimentally Induced Mesothelioma in White Rats in Response to Intraperitoneal Administration of Amorphous Crocidolite Asbestos: Preliminary Report, *Environ. Res.*, **29**, 123–133.
29. Brown, R.C., Chamberlain, M., Griffiths, D.M. and Timbrell, V. (1978) The Effect of Fibre Size on the *in vitro* Biological Activity of Three Types of Amphibole Asbestos, *Int. J. Cancer*, **22**, 721–727.
30. Wagner, J.C., Berry, G., Skidmore, J.W. and Timbrell, V. (1974) The effect of the inhalation of asbestos in rats, *Br. J. Cancer*, **29**, 252–269.
31. Schnitzer, R.J. and Pundsack, F.L. (1970) Asbestos Hemolysis, *Environ. Res.*, **3**, 1013.
32. Pott, F., Roller, M., Ziem, U., Reiffer, F.J., Bellmann, B., Rosenbruch, M. and Huth, F. (1989) Carcinogenicity studies on natural and man-made fibres with intraperitoneal test in rats. In, 'Non-Occupational Exposure to Mineral Fibres', Bignon, J., Peto, J., Saracci, R., eds., IARC Scientific Publication, **90**, IARC, Lyon, pp. 173–179.
33. Wagner, J.C., Skidmore, J.W., Hill, R.J. and Griffiths, D.M. (1985) Erionite Exposure and Mesothelioma in Rats, *Br. J. Cancer*, **51**, 727–730.
34. Johnson, N.F. and Wagner, J.C. (1989) Effect of Erionite Inhalation on the Lungs of Rats. In, 'Biological Interaction of Inhaled Mineral Fibres and Cigarette Smoker', Wehner, A.P., Felton, D.L., eds., Battelle Press, Columbus, OH, pp. 325–345.
35. Johnson, N.F., Edward, R.E., Munday, D.E., Rowe, N. and Wagner, J.C. (1984) Pleuripotential Nature of Mesotheliomata Induced by Inhalation of Erionite in Rats, *Br. J. Exp. Pathol.*, **65**, 377–388.

36. Hill, R.J., Edwards, R.E. and Carthew, P. (1990) Early changes in the pleural mesothelium following intrapleural inoculation of the mineral fibre erionite and subsequent development of mesothelioma, *J. Exp. Pathol.*, **71**, 105–118.
37. Baris, Y.I., Simonato, L., Artvinli, M., et al. (1987) Epidemiological and Environmental Evidence of Health Effects of Exposure to Erionite Fibres: A Four Year Study of the Cappadocian Region of Turkey, *Int. J. Cancer*, **39**, 10–17.
38. Baris, Y.I. (1991) Fibrous zeolite (erionite)-related disease in Turkey, *Am. J. Ind. Med.*, **19**, 303–316.
39. McConnell, E.E., Mast, R.W., Hesterberg, T.W., Chevalier, J., Kotin, P., Bernstein, D.M., Thevenaz, P., Glass, L.R. and Anderson, R. (1995) Chronic Inhalation Toxocity of a Kaolin-based Refractory Ceramic Fibre in Syrian Golden Hamsters. *Inhal. Toxicol.*, **7**, 503–532.
40. Rice, C., Lockey, J., Lemaster, S.G., Dimas, J. and Gartside, P. (1994) Assessment of Current Fibre and Silica Exposure in the U.S. Refractory Ceramic Fibre Manufacturing Industry, *Ann. Occup. Hyg.*, **38** (Suppl. 1), 739–744.
41. Glass, L.R., Brown, R.C. and Hoskins, J.A. (1995) Health effects of refractory ceramic fibres: scientific issues and policy considerations, *Occup. Environ. Med.*, **52**, 433–440.
42. Lemaster, S.G., Lockey, J., Rice, C., McKay, R., Hansen, K., Lu, J., Levin, L. and Gartside, P. (1994). Radiographic Changes Among Workers Manufacturing Refractory Ceramic Fibres and Products, *Ann. Occup. Hyg.*, **38** (Suppl. 1), 745–751.
43. Dement, J.M., Harris, R.L., Symons, M.J. and Shy, C.M. (1982a) Exposures and Mortality among Chrysotile Asbestos Workers, Part I: Exposure Estimates, *Am. J. Ind. Med.*, **4**, 399–419.
44. Dement, J.M., Harris, R.L., Symons, M.J. and Shy, C.M. (1982b). Exposures and Mortality among Chrysotile Asbestos Workers, Part II: Mortality, *Am. J. Ind. Med.*, **4**, 421–433.
45. McDonald, A.D., Fry, J.S., Woolley, A.J. and McDonald, J.C. (1983) Dust exposure in an American chrysotile textile plant, *Br. J. Ind. Med.*, **40**, 361–366.
46. Berry, G. and Newhouse, M.L. (1983) Mortality of workers manufacturing friction materials using asbestos, *Br. J. Ind. Med.*, **40**, 1–7.
47. Newhouse, M.L. and Sullivan, K.R. (1989) A mortality study of workers manufacturing friction materials, *Br. J. Ind. Med.*, **46**, 176–179.
48. Harington, J.S. (1991) The carcinogencity of chrysotile asbestos. In, 'The Third Wave of Asbestos Disease: Exposure to Asbestos in Place', Landrigan, P.J., Kazemi, H., eds., *Annal. NYAS*, **643**, 465–472.
49. Sébastien, P., McDonald, J.C., McDonald, A.D., Case, B. and Harley (1989) Respiratory cancer in chrysotile textile and mining industries: Exposure inferences from lung analysis, *Br. J. Ind. Med.*, **46**, 180–187.
50. Wagner, J.C. (1983) Mesothelioma and Mineral Fibres. In, 'Accomplishments in Cancer Research', Fortner, J.G., Rhoads, J.E., eds., J.B. Lippincott Company, Philadelphia, pp. 60–72.
51. Timbrell, V., Griffiths, D.M. and Pooley, F.D. (1971) Possible biological importance of fibre diameters of South African amphiboles, *Nature (London)*, **232**, 55–56.
52. Wagner, J.C. and Berry, G. (1969) Mesotheliomas in Rats following Inoculation with Asbestos, *Br. J. Cancer*, **23**, 567–581.
53. Meurman, L.O., Kiviluoto, R. and Hakama, M. (1974) Mortality and Morbidity among the Working Population of Anthophyllite Asbestos Miners in Finland, *Br. J. Ind. Med.*, **31**, 105–112.
54. Nurminen, M. (1975) The Epidemiological Relationship between Pleural Mesothelioma and Asbestos Exposure, *Scand. J. Work Environ. Health*, **1**, 128–137.

55. Huuskonen, M.S., Ahlman, K., Mattson, T. and Tossivainen, A. (1980) Asbestos Disease in Finland, *J. Occup. Med.*, **22**, 751–754.
56. Tuomi, T., Segerberg-Kontinnen, M., Tammilehto, L., Tossavainen, A. and Vanhala, E. (1989) Miner fibre concentration in lung tissue of mesothelioma patients in Finland, *Am. J. Ind. Med.*, **16**, 247–254.
57. Timbrell, V. (1989) Review of the significance of fibre size in fibre-related lung disease: A centrifuge cell for preparing accurate microscope-evaluation specimens from slurries used in sonification studies, *Ann. Occup. Hyg.*, **53**, 483–505.
58. Peterson, J.T., Greenberg, S.D. and Buffler, P.A. (1984) Non-asbestos-related malignant mesothelioma – a review, *Cancer*, **54**, 951–960.
59. Carbone, M., Pass, H.I., Rizzo, P., Marinetti, M.R., DiMuzio, M., Mew, D.J.Y., Levine, A.S. and Procopio, A. (1994) Simian Virus 40-like DNA Sequences in Human Pleural Mesothelioma, *Oncogene*, **9**, 1781–1790.
60. Carbone, M., Rizzo, P. and Pass, H.I. Association of Simian Virus 40 with Rodent and Human Mesotheliomas. In, 'DNA Tumor Viruses: Oncogene Mechanisms', Borbent-Brodono and Friedman, eds., Plenum Press (in press).
61. Okada, S., Hamazaki, S., Toyokuni, S. and Midorikawa, O. (1980) Induction of mesothelioma by intraperitoneal injections of ferric saccharate in male Wistar rats, *Br. J. Cancer*, **60**, 708–711.
62. Kennedy, T.P., Dodson, R., Rao, N.V., Ky, H., Hopkins, C., Baser, M., Tolley, E. and Hoidal, J.R. (1989) Dust Causing Pneumoconiosis Generate × OH and Produce Hemolysis by Acting as Fenton Catalysts, *Arch. Biochem. Biophys.*, **209**, 359–364.
63. McDonald, J.C., Liddell, D., Gibbs, G.W., Eyssen, G.E. and McDonald, A.D. (1980) Dust exposure and mortality in chrysotile mining 1910–75, *Br. J. Ind. Med.*, **37**, 11–24.
64. McDonald, J.C., Liddell, D., Dufresne, A. and McDonald, A.D. (1993) The 1891–1920 birth cohort of Québec chrysotile miners and millers: mortality 1976–88, *Br. J. Ind. Med.*, **50**, 1073–1081.

Mineralogical Factors and the Relationship of Fibres and Dusts in Humans to Disease*

A. CHURG

DEPARTMENT OF PATHOLOGY, UNIVERSITY OF BRITISH COLUMBIA AND VANCOUVER, HOSPITAL AND HEALTH SCIENCES CENTRE, VANCOUVER, CANADA

1 Introduction

This paper reviews the relationship of pulmonary inorganic particle burden and pleuropulmonary malignancies using data derived from analytical electron microscopy (*i.e.* electron microscopy that allows identification of specific types, numbers and sizes of particle) of human lung tissues. Detailed information about the mineral particle content of human lungs is available for only two groups: the general population; and workers with asbestos exposure. For most other types of inorganic carcinogens there is either no information in the literature, or the substance in question is highly soluble and cannot be detected by electron microscopic techniques. Even other types of putatively carcinogenic fibres, for example fibrous glass, recently classified as a possible carcinogen by the US National Toxicology Program,[1] are apparently cleared (in large part through dissolution) at a very rapid rate, so that electron microscopic studies are remarkable mostly for the failure to find any fibres in the lungs of those with known occupational exposure.[2]

When evaluating dust data generated by analytical electron microscopy, it is important for the reader to know that there are marked interlaboratory variations in preparative methods, instruments used and counting rules, and that these differences result in quite substantial differences in the concentrations of particles recorded, even when examining the same specimen.[3] Experience has shown that comparison of absolute concentration values from laboratory to laboratory is treacherous, and much more useful information is obtained by looking for consistent relationships between particle burden and disease in each reporting

---

* Supported by grants MA7820 and MA6907 from the Medical Research Council of Canada.

laboratory. Even something as apparently obvious as reported mean particle size cannot be readily compared among laboratories, because some count all particles greater than a certain size, some count only relatively large particles (or long fibres) and some count various numbers of particles or fibres in different size categories. Again, comparisons within the data generated by a given laboratory are necessary to interpret the effects of fibre size. Detailed reviews of the preparative methods and limitations of analytical electron microscopy can be found in references.[4]

2 Particles in the General Population

2.1 Particles in the Parenchyma and Airways

Everyone in the population is exposed to mineral particles from ambient air; these are derived in large part from rock and soil weathering and also from human activity.

Only a few reports have been published on the particle content of parenchymal tissue.[5-8] Three of these reports[6-8] document mean particle concentrations on the order of 500 to 1000 × 10^6 particles gm^{-1} dry lung, while the fourth[5] finds values an order of magnitude lower. Within each study, the patient to patient variation in particle content is quite striking, ranging up to one order of magnitude. Where information on particle type is supplied [6-8] there is reasonably good agreement that aluminium silicates (micas, feldspars, kaolin, *etc.*) represent the largest group (up to 60% of particles), followed by crystalline silica, talc, and aluminium and titanium oxides. We examined particle concentration in different portions of the lung but were unable to find any consistent differences in random samples of peripheral upper and lower and central upper and lower lobes.[6] Particle size has been examined by Stettler *et al.*[7] and Churg and Wiggs;[6] both groups found that the geometric mean case to case particle size was 0.4 to 0.7 µm.

It is believed from animal and explant studies, and from limited experimental human data using inert test particles, that smoking slows particle clearance.[9-15] However, attempts to show that the parenchyma of smokers contains higher particle concentrations than the parenchyma of non-smokers have been largely unsuccessful,[5-8] although Vallyathan *et al.*[16] claimed that the concentration of aluminium and silicon correlated with the amount of smoking, and Churg and Wiggs[6] found a correlation between pack years of smoking and particle concentration in the upper lobe but not the lower lobe. Attempts to show an effect of smoking are probably confounded by the very large person to person variations in retained particle load, even in non-smokers.[5-8]

Most inhaled particles that land on tracheobronchial epithelial cells are removed by mucociliary clearance and/or phagocytosis by macrophages. Nonetheless, a small fraction (estimated by Gore and Patrick[17] at 1%) of such particles are translocated through the epithelium and come to rest in the connective tissue and interstitial macrophages of the airway wall. Particle removal from this compartment is apparently very slow.

Using human autopsy lungs, it is possible to dissect off the bronchial mucosa (*i.e.* all tissues between the cartilage and the lumen) from the larger airways and then to digest this tissue and determine its mineral content in much the same fashion as is done for parenchymal tissue. Churg and Stevens[18] examined the particle burden in airway generations two to four from a series of never-smokers and current smokers. Perhaps the most striking finding was the relatively large particle burden in the airway wall; per gram of dry tissue, this was often as much as one half of the parenchymal burden. In never-smokers there was a consistent increase in particle concentration as the airways narrowed, a finding that matches mathematical models of particle deposition, but in about half of the smokers this pattern was disrupted, perhaps indicating that some smokers are highly resistant to the effects of smoke and some much more susceptible. In contrast to the parenchyma, crystalline silica was found to be the predominant particle type in the airway mucosa in both groups.

An additional finding of potential usefulness was the observation of particles analysing as calcium carbonate in fairly large numbers in the airways and parenchyma of smokers but not non-smokers.[18] Calcium carbonate has been reported by Langer and Nolan[19] as one of the major mineral constituents of cigarette smoke, and the distribution of the calcium carbonate particles in these smokers' airways was a good match to a recently published model of smoke deposition.[20] These observations suggest that such particles can be used as a tracer of smoke deposition in the human lung.

2.2 Correlations of Particle Burden and Disease

Probably the most important conclusion from the studies just summarized is the very large numerical burden of particles that is found in both the parenchyma and airways in the general population. As a rule, this burden appears to be tolerated without the production of obvious disease. However, this conclusion must be tempered by the growing number of reports of increased morbidity and mortality from increased levels of fine atmospheric particulate (PM_{10})[21–23]. Most of the particles found by electron microscopy in the lungs of the general population would be counted in PM_{10}. The pathogenetic mechanisms involved in PM_{10} toxicity are matters of intense speculation,[24] and it would be useful to know whether correlations could be found between particle burden in the lung and measured atmospheric PM_{10} levels.

The issue of atmospheric particulate pollution and lung cancer has been examined for many years, and in some epidemiological studies there are correlations between lung cancer rates and concentrations of particulate air pollutants.[25–27] There is experimental evidence that a crude particulate fraction isolated from ambient air or even so-called 'inert' particles of the type found in ambient air, for example iron oxide, increase the incidence of lung cancers when administered along with organic chemical carcinogens to test animals.[28–31] Cigarette smoke itself contains both chemical carcinogens and mineral particles that may adsorb these carcinogens, and thus smoke may behave in a similar fashion.[30] As well, small atmospheric particles, and presumably smoke particles, have been shown to adsorb a variety of toxic trace elements,[32] and smoke-derived particles may also adsorb radioactive species from the smoke, especially ^{210}Po.[33] Thus, there is a variety of mechanisms by which particles might act as carcinogens.

There are some (limited) data that directly support the association of local particle deposition and lung cancer. Schlesinger and Lippmann[34] used a hollow cast of the human bronchial tree and found that there was a good correlation between the number of particles deposited in various lobar bronchi and the known relative incidence of lung cancer in those same lobes. Churg and Stevens examined lungs from 15 cigarette smokers with nonresected lung cancers[35] and found that particle concentrations were significantly higher in the bronchial mucosa of upper or lower lobe bronchi leading to the lobes containing the cancers compared with the lobes without tumours, but no such difference was found for the particle concentrations in the lobar parenchyma. These data provide additional evidence that the distribution of particles within the bronchial mucosa is important in the genesis of carcinoma. However, whether the particles themselves are playing a role in carcinogenesis, as implied by the observations detailed above, or are merely markers of some other, actual carcinogen, is unknown.

3 Fibre Burden and Disease Caused by Asbestos

3.1 Types of Asbestos and their Persistence in Lung Tissue.

Six different mineral fibres are classified as asbestos. Five of these are members of the amphibole family of minerals: amosite, crocidolite, tremolite, actinolite and anthophyllite, while the remaining fibre type, chrysotile, is mineralogically quite distinct. Amosite and crocidolite have seen fairly extensive commercial usage (and are the fibres meant in this paper by the term 'commercial amphibole'), whereas tremolite, actinolite and anthophyllite are mainly encountered as natural constituents of other mineral ores; in particular, tremolite and actinolite are found to a greater or lesser extent in virtually all chrysotile ores. It should be remembered that, historically, chrysotile has constituted more than 90% of the asbestos used world-wide.[36,37]

While the mineralogical distinction between chrysotile and amphiboles is based on physical and chemical properties, it has become clear that there are major differences in biological properties between these two broad classes of fibre. As shown in Table 1, no matter what the claimed exposure might be, analytical studies of human lung always reveal a much greater proportion of amphiboles and a much lesser proportion of chrysotile than are present in the inhaled dust.[38,39]

There has been extensive discussion of the reasons for this phenomenon and the topic is reviewed at length in the references.[38-41] Suffice it to say here that the effect appears to be one of preferential clearance, by uncertain mechanisms, of deposited chrysotile compared with deposited amphibole fibres rather than a failure of chrysotile deposition. As a result, the estimated half life of chrysotile in the lung is of the order of only a few months,[39,42] whereas the estimated half life of amphiboles is in the range of years to decades.[38] It is generally believed that the much greater biodurability of amphibole compared with chrysotile fibres in the

Table 1 *Relative proportion of chrysotile and amphiboles in various exposure groups (from Ref. 39). Values as mean (range) percent of total fibres*

	No. of Reports	Chrysotile (%)	Tremolite (%)	Amosite/Crocidolite (%)
General population	7	72 (31–98)	18 (2–50)	14 (0–46)
Workers in industries using predominantly chrysotile	8	47 (19–80)	46 (10–81)	15 (0.3–51)
Workers with mixed amphibole and chrysotile exposure	5	40 (6–89)	4 (2–5)	58 (9–89)

lung is the reason that amphiboles are much more potent mesothelial carcinogens in man than is chrysotile (see below).

It should be noted that preferential chrysotile clearance is also seen in experimental animals[38,39]; however, it is likely that the relatively short lifetime of small laboratory rodents vitiates the differences between chrysotile and amphibole in terms of mesothelioma induction since, in such models, chrysotile asbestos is just as efficient a mesothelial carcinogen as is amphibole asbestos.[41]

3.2 Asbestos fibres in the general population

Asbestos fibres are present in urban air (estimated levels in North America on the order of 0.0001 f cm^{-3}),[43] and can be found in the lungs of everyone in the population.[36,37] Some of the ambient fibre is of commercial product origin, but some fraction, particularly some fraction of chrysotile fibres, is undoubtedly a result of natural rock weathering, particularly in the western portions of North America, where there is extensive serpentine (the parent rock of chrysotile) outcrop.[36,37]

Table 2 shows the typical concentrations of different types of asbestos in the lungs of the general population.[44,45] Values from two different laboratories are presented. We count all fibres longer than 0.5 μm, and hence our data suggest that the bulk of fibres are short fragments of chrysotile and that the most common amphibole is tremolite derived from the chrysotile. Levels of amosite and crocidolite are extremely low. Case, Sebastien and McDonald[45] counted only fibres longer than 5 μm in a series of accident victims from across Canada, and in their data there is a much greater relative fraction of amosite and crocidolite because these fibres do not readily fragment into short lengths. But despite these differences, the conclusion to be drawn from Table 2 is that everyone in the population carries a numerically substantial burden of asbestos fibres.

In recent years there has been considerable scientific and public debate about the possible adverse effects of asbestos fireproofing and insulation in public buildings.[43,46,47] A survey of such buildings found a mean level of 0.0002 f cm^{-3},[43] a value quite similar to that of ambient air. Virtually all of the asbestos in question is chrysotile. It should be appreciated that when asbestos-induced disease has been found in persons with occupational exposure, the typical exposure levels have been upwards of 5 to 10 f cm^{-3}, and sometimes much higher. Since epidemiological demonstration of an adverse effect at levels of 0.0002 f cm^{-3} is difficult, even as a theoretical proposition, most of the arguments about this issue have been based on mathematical extrapolations from high to low exposure levels

Table 2 Concentration of asbestos fibres in the general population. Values as fibres/g/g dry lung

Report/Fibre	Mean	Median	95th Percentile
Case & Sebastien (Ref. 45)(Fibres >5 μm)[1]			
Chrysotile	62 000		
Tremolite	14 000		
Amosite/Crocidolite	10 000		
Churg & Wiggs (Ref. 44)(Fibres >0.5 μm)[2]			
Chrysotile	300 000	200 000	1 100 000
Tremolite	400 000	200 000	1 200 000
Amosite/Crocidolite	1000	0	10 000

[1] Accident victims aged 61 or greater.
[2] General hospital autopsy population.

over 5 to 7 orders of magnitude, and many believe that such extrapolations are of questionable scientific validity.[46]

It is particularly instructive in this regard to consider the general population of the mining townships of Eastern Québec. Because of the presence of mine tailings, open pit mines and chrysotile/tremolite bearing rock and soil, the atmospheric levels of chrysotile and tremolite in these townships are several-hundred-fold greater than those found in the air of most cities in North America:[48,49] Sebastien et al. measured an average ambient chrysotile level in the mining townships of 0.01 f cm^{-3},[48] a value within one order of magnitude of the current US OSHA standard. Increased mean pulmonary concentrations of chrysotile and tremolite, concentrations some 5- to 10-fold higher than are seen in residents of Vancouver or Montreal,[50–52] can be found in the lungs of those who have lived in or downwind from the mining townships, even if they have never worked in the asbestos industry. As well, the fibres in the lungs of such individuals are distinctly longer than those in urban areas,[44,51,52] a potentially important observation if theories about the greater pathogenicity of long fibres are correct (see below). Nonetheless, epidemiological studies of persons living for whole lifetimes in the mining townships and who have never worked in the mining or milling industry have, thus far, failed to show an increased incidence of either lung cancer or mesothelioma (reviewed in Refs. 46, 49, 51).

These are important observations because they indicate that lifetime chrysotile and tremolite burdens considerably greater than those in the general population of North American cities do not produce detectable asbestos-related malignancies. By the same token, these findings provide support for the conclusion that the much lower urban ambient or asbestos insulated building levels of chrysotile to which every one in the population is exposed do not cause adverse effects.

3.3 Asbestos Fibre Concentration and Mesothelioma

As noted above, chrysotile does not persist in lung tissue, and this phenomenon complicates interpretation of human lung burden studies, the more so as most

Table 3 *Asbestos fibre concentrations by disease in chrysotile miners and millers and in shipyard workers and insulators with heavy amosite exposure. Values are geometric mean (GSD) in millions/g dry lung*

Disease	Chrysotile miners & millers		Shipyard/insulators
	Chrysotile[a]	Tremolite[a]	Amosite[b]
Asbestosis	30 (5) [$p < 0.01$]	140 (6) [$p<0.01$]	10 (60.6) [$p<0.001$]
Airway fibrosis	27 (4) [$p<0.01$]	120 (3) [$p<0.01$]	4.3 (12) [$p=0.001$]
Mesothelioma	34 (5) [$p<0.10$]	180 (4) [$p<0.01$]	0.9 (8)
Pleural plaques	15 (5)	75 (6)	1.4 (8)
Lung cancer	13(6)	49 (5)	1.1 (8)
No asbestos disease	2 (20)	9 (10)	0.7 (5)

[a] Data on 94 workers, modified from Ref. 54.
[b] Data on 144 workers, modified from Ref. 55.
Note: For both groups, significance values in brackets are comparisons with the no disease group using a multiple regression model that accounts for the presence of more than one asbestos disease in many subjects.

working populations have been exposed to both chrysotile and commercial amphiboles. To a certain extent, the tremolite in the chrysotile ore may serve as a substitute measure for the chrysotile itself[53,54] (see also Table 3). However, while chrysotile miners and millers and chrysotile textile workers accumulate substantial tremolite burdens (in the case of miners and millers, usually much greater than the residual chrysotile burden by the time of autopsy), analysis of human lungs suggests that the amount of tremolite in other end products is generally less, but quite variable.[44,55]

When populations with heavy commercial amphibole, as opposed to only chrysotile exposure, are examined, quite dramatically different fibre–disease relationships are observed. References 36, 37, 55 and 56 provide extensive detailed tabulations of the numbers of commercial amphibole fibres found in populations with occupational asbestos exposure and various asbestos-related diseases. These studies indicate quite clearly that, for exposure to amosite or crocidolite, pleural mesothelioma appears at fibre levels considerably greater than are found in the general population, but much lower than those seen in workers with asbestosis. Representative data from our laboratory,[55] derived from a large group of shipyard workers and insulators with heavy amosite exposure, are shown in Table 3.

Few data are available for chrysotile-induced mesothelioma. We analysed a group of 94 chrysotile miners and millers from Thetford Mines in Québec[54] and concluded that, in contrast to the situation for amosite, chrysotile-induced mesothelioma appeared at mean fibre levels equal to those associated with asbestosis (Table 3). These observations are entirely consistent with a large body of epidemiological evidence that concludes that chrysotile asbestos is a much weaker mesothelial carcinogen in man than is amphibole asbestos.[57]

These findings are also of considerable importance for regulating occupational exposures and for estimating the risks of mesothelioma to the general population from ambient and building exposure, since they imply that avoiding exposures sufficient to produce asbestosis (minimal cumulative exposure estimated at upwards of 25 f cm-yrs^{-1},[58] – a level that is impossible to achieve with current workplace or ambient exposures) will avoid producing mesothelioma from chrysotile exposure. On the other hand, they also indicate that exposure to amosite or crocidolite is potentially much more dangerous in terms of mesothelioma development, although it must be repeated that the mean amphibole fibre concentrations shown for mesothelioma cases in Table 3 are several orders of magnitude greater than the levels found in the general population (compare Table 2).

At first glance, Table 3 appears to indicate that both tremolite and chrysotile show equally good associations with disease, particularly mesothelioma. In fact this is misleading, because regressions that are run, using the chrysotile miner data for tremolite with adjustment for chrysotile, yield persisting correlations, whereas regressions run for chrysotile with adjustment for tremolite produce no correlations at all.[54] Further support for the idea that tremolite and not chrysotile is responsible for 'chrysotile-induced' mesothelioma comes from a study by McDonald et al.[59] comparing the fibre burden in autopsy cases of mesothelioma across Canada and general population controls: McDonald et al. concluded that the presence of amphibole asbestos fibres, including tremolite, could explain most cases of mesothelioma, and the presence of chrysotile fibres, very few. Additional evidence comes from the observation that, within the chrysotile mining region of Québec, there are differences in the tremolite content of the ore, and the incidence of mesothelioma is higher in the Thetford Mines region, an area with relatively high tremolite levels in the ore, than the Asbestos region, where tremolite levels are lower.[60] More recently McDonald and McDonald[60] re-evaluated 24 mesothelioma cases from the Thetford Mines region and showed that the mesothelioma cases were concentrated in 5 of 15 mines; the lungs of workers from these five mines contained significantly higher concentrations of tremolite than the lungs of workers from the other ten mines. No such differences were found for chrysotile.

These findings are also complemented by two studies by Rogers et al.[61,62] using Australian mesothelioma cases and general population controls. In the first study,[61] Rogers et al. concluded that, for a 10-fold increase in retained fibre concentration, the risk of mesothelioma was greatest for crocidolite, less for chrysotile and even less for amosite. However, in the second study[62] these data were combined with estimates of fibre retention by fibre type, leading the authors to conclude that, overall, the risk of mesothelioma from commercial amphibole exposure far exceeds that from chrysotile exposure.

Lastly, additional support for the association of commercial amphiboles with mesothelioma comes from the series of amosite-exposed workers shown in Table 3.[55] These workers in fact had historic heavy exposure to both amosite and chrysotile, but by the time of autopsy very little chrysotile (or tremolite) was present, and neither chrysotile nor tremolite concentrations correlated with the presence of any disease.

3.4 Asbestos Fibre Size and Mesothelioma

Examination of Table 3 shows that fibre concentration alone fails to explain the presence of some diseases; for example, in the amosite-exposed workers, those with plaques, lung cancers and mesotheliomas, and workers with exposure but no disease, have about the same mean fibre burden.

There is a considerable body of experimental animal data, which indicate that long fibres are much more dangerous in terms of mesothelioma induction (and also asbestosis induction) than are short fibres.[63-67] Unfortunately, attempts to demonstrate such an association in man have been problematic. Table 4 shows data on fibre length and disease from the same group of amosite-exposed workers mentioned above.[55] There was no association of fibre length and any disease, although in fact all subjects with disease had longer fibres than subjects without. Similar analyses were run for fibre width, aspect ratio (length to width ratio), surface area and mass. The only size measure that showed an association with disease was high fibre aspect ratio with pleural plaques; mesothelioma subjects in fact had lower aspect ratio fibres than the subjects with exposure but no disease. The study of chrysotile miners and millers[54] shown in Table 3 also concluded that pleural plaques were associated with high aspect ratio fibres (but only tremolite and not chrysotile fibres); mesothelioma was not associated with either long or high aspect ratio fibres.

Table 4 Mean amosite fibre length by disease in shipyard workers and insulators (from Ref. 55). Length values in μm

Disease	Cases with Disease		t-Test[a]	Multiple regression model[b]
	No.	Length		
Asbestosis	23	6.4	$p=0.18$	$p=0.33$
Airway fibrosis	16	6.0	$p=0.29$	$p=0.64$
Mesothelioma	83	5.5	$p=0.48$	$p=0.32$
Lung cancer	32	5.7	$p=0.39$	$p=0.73$
Plaque	103	5.9	$p=0.32$	$p=0.16$
No disease	8	4.6		

[a]Comparison of cases with specified disease to those with no disease by t-test using square root transformed length values.
[b]Multiple regression model accounting for the presence of more than one disease in most subjects and using square root transformed values.

The case control studies of McDonald et al.[59] and Rogers et al.[61,62] have been described above. McDonald et al.[59] concluded that the concentration of fibres longer than 8 μm was best related to the relative risk of mesothelioma, and that short fibre concentration provided no additional information. Rogers et al.[61] concluded that mesothelioma risk related to the concentration of crocidolite longer than 10 μm, and amosite and chrysotile shorter than 10 μm. Aside from the obvious contradiction between these reports, both are probably confounded by the fact that most mesotheliomas occur in those with occupational asbestos exposure, but their control groups are general population autopsies, and fibres in the general population as a rule are shorter than those in exposed workers.[44,51] Thus, these studies would probably have come to the same conclusions whether they had examined workers with asbestosis or even workers with heavy exposure but no disease.

One piece of indirect evidence from human lungs that does support the association of mesothelioma and long fibres is the relatively short size and low aspect ratio of tremolite fibres in the lungs of the Québec chrysotile miners and millers: we found that the geometric mean tremolite in these workers was 2.2 μm and the aspect ratio 11, compared with shipyard and insulation workers with heavy amosite exposure, where the geometric mean amosite length was 3.9 μm and the aspect ratio 22.[44,54] These size differences are consistent with the much lower incidence of mesothelioma in chrysotile miners and millers compared with workers with amosite exposure.[57]

At this point, the issue of whether mesothelioma is really associated with long high aspect ratio fibres in man is unresolved. This is an important mechanistic and regulatory question that needs to be addressed.

3.5 Asbestos Fibre Location and Mesothelioma

It appears logical to assume that mesothelioma is caused by fibres that interact with mesothelial cells, and, indeed, intrapleural inoculation of fibres in animals is an excellent way to induce mesothelioma.[69,70] However, even in experimental animals, there is some suggestion that asbestos that reaches the lung via the trachea produces mesothelial cell proliferation by remote (?cytokine) effects rather than by direct contact with the pleura.[68]

While it is easy to demonstrate the presence of both amphiboles and chrysotile in the subpleural lung tissue,[69] studies of the pleura (specifically parietal pleura) itself in man have generally failed to show significant accumulation of asbestos fibres, or have shown that short chrysotile fibres greatly outnumber amphibole fibres,[70,71] even though amphiboles are present in much higher concentration in the parenchyma. It is possible that very local accumulations of asbestos fibres occur in 'hot spots'; there is some evidence for such fibre redistribution in rats over the course of several months,[72] although this phenomenon has not been observed by others.[73] Deposition patterns may also favour accumulation of long fibres in localized subpleural locations; for example, we found that the posterior upper lobe tended to accumulate long amphibole fibres.[74] But at this point, a clear and consistent demonstration of fibres in human pleural tissues in types and sizes that match epidemiological and experimental animal predictions is lacking.

3.6 Asbestos Fibre Burden and Lung Cancer

Any attempt at examining the relationship between asbestos fibre burden and lung cancer must deal with the major problem of confounding by cigarette smoke. This problem is even more difficult to approach using human autopsy tissue than it is epidemiologically, because autopsies by definition suffer from (generally unknown) types of selection bias, and autopsies from cohorts with asbestos exposure have even greater selection bias, since many such autopsies are performed in the hope of demonstrating a compensable disease.

A further complication is the clear and strong association of lung cancer with asbestosis; indeed, in my opinion, there is extremely good evidence to support the idea that an increased risk of lung cancer from asbestos exposure exists only in those who have asbestosis (reviewed in Ref. 75). Analytical studies show that such cancer cases have the typical, very high fibre load seen in asbestosis. For example, Wagner et al.[56] examined the fibre content of the lungs of 36 east London factory workers. All 14 cases of lung cancer had asbestosis, and the mean fibre content of the lungs, which was about the same as the subjects with asbestosis without lung cancer, was much above that of subjects with mesothelioma. A similar result was reported by Roggli et al.[76]

Analytical studies on workers without asbestosis are probably really studies of cigarette smoke induced lung cancer in persons with asbestos exposure, and shed little light on the relationship between fibre burden and lung cancer.[77] Support for this idea comes from data of Roggli et al.[76] who noted that, after excluding lung cancers associated with asbestosis, the remaining lung cancer cases in his collection had fibre burdens very similar to individuals with exposure and pleural plaques. Roggli et al. point out that the majority epidemiological opinion is that pleural plaques are not associated with lung cancers, and this implies that the fibre burden seen in pleural plaques also is not associated with lung cancer. As shown in Table 3, we also found that, after excluding cases of asbestosis, lung cancer cases and pleural plaque cases had statistically identical fibre burdens, and this was true for both the amosite-exposed[55] and chrysotile-exposed[54] cohorts. In fact, in the amosite cohort, subjects with lung cancer and no asbestosis did not have a mean fibre concentration any greater than subjects with exposure but no disease. Given this situation, analysis of fibre burden appears to be of no help in establishing the aetiology of a lung cancer.

Two recent case–control studies have, nonetheless, attempted to determine whether there is a relationship between pulmonary fibre burden and lung cancer in a general autopsy population. Mollo et al.[78] concluded that increased numbers of asbestos bodies in lung tissue were associated with the presence of adenocarcinomas; Karjalainen et al.[79] reported a similar association with adenocarcinoma, and also claimed that the risk was higher with higher fibre concentrations for lower lobe than upper lobe tumours. These studies suffer from a number of methodological problems, not the least of which is the general nonreproducibility of pathological diagnoses of the specific histological type of lung cancer; indeed, the literature on the association of asbestos exposure and specific histological types of lung cancer is a mass of contradictions, and two detailed reviews of this issue have concluded that there are no consistent associations.[80,81] The issue of upper vs. lower lobe tumours is very similar to that of cell type: the aggregate body of reports appears to give random results. Both Mollo et al. and Karjalainen et al. also failed to show an effect of smoking in increasing the risk of lung cancer, a problem that raises serious doubts about the associations with asbestos. Overall, I find these reports unconvincing.

Lippmann[82] has reviewed a variety of animal inhalation studies and concluded that lung cancer is particularly associated with fibres 10 μm and longer. It should be noted, however, that there is considerable evidence to suggest that lung cancer in such animal models only develops in the presence of asbestosis (reviewed in Ref. 83), and such size predictions may actually be predictions about the sizes of fibres associated with asbestosis. In man, there is little information on fibre size and lung cancer. In the two studies mentioned above[54,55] and shown in Tables 3 and 4, we found no association of any fibre size measure and lung cancer.

As indicated previously, a number of experimental studies using both whole animal and explant systems have concluded that cigarette smoke increases the retention of many types of mineral dust, including asbestos,[12-15] but no data exist on the question of the effects of cigarette smoke and asbestos retention in human smokers. This issue is of considerable interest, since a large number of radiographic studies have concluded that cigarette smoking increases the incidence and perhaps the progression rate of asbestosis,[84-89] and also because, epidemiologically, the combination of cigarette smoke and asbestos exposure produces more than additive increases in lung cancer incidence rates in heavily exposed populations.[90]

These smoke effects could be explained by enhanced retention of asbestos fibres in cigarette smokers, leading to an effectively increased dose of fibres, and thus an increased incidence of asbestosis (which is quite clearly dose related [36,37]) and finally an increased incidence of lung cancer. However, we were unable to show that cigarette smoke increased parenchymal asbestos retention in the two cohorts shown in Table 2,[54,55] probably because of the lack of sufficiently detailed data on asbestos exposure.

To attempt to examine this question in another fashion, we compared the airway mucosal asbestos burden in six smokers and six never-smokers from our amosite-exposed shipyard and insulation worker cohort.[91] To overcome the problem of unknown actual exposure levels, we matched test and control cases by age, sex, years of exposure and mean parenchymal amosite burden. As shown in Table 5, non-smokers had very low levels of fibres of any kind in their airway tissues, while the smokers had about a 6-fold greater concentration of amosite and a 50-fold greater concentration of chrysotile. These observations indicate that smoking does enhance uptake of asbestos fibres by airway mucosa. Of additional interest was the observation that fibres of all types were significantly shorter in both the airways and the parenchyma in the smokers. This finding is consistent with experimental data indicating that cigarette smoke acts to increase asbestos retention by interfering with the removal of shorter fibres that would normally be cleared.[92] It also implies that smoking probably does increase the parenchyma asbestos fibre concentration in man.

Table 5 *Comparison of airway mucosal fibre burdens in smokers and non-smokers. Values are thousands of fibres/g dry tissue. Mean ± SD [Range] (Geometric Mean) (from Ref. 91)*

	Non-smokers	Smokers	P Value
Amosite	10 ± 26 [0–120] (0)	62 ± 120 [0–610] (0)	0.02
Tremolite	21 ± 67 [0–420] (0)	170 ± 430 [0–2100] (0)	NS
Chrysotile	2 ± 5 [0–23] (0)	110 ± 310 [0–1500] (0)	0.006

References

1. US Department of Health and Human Services (1994) National Toxicology Program, Seventh Annual Report on Carcinogens.
2. Sebastien, P. (1994) Biopersistence of man-made vitreous silicate fibres in the human lung, *Environ. Health Perspect.*, **102** (Suppl. 5), 225–228.
3. Gylseth, B., Churg, A., Davis, J.M.G., et al. (1985) Analysis of asbestos fibres and asbestos bodies in human lung tissue samples. An international laboratory trial, *Scand. J. Work Environ. Health*, **11**, 107–110.
4. Ingram, P., Shelburne, J.D. and Roggli, V.L., eds., (1989) 'Microprobe Analysis in Medicine', Hemisphere Press, New York.
5. Paoletti, L., Eibenschutz, L., Cassano, A.M., et al. (1989) Mineral fibres and dust in lungs of subjects living in an urban environment. In, 'Non-occupational Exposure to Mineral Fibres', Bignon, J., Peto, J., Saracci, R., eds., IARC, Lyon, pp. 354–360.
6. Churg, A. and Wiggs, B. (1987) Types, numbers, sizes, and distribution of mineral particles in the lungs of urban male cigarette smokers, *Environ. Res.*, **42**, 121–129.
7. Stettler, L.E., Growth, D.H., Platek, S.F. and Burg, J.R. (1989) Particulate concentrations in urban lungs. In, 'Microprobe Analysis in Medicine', Ingram, P., Shelburne, J.D., Roggli, V.L., eds., Hemisphere Press, New York, pp. 133–146.
8. Kalliomaki, P.L., Taikina-aho, O., Paakko, P., et al. (1989) Smoking and the pulmonary mineral particle burden. In, 'Non-occupational Exposure to Mineral Fibres', Bignon, J., Peto, J., Saracci, R., eds., IARC, Lyon, pp. 323–329.
9. Cohen, D., Arai, S.F. and Brain, J.D. (1979) Smoking impairs long-term clearance from the lung, *Science*, **204**, 514–516.
10. Bohning, D.E., Atkins, H.L. and Cohn, S.H. (1982) Long-term particle clearance in man: Normal and impaired, *Ann. Occup. Hyg.*, **26**, 259–271.
11. Friedman, A.P., Robsinson, S.E. and Street, M.R. (1988) Magnetopneumographic study of human alveolar clearance in health and disease, *Ann. Occup. Hyg.*, **32** (Suppl. 1), 809–820.
12. Mauderly, J.M., Chen, B.T., Hahn, F.F., Lundgren, D.L. and Cuddihy, R.G. (1989) The effect of chronic cigarette smoke inhalation on the long-term pulmonary clearance of inhaled particles in the rat. In, 'Biological Interactions of Inhaled Mineral Fibres and Cigarette Smoke', Wehner, A.P., ed., Battelle Press, Columbus, OH, pp. 223–240.
13. McFadden, D., Wright J.L., Wiggs, B. and Churg, A. (1986) Smoking inhibits asbestos clearance, *Am. Rev. Respir. Dis.*, **133**, 372–374.
14. Gilks, B., Wright, J.L. and Churg, A. (1988) Effects of cigarette smoke on tissue uptake and retention of iron oxide in the guinea pig, *Am. Rev. Respir. Dis.*, **137**, 1382–1384.
15. Keeling, B., Hobson, J. and Churg, A. (1993) Effects of cigarette smoke on tracheal epithelial uptake of non-asbestos mineral particles in organ culture, *Am. J. Respir. Cell Mol. Biol.*, **9**, 335–340.
16. Vallyathan, V. and Hahn, L.H. (1985) Cigarette smoking and inorganic dust in human lungs, *Arch. Environ. Health*, **40**, 69–73.
17. Gore, D.J. and Patrick, G. (1982) A quantitative study of the penetration of insoluble particles into the tissue of the conducting airways, *Ann. Occup. Hyg.*, **26**, 149–161.
18. Churg, A. and Stevens, B. (1992) Mineral particles in the human bronchial mucosa. II. Cigarette smokers without emphysema, *Exp. Lung Res.*, **18**, 687–714.
19. Langer, A.M., Nolan, R.P., Bowes, D.R. and Shirey, S.B. (1989) Inorganic particles found in cigarette tobacco, cigarette ash, and cigarette smoke. In, 'Biological Interactions of Inhaled Mineral Fibres and Cigarette Smoke', Wehner, A.P., Felton, D., eds., Battelle Press, Columbus, OH, pp. 421–440.

20. Muller, W.J., Hess, G.D. and Scherer, P.W. (1990) A model of cigarette smoke particle deposition, *Am. Ind. Hyg. Assoc. J.*, **51**, 245.
21. Schwartz, J. (1983) Particulate air pollution and chronic respiratory disease, *Environ. Res.*, **62**, 7–13.
22. Utell, M.J. and Samet, J.M. (1993) Particulate air pollution and health, *Am. Rev. Resp. Dis.*, **147**, 1334–1335.
23. Dockery, D.W. and Pope, C.A., III. (1994) Acute respiratory effects of particulate air pollution, *Ann. Rev. Public Health*, **15**, 107–132.
24. Seaton, A., MacNee, W., Donaldson, K. and Goddens, D., (1995). Particulate air pollution and acute health effects, *Lancet*, **345**, 176–178.
25. Doll, R. (1978) Atmospheric pollution and lung cancer, *Environ. Health Perspect.*, **22**, 23–31.
26. Vena, J.E. (1982) Air pollution as a risk factor in lung cancer, *Am. J. Epidemiol.*, **116**, 42–56.
27. Hitosugi, M. (1968) Epidemiological study of lung cancer with special reference to the effect of air pollution and smoking habit, *Inst. Public Health Bull. (Jpn)*. **17**, 237–256.
28. Saffiotti, U., Montesano, R., Sellakumar, A.R., Cetis, F. and Kaufman, D.G. (1972) Respiratory tract carcinogenesis in hamsters induced by benzo[*a*]pyrene and ferric oxide, *Cancer Res.*, **32**, 1073–1081.
29. Stenback, F., Rowland, J. and Sellakumar, A. (1976) Carcinogenicity of benzo[*a*]pyrene and dusts in the hamster lung, *Oncology*, **33**, 29–34.
30. Lubawy, W.C. and Isaac, R.S. (1980) Acute tobacco smoke exposure alters the profile of metabolites produced by benzo[*a*]pyrene by the isolated perfused rabbit lung, *Toxicology*, **18**, 37–47.
31. Warshawsky, D., Bingham, E. and Niemeier, R.W. (1983) Influence of airborne particulate on the metabolism of benzo[*a*]pyrene in the isolated perfused lung, *J. Toxicol. Environ. Health*, **11**, 503–517.
32. Natusch, D.F.S., Wallace, J.R. and Evans, C.A. (1974) Toxic trace elements: Preferential concentration in respirable particles, *Science*, **183**, 202–204.
33. Martell, E.A. (1983) Radiation at bronchial bifurcations of smokers from indoor exposure to radon progeny, *Proc. Natl. Acad. Sci. U.S.A.*, **80**, 1285–1289.
34. Schlesinger, R.B. and Lippmann, M. (1978) Selective particle deposition and bronchogenic carcinoma, *Environ. Res.*, **15**, 424–431.
35. Churg, A. and Stevens, B. (1988) Association of lung cancer and airway particle concentration, *Environ. Res.*, **45**, 58–63.
36. Churg, A., and Green, F.H.Y (1988) 'Pathology of Occupational Lung Disease', Igaku-Shoin, New York.
37. Roggli, V.L., Greenberg, S.D. and Pratt, P.C. (1992) 'Pathology of Asbestos-Related Diseases', Little, Brown and Company, Boston.
38. Churg, A. and Wright, J. (1994) Persistence of natural fibres in lung tissue, *Environ. Health Perspect.*, **102** (Suppl. 5), 229–233.
39. Churg, A. (1994) Deposition and clearance of chrysotile asbestos, *Ann. Occup. Hyg.*, **38**, 625–633.
40. Bignon, J., Saracci, R. and Tournay, J.-C., eds., (1994) Biopersistence of respirable synthetic fibres and minerals, *Environ. Health Perspect.*, **102**, (Suppl. 5), 3–284.
41. Gibbs, G.W., Valic, F. and Browne, K., eds., (1994) Health risks associated with chrysotile asbestos, *Ann. Occup. Hyg.*, **38**, 399–646.
42. Oberdorster, G. (1994) Macrophage associated responses to chrysotile, *Ann. Occup. Hyg.*, **38**, 601–615.

43. Health Effect Institute-Asbestos Research (1991) 'Asbestos in public and commercial buildings: A literature review and synthesis of current knowledge', Cambridge, MA, Health Effects Institute.
44. Churg, A. and Wiggs, B. (1986) Fibre size and number in users of processed chrysotile ore, chrysotile miners, and members of the general population, *Am. J. Ind. Med.*, **9**, 143–152.
45. Case, B.W., Sebastien, P. and McDonald, J.C. (1988) Lung fibre analysis in accident victims: A biological assessment of general environmental exposure, *Arch. Environ. Health*, **43**, 178–179.
46. McDonald, J.C. (1985) Health implications of environmental exposure to asbestos, *Environ. Health Perspect.*, **62**, 328–329.
47. Hughes, J.M. and Weill, H. (1986) Asbestos exposure–quantitative assessment of risk, *Am. Rev. Respir. Dis.*, **133**, 5–13.
48. Sebastien, P., Plourde, M., Robb, R., *et al.* (1986) Ambient air asbestos survey in Québec mining towns. Part 2–Main study. Environment Canada Report 5/AP/RQ-2E.
49. Siemiatycki, J. (1982) Health effects on the general population (Mortality in the general population in asbestos mining areas). Proc. World Symposium on Asbestos, The Asbestos Institute, Montreal, 337–348.
50. Case, B.W. and Sebastien, P., Fibre levels in lung and correlation with air samples. In, (1989) 'Non-occupational Exposure to Mineral Fibres', Bignon, J., Peto, J., Saracci, R., eds., IARC, Lyon, pp. 207–219.
51. Churg, A. (1986) Lung asbestos content in long-term residents of a chrysotile mining town, *Am. Rev. Respir. Dis.*, **134**, 125–127.
52. Case, B.W. and Sebastien, P. (1987) Environmental and occupational exposures to chrysotile asbestos: A comparative microanalytic study, *Arch. Environ. Health*, **42**, 85–191.
53. Churg, A. (1988) Chrysotile, tremolite, and malignant mesothelioma in man, *Chest*, **93**, 621–628.
54. Churg, A., Wright, J. and Vedal, S. (1993) Fibre burden and patterns of asbestos-related disease in chrysotile miners and millers, *Am. Rev. Respir. Dis.*, **148**, 25–31.
55. Churg, A. and Vedal. S. (1994) Fibre burden and patterns of asbestos-related disease in workers with heavy mixed amosite and chrysotile exposure, *Am. J. Respir. Crit. Care Med.*, **150**, 663–669.
56. Wagner, J.C., Newhouse, M.L., Corrin, B., Rossiter, C.E.R. and Griffiths, D.M. (1988) Correlation between fibre content of the lung and disease in East London asbestos factory workers, *Br. J. Ind. Med.*, **45**, 305–308.
57. McDonald, J.C. and McDonald, A.D. (1986) Epidemiology of malignant mesothelioma. In, 'Asbestos-Related Malignancy', Antman, K. and Aisner, J., eds., Grune and Stratton, New York, pp. 31–35
58. Report of the Royal Commission (1984) On Matters of Health and Safety Arising from the Use of Asbestos in Ontario, Queen's Printer for Ontario.
59. McDonald, J.C., Armstrong, B., Case, B., Doell, D., McCaughey, W.T.E., McDonald, A.D. and Sebastien, P. (1989) Mesothelioma and asbestos fibre type. Evidence from lung tissue analysis, *Cancer*, **63**, 1544–1547.
60. McDonald, J.C. and McDonald, A.D. (1995) **Chrysotile, tremalite,** and mesothelioma, *Science*, **267**, 775–776.
61. Rogers, A.J., Leigh, J., Berry, G., Ferugson, D.A., **Mulder, H.B.** and Ackad, M. (1991) Relationship between lung asbestos fibre type and concentration and relative risk of mesothelioma, *Cancer*, **67**, 1912–1920.

62. Rogers, A.J., Leigh, J., Berry, G., Ferugson, D.A., Mulder, H.B., Ackad, M. and Morgan, G.G. (1994) Dose-response relationship between airborne and lung asbestos fibre type, length, and concentration, and relative risk of mesothelioma, *Ann. Occup. Hyg.*, **38** (Suppl. 1), 631–638.
63. Davis, J.M.G. and Jones, A.D. (1986) Comparisons of the pathogenicity of long and short fibre chrysotile asbestos in rats, *Br. J. Exp. Pathol.*, **69**, 717–737.
64. Davis, J.M.G., Addision, J., Bolton, R.E., Donaldson, K., Jones, A.D. and Smith, T. (1986) The pathogenicity of long versus short fibre samples of amosite asbestos administered to rats by inhalation and intraperitoneal injection, *Br. J. Exp. Pathol.*, **67**, 415–430.
65. Stanton, M.F., Layard, M., Tegeris, A., *et al.* (1981) Relation of particle dimension to carcinogenicity in amphibole asbestoses and other fibrous minerals, *JNCI*, **67**, 965–975.
66. Adamson, I.Y.R. and Bowden, D.H. (1987) Response of mouse lung to crocidolite asbestos. I. Minimal fibrotic reaction to short fibres, *J. Pathol.*, **152**, 99–107.
67. Adamson, I.Y.R. and Bowden, D.H. (1987) Response of mouse lung to crocidolite asbestos. II. Pulmonary fibrosis after long fibres, *J. Pathol.*, **152**, 109–117.
68. Adamson, I.Y.R., Bakowska, J. and Bowden, D.H. (1993) Mesothelial cell proliferation after instillation of long or short asbestos fibres into mouse lung, *Am. J. Pathol.*, **142**, 1209–1216.
69. Churg, A., Wiggs, B., DiPaoli, L., Kempe, B. and Stevens B. (1984) Lung asbestos content in chrysotile workers with mesothelioma, *Am. Rev. Respir. Dis.*, **130**, 1042–1045.
70. Sebastien, P., Janson, X., Gaudichet, A., Hirsch, A. and Bignon, J. (1980) Asbestos retention in human respiratory tissues: comparative measurements in lung parenchyma and in parietal pleura. 'Biological Effects of Mineral Fibres', Wagner, J.C., ed., IARC, Lyon, pp. 237–246.
71. Dodson, R.F., Williams, M.G., Corn, C.J., Rollo, A. and Bianchi, C. (1990) Asbestos content of lung tissue, lymph nodes, and pleural plaques from former shipyard workers, *Am. Rev. Respir. Dis.*, **142**, 843–847.
72. Morgan, A., Evans, J.C. and Holmes, A. (1977) Deposition and clearance of inhaled fibrous minerals in the rat. Studies using radioactive tracer technique. In, 'Inhaled Particles IV', Walton, W.H. and McGovern, B., eds., Pergamon Press, New York, pp. 259–274.
73. Coin, P.G., Roggli, V.L. and Brody, A.R. (1992) Deposition, clearance, and translocation of chrysotile asbestos from peripheral and central regions of the lung, *Environ. Res.*, **58**, 97–116.
74. Churg, A. and Wiggs, B. (1987) Accumulation of long asbestos fibres in the peripheral upper lobe in cases of mesothelioma in man, *Am. J. Ind. Med.*, **11**, 563–570.
75. Churg, A. (1993) Asbestos, asbestosis, and lung cancer, *Mod. Pathol.*, **6**, 509–511.
76. Roggli, V.L., Pratt, P.C. and Brody, A.R. (1992) Analysis of tissue mineral fibre content. In, 'Pathology of Asbestos-Related Diseases', Roggli, V.L., Greenberg, S.D. and Pratt, P.C., Little, Brown and Company, Boston, pp. 299–346.
77. Weiss, W. (1993) Asbestos-related pleural plaques and lung cancer, *Chest*, **103**, 1854–1859.
78. Mollo, F., Pira, E., Piolatto, G., Bellis, D., Burlo, P., Andreozzi, A., Bontempi, S. and Negri, E. (1995) Lung adenocarcinoma and indicators of asbestos exposure, *Int. J. Cancer*, **60**, 289–293.
79. Karjalainein, A., Anttila, S., Vanhala, E. and Vainio, H. (1994) Asbestos exposure and the risk of lung cancer in a general urban population, *Scand. J. Work Environ. Health*, **20**, 243–230.

80. Ives, J.C., Buffler, P.A. and Greenberg, S.D. (1983) Enviromental associations and histopathologic patterns of carcinoma of the lung, *Am. Rev. Respir. Dis.*, **128**, 195–209.
81. Churg, A. (1994) Lung cancer cell type and occupational exposure. In, 'Epidemiology of Lung Cancer', Samet, J.M., ed., Marcel Dekker, New York, pp. 413–436.
82. Lippmann, M. (1994) Deposition and retention of inhaled fibres: effects on incidence of lung cancer and mesothelioma, *Occup. Environ. Med.*, **51**, 793–798.
83. Davis, J.M.G. and Cowie, H.A. (1990) The relationship between fibrosis and cancer in experimental animals exposed to asbestos and other fibres, *Environ. Health Perspect.*, **88**, 305–309.
84. Weiss, W. (1984) Cigarette smoke, asbestos, and small irregular opacities, *Am. Rev. Respir. Dis.*, **130**, 293–301.
85. Barnhart, S., Thornquist, M., Omenn, G.S., Goodman, G., Feigl, P. and Rosenstock, L. (1990) The degree of roentgenographic parenchymal opacities attributable to smoking among asbestos exposed subjects, *Am. Rev. Respir. Dis.*, **141**, 1102–1106.
86. Sluis-Cremer, G.K. and Hnizdo, E. (1989) Progression of irregular opacities in asbestos miners, *Br. J. Ind. Med.*, **46**, 846–852.
87. Kilburn, K.H., Lilis, R., Anderson, H.A., Miller, A. and Warshowa, R.H. (1986) Interaction of asbestos, age, and cigarette smoking in producing radiographic evidence of diffuse pulmonary fibrosis, *Am. J. Med.*, **80**, 377–381.
88. Lilis, R., Selikoff, I.J., Lerman, Y., Seidman, H. and Gelb, S.K. (1986) Asbestosis: interstitial pulmonary fibrosis in a cohort of asbestos insulation workers: influence of cigarette smoking, *Am. J. Ind. Med.*, **10**, 459–470.
89. Blanc, P.D., Golden, J.A., Gamsu, G., Aberle, D.R. and Gold, W.M. (1988) Asbestos exposure-cigarette smoking interactions among shipyard workers, *JAMA*, **259**, 370–373.
89. Ducatman, A.M., Withers, B.F. and Yang, W.N. (1990) Smoking and roentgenographic opacities in US asbestos workers, *Chest*, **97**, 810–813.
90. McDonald, J.C. and McDonald, A.D. (1986) Epidemiology of asbestos-related lung cancer. In, 'Asbestos-Related Malignancy', Antman, K., Aisner, J., eds., Grune and Stratton, New York, pp. 57–79.
91. Churg, A. and Stevens, B. (1995). Enhanced retention of asbestos fibres in the airways of human smokers, *Am. J. Resp. Crit. Care Med.*, **151**, 1409–1413.
92. Churg, A., Wright, J.L., Hobson, J. and Stevens, B. (1992) Effects of cigarette smoke on the clearance of long and short asbestos fibres from the lung, *Int. J. Exp. Pathol.*, **73**, 287–297.

In Vitro Studies of Genotoxicity and their Significance

M.-C. JAURAND

INSERM UNITE 139, INSTITUT MONDOR DE MEDECINE MOLECULAIRE
FACULTE DE MEDECINE, FRANCE

1 Introduction

Several methodologies have been developed to investigate the carcinogenicity of inorganic substances. As far as the risk of exposure by inhalation is concerned, *i.e.* for lung cancer and malignant mesothelioma, inhalation procedures and intracavitary (intrapleural and intraperitoneal) routes have been employed to expose animals to inorganic compounds. *In vitro* cell systems have been also used to investigate the carcinogenic potency. Every type of investigation has advantages and disadvantages, as summarized in Tables 1 and 2.

In vitro studies are of particular interest to investigate the mechanisms of toxicity of inorganic substances; they allow determination of the nature of the interactions with the cells, they permit the study of, on one hand, of interactions between specific cell types relevant to the type of exposure and the substance of interest; and on the other hand, specific types of interactions relevant to the carcinogenic processes (Table 2).

Since short-term assays using cells in culture are both cheaper and more rapid than long-term inhalation studies, they are also of interest to evaluate the toxic,

Table 1 *Methodologies used to investigate the carcinogenic potency of inhaled particles*

• **Inhalation**	• **Intracavitary**
— For	**inoculation**
• route of exposure similar to the human situation	— For
— Against	• higher tumour rates
• size selection by rodent airways	• lower cost (expenses)
• duration of exposure (low tumour yield)	— Against
• species specificity	• different route of exposure compared with the human situation
• high cost (animal, expenses)	

Table 2 *Methodologies used to investigate the carcinogenic potency of inhaled particles*

- ***In vitro* cell systems**
  — For
    - study of cell/particle interactions => assessment of cellular changes, especially related to carcinogenesis
    - specific and unspecific cell models
    - lower cost (animal, expenses)
  — Against
    - different routes of 'exposure' compared with the human situation
    - isolated systems (but co-cultures available)

especially carcinogenic, potency of substances. The validation of the short term *in vitro* tests should be ideally made by comparison with the carcinogenic potential in humans; however, the lack of human data concerning exact qualitative and quantitative past exposure producing the disease, and the occurrence of numerous uncontrolled factors (exposure to several types of agents, cofactors, individual factors...) render it difficult to establish correlation. Therefore, comparisons between animal and *in vitro* data can be helpful to investigate the advantages and limitations of both sorts of methods *in vitro* and *in vivo*.

In vitro assays are based on the detection of cellular changes associated with the carcinogenic process. Since neoplastic transformation is a multistep process, it is possible to study different markers of events involved in the cell progression (Figure 1). They include determination of DNA damage, repair of damage, chromosome abnormalities and cell transformation. In the following, a genotoxic agent will be considered as a compound for which the biological activity is to alter the information encoded in the DNA. This can occur by point mutations as well as insertions, deletions or changes in chromosome number and/or structure.

The relationship between genotoxicity and carcinogenesis should be considered to realise the limits and/or relevance of genotoxicity assays. Genotoxicity is not necessarily associated with carcinogenesis; for instance, if DNA damage is correctly repaired or if the genetic changes occur in genes other than the critical genes mentioned above. Carcinogenesis is not only dependent on genotoxicity since other changes can occur in the progression of neoplastic transformation. A genetic instability can follow the alteration of critical genes; moreover, other processes (deregulation of cell proliferation) are of importance in neoplastic progression. However, carcinogenesis is always dependent, directly or indirectly, on genotoxicity, since cancer cells always have genetic abnormalities. On a molecular basis, neoplastic transformation results from the activation of oncogenes and inactivation of tumour suppressor genes associated with a modification of the cell cycle regulation. The analysis of molecular changes in cells exposed to carcinogenic agents in order to assess the carcinogenic risk is not yet developed, but one can assume that this could be or will be the aim of further experiments.

In the past years, *in vitro* studies have been developed to investigate the carcinogenic risk of different substances, including the study of effects of mineral and synthetic fibres that represent an important problem in respiratory carcino-

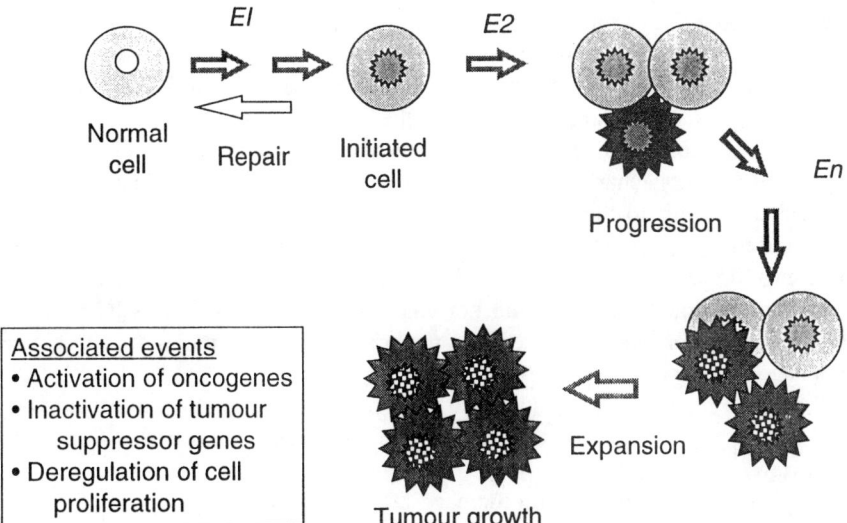

Figure 1 *Schematic representation of multistep transformation. The process of neoplastic transformation is complex; cells undergo several mutational changes due to the occurrence of different events (E_1, E_2 ... E_n) (DNA damage, misrepair or damage, loss of control of cell homeostasis, chronic inflammation...). Associated molecular events are activation of oncogenes, inactivation of tumour suppressor genes and deregulation of cell proliferation. These processes produce the expansion of neoplastic cells and tumour growth.*

genesis. In the following, results will mainly refer to experiments carried out with asbestos fibres and some sorts of synthetic fibres. Several reviews have been devoted to the interpretation of the mechanisms of action of fibres on the basis of both *in vitro* and *in vivo* findings.[1–8] The reader will find in these reviews the references of the studies that permitted the assumptions and conclusions discussed here.

2 Genotoxicity Studies

2.1 DNA Damage

The assessment of DNA damage can be made by the detection of point mutations or, indirectly, by the determination of DNA repair, a process that follows DNA damage in most circumstances.

Point mutations are generally studied in bacteria, but also in some eukaryotic cells. While the first experiments using bacteria failed to demonstrate mutations by asbestos fibres, more recent studies using other cell types demonstrated that asbestos induces mutations (see Ref. 4 for review). Following treatment with asbestos A_L cells, a human–hamster hybrid cell line, previously developed to investigate the effects of radiation[9] and lymphocytes,[10] exhibited mutations characterised by the occurrence of large multifocus deletions and loss of heterozygocity, respectively. The discrepancies between previous and present reports may be due to the nature of the cells and to the different methods used to detect mutations. The selection of revertant bacteria and the point mutation assays in eukaryotic cells need the proliferation of viable cells. If large mutations have occurred, mutants may not be selected. In contrast, the hamster–human hybrids can grow even in the presence of large mutations in the selection gene.

Another reason for discrepancies between results in bacteria and eukaryotes may stand in the mechanism of DNA damage. Several studies indicate that DNA damage can be due to the formation of oxygen derivatives. These molecules may be produced by phagocytic cells in the early steps in the process of phagocytosis. Fibre engulfment has generally been observed in cells in culture. As far as lymphocytes are concerned these cells do not ingest asbestos fibres, but as lymphocytes are obtained from blood samples, the preparations contain monocytes that carry out phagocytosis.

It may be noted here that some earlier reports have, however, indicated some weak mutagenic effect in bacteria and lung cells treated with asbestos (see Ref. 4 for review).

The ability of a carcinogen to damage DNA can also be assessed by the determination of DNA damage (DNA breakage) and DNA repair. DNA strand breaks can be shown by alkaline elution or separation of DNA fragments. DNA repair is generally detected by autoradiography, where labelled nuclei are observed after incorporation of [^{3}H] thymidine in growth arrested (G0/G1) cells.

Asbestos fibres failed to induce strand breaks in several test systems and DNA repair was not detected by autoradiography in hepatocytes.[11] However, using other methods of investigation, DNA breakage can be suggested[12,13] by the activation of poly(ADP)ribose polymerase; moreover, DNA repair was found in rat pleural mesothelial cells (RPMC). In this latter assay, DNA repair has been studied by the measurement of unscheduled DNA synthesis (UDS).[14] Both DNA breakage and DNA repair seem partly related to the production of oxygen derivatives by the asbestos-treated cells; this production is more likely dependent on fibre–cell interactions than on fibre uptake (unpublished data from our laboratory). All fibre samples, MMF and asbestos, were efficient in producing DNA repair in rat pleural mesothelial cells. Several samples (including some sample of talc and attapulgite) tested in the UDS assay have been studied for their carcinogenic potency by intrapleural inoculation in rats. A good correlation was obtained between UDS enhancement by mineral samples and tumourigenicity. When MMF were taken into consideration, the correlation between *in vivo* and *in vitro* findings were less evident, but these fibres have been tested *in vivo* by inhalation, a method where the tumour yield is often at the limit of statistical significance, rendering difficult correlation studies. The results must be interpreted regarding the nature of the agent tested. In our hands, DNA damage by inorganic particles is difficult to assess using methods based on centrifugation or separation procedures by chromatography, because of interactions between DNA and fibres; therefore, indirect methods may be more appropriate to search for DNA damage. Similarly, autoradiographic methods may be difficult to apply because of the microscopic observation step, especially when numerous or large fibres are used. Other more recent methods based on electrophoresis could be applied.

The observation of some forms of DNA damage *in vitro* following treatment of cells with asbestos demonstrates the fibre's ability to be responsible for events involved in the carcinogenic process. According to the authors knowledge, DNA damage and repair by asbestos has not been studied *in vivo*. A recent report has demonstrated that DNA double-strand breaks and anti-ds DNA antibodies were detected in the blood of workers occupationally exposed to asbestos;[15] therefore, some identity between *in vitro* and *in vivo* effects can be suggested.

If DNA damage and repair induced by a substance is a mark of the substance's potency to produce some changes associated with the carcinogenic process, this does not mean that these changes will inevitably generate neoplasia. While some forms of repair lead to mutations, others may be error free. This must be kept in mind if one wants to extrapolate *in vitro* results to the *in vivo* situation.

Short-term *in vitro* tests were originally developed to study the mechanisms of DNA damage by chemicals. They have been used to assess the potential genotoxicity of chemicals to humans. These tests have originally focused on the detection of point mutations. However, not all carcinogens are positive in these assays and 'nongenotoxic' carcinogens have been identified.[16] Recently, an evaluation based on EPA/IARC databases has demonstrated that many of the putative nongenotoxic carcinogens induce gene or chromosomal mutations or aneuploidy in short-term tests.[17] This indicates that agents inducing DNA repair and cytogenetic effects may also be considered as 'genotoxic' in a large sense. Therefore, asbestos fibres may be considered amongst these compounds.

2.2 Chromosome Damage

There are several ways to assess chromosome damage. Two types of chromosome abnormalities can be detected, namely structural chromosome damage (chromosome aberrations, micronucleus), and numerical chromosome damage (aneuploidy, polyploidy). Moreover, studies of mitotic cells at the anaphase stage allow investigation of chromosome mis-segregation.

The results of chromosome damage in asbestos- or fibre-treated cells have been discussed earlier.[4] Briefly, low but significant levels of chromosome breaks and other structural abnormalities are found in different cell types as the result of the formation of clastogenic factors possibly associated with oxyradicals.[18,19] Similar to what has been mentioned above for mutations, lymphocytes do not appear to form chromosome aberrations in the absence of monocytes, an observation in agreement with the hypothesis of a clastogenicity due to mediators released by phagocytes, where oxyradicals and oxygen derivatives can be involved.

Aneuploidy has been generally found after treatment of different cell types with asbestos.[4] The formation of aneuploid cells provokes the development of cell populations with different gene dosage and may favour the emergence of altered cells with genetic changes related to neoplasia (mutations, lack of growth regulation).[20] Aneuploidy has also been detected in *Drosophila melanogaster* using the ZESTE genotoxic test systems.[21]

Aneuploidy can be produced by substances that interact with microtubules or kinetochores, the motors that drive the chromosomes during mitosis. This can occur because of physical interaction and/or affinity between the cell components and the toxic agents. The chromosome damage could be due to the impairment of chromosome movement by fibres strongly anchored to the cellular matrix.[22] *In vitro*, it is possible to detect anaphase aberrations in log phase cells treated with different substances, including asbestos. The abnormalities are defined as bridges between chromosomes, lagging chromatin and fragments that correspond to delayed migration of some chromosome material or chromosome breakage.[23–25]

Many human and rodent tumours exhibit specific cytogenetic changes. In malignant mesothelioma, the tumour specifically related to asbestos exposure, a great range of chromosome alterations has been found, but no specific changes in spite of some non-random changes located on chromosomes 1, 3, 4, 5, 7, 9, 20 and 22.[26–28] While chromosome aberrations may be related to the production of oxygen derivatives and the formation of more stable clastogenic factors, anaphase abnormalities are more likely dependent on fibre dimensions. In a recent study where several samples of mineral fibres were studied for their ability to produce anaphase abnormalities in rat pleural mesothelial cells, no anaphase abnormalities were found in cells exposed to short fibre samples, in contrast with other fibre samples containing long and thin fibres. However, a sample of asbestos (amosite) did not form anaphase abnormalities.[29] Comparison between effects of different samples suggested that a threshold amount of critical fibres was necessary to detect an effect. Interestingly, a correlation was found between the fibre potency to produce abnormal anaphases and the carcinogenic potency by intrapleural inoculation. Therefore, in this test system, the fibre geometry strongly accounts for the effects, but the results do not exclude the role of other parameters. These results emphasize the interest for understanding the mechanisms whereby the fibres produce the effects.

2.3 Neoplastic Cell Transformation

The observation of neoplastic cell transformation after exposure to test substances is an indication that the cells have undergone multiple changes following application of the agent. It suggests that the substance may operate at different steps of the carcinogenic process. The criteria for the evaluation of cell transformation may depend on the cell type; they are generally based on observation of morphological changes (loss of contact inhibition), modification of growth properties (growth in semi-solid medium) and tumour growth in nude mice. Morphologically transformed colonies may represent the signature of the occurrence of early steps in a progressive, multistep process leading to neoplastic transformation. In some specific cell types where cell progression is better known, some specific changes can be studied. Transformation of C3H 10 T1/2 mice cells, Syrian hamster embryo (SHE) cells, Balb 3T3 and rat pleural mesothelial cells have been observed following treatment with asbestos (see Ref. 4 for review) and with glass microfibres and other dusts in SHE cells. Aneuploidy has been identified as a change in neoplastic progression of SHE and rat pleural mesothelial cells and detected as an early change after treatment with asbestos. In order to explore the mechanism of fibre-induced SHE cells transformation, fibres of different lengths have been studied. The transformation was dependent on fibre length, short fibres being less active than long fibres. Two factors appear to account for fibre size dependence of SHE cell transformation: the phagocytosis selects long fibres, and chromosome mutations induced by the fibres are fibre length dependent.[30,31] So far, it remains unknown whether oxygen derivatives produced in asbestos–cell interactions are involved in this cell transformation.

3 Cytotoxicity, Cell Viability

The study of cell viability is currently made to determine the lethal concentrations of the test substances and to compare the activity of different samples at equitoxic doses. Large numbers of methods are applied, based on the assessment of metabolic changes (ATP production, oxygen consumption, mitochrondrial integrity...) or cell death; other specific techniques have been developed. If cytotoxicity tests are useful for *in vitro* testing, the cytotoxicity is not generally directly associated to the carcinogenicity. However, as far as fibres are concerned, a good correlation was obtained between cytotoxicity and fibre length in several cell systems.[29,32,33]

4 Present and Further Research Areas

From the *in vitro* studies carried out with several sorts of mammalian cells and inorganic substances such as asbestos, it appears that DNA damage and repair of damage can be detected by different test systems and likely result, at least partly, from cell–fibre interactions involving oxygen derivatives. The occurrence of structural chromosome aberrations is dependent on the cell system, but the formation of numerical chromosome abnormalities and associated changes (abnormal anaphases) are more currently observed. The extent of these latter abnormalities seems dependent on the number of critical fibres present in the sample and suggests that the fibre geometry is of importance in the development of such abnormalities. Long and thin fibres are also more efficient in inducing cell transformation, in agreement with the relationship between chromosome changes and cell transformation demonstrated in some cell systems.

An extended discussion of the relationship between *in vitro* assessment of cell damage in mesothelial cells and the carcinogenic potential of different fibres in animals has been published elsewhere.[33]

In vitro, the cell response is dependent on fibre parameters such as fibre dimensions, demonstrated as playing a role *in vivo* in carcinogenesis. With asbestos, fibre dimensions may be important both in the generation of active oxygen species by asbestos-treated cells[34] and in the formation of chromosome mutations.

As neoplastic transformation results from molecular changes involving critical genes, the molecular mechanisms whereby fibres produce neoplastic transformation are under investigation. It is likely that general mechanisms will continue to be identified that are independent of the cell type but that some cell type specificities and species specificities will be discovered.

Numerous processes can be investigated to determine a carcinogenic potential (Figure 2). Carcinogens produce DNA damage, the type of which depends on the

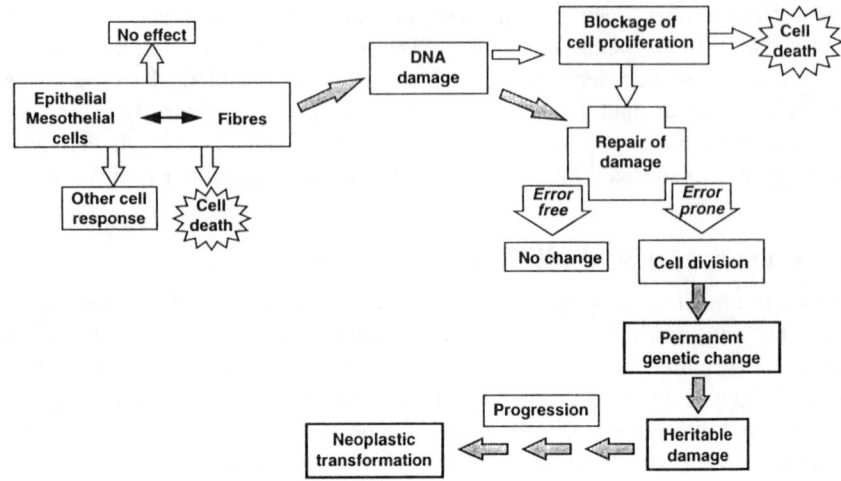

Figure 2 *Schematic representation of the possible pathways resulting from interactions between fibres and target cells, and that eventually could produce neoplastic transformation. The figure illustrates the numerous processes that can be assessed* in vitro.

nature of the agent. The DNA integrity is controlled in a normal cell and its impairment is associated with the activation of genes of DNA repair, growth arrest, or even death commitment such as apoptosis.[35] Therefore, DNA damage and repair and the balance between cell life and cell death are basic concepts in neoplastic cell transformation and tumour growth. What will happen to a damaged cell in terms of survival will depend on the extent of damage, ability to repair and level of survival factors. Tumour growth will develop when cell birth is greater than cell death.

DNA and chromosome damage are key events in the initiation of neoplastic cell transformation, and may well be involved in further steps allowing its achievement. What will happen to the damaged cell, in terms of cell pathology will depend on the fidelity of DNA repair; misrepair of damage will produce abnormal cells prone to progress further additional damages. Alternatively, DNA-damaged cells may undergo a suicide process (apoptosis) to avoid to emergence of pathological cells.[36]

The different pathways that control the cell evolution following damage due to external causes is regulated by genes and proteins that play a role in the control of the cell progression in the cell cycle.[37,38] The regulation is exerted by transcription factors, themselves regulated by post-translational mechanisms. The proteins encoded by the genes *p53, ECCR45, c-FOS, c-MYC, BCL2* and the CDKs are important factors that may be induced or activated following damage. The poly(ADP) ribosylation of proteins is another event that controls the protein activity and stability after DNA damage.[39] Therefore, the effects of inorganic substances on the genes involved in cell cycle regulation are under investigation. So far, induction of *c-FOS* and *c-JUN* has been reported in mesothelial cells.[40] With early passages of rat pleural mesothelial cells, little apoptosis was detected in asbestos-treated cells[41] in agreement with previous findings on the activation of repair pathways. To study these processes, *in vitro* cell systems are of great interest.

5 Conclusions

The *in vitro* tests allow the study of individual steps involved in the carcinogenic process. The performance of different assays is useful to determine whether the activity of the substance under investigation is limited to one specific effect or characterised by the occurrence of several effects. The comparison between data obtained in different cell systems is certainly useful to identify a carcinogenic substance. It is also important to determine the basic mechanisms explaining the effects. As emphasized in the paragraph related to the study of anaphase abnormalities, an absence of effect does not always mean that the substance has no potential; this may be only due to dose–response considerations, if the amount of the active part of the agent tested is under the non-observable effect levels.

The extrapolation between results obtained *in vitro* and the human situation is a critical point. As far as asbestos is concerned, a good correlation is obtained between *in vitro* results with cells systems and animal findings, but the epidemiological evidence is sometimes different.[42] The differences can be due to some uncertainties in the knowledge of human past asbestos exposure and to differences in the biopersistence of the fibres in human, animal and *in vitro* situations.

References

1. Barrett, J.C., Lamb, P. and Wiseman, R.W. (1989) Multiple mechanisms for the carcinogenic effects of asbestos and other mineral fibres, *Environ. Health Persp.*, **81**, 81–92.
2. Jaurand, M.C. (1989) Particulate-state carcinogenesis. A survey of recent studies on the mechanism of action of fibres. In, 'Non-occupational Exposure to Mineral Fibres', Bignon, J., Peto, J. and Saracci, R., eds., IARC Scientific Publication, **90**, pp. 54–73.
3. Mossman, B.T., Bignon, J., Corn, M. and Seaton, A. (1990) Asbestos: Scientific developments and implications for public policy, *Science*, **247**, 294–301.
4. Jaurand, M.C. (1991) Mechanisms of fibre genotoxicity. In, 'Mechanisms in Fibre Carcinogenesis', Brown, R.C., Hoskins, J.A., Johnson, N.F., eds., NATO Advanced Workshop, Plenum Press, New York and London, pp. 287–307.
5. Barrett, J.C. (1992) Mechanisms of action of known carcinogens. Mechanisms of Carcinogenesis in Risk Identification, International Agency for Research on Cancer, 115–134.
6. Walker, C., Everitt, J. and Barrett, J.C. (1992) Possible cellular and molecular mechanisms for asbestos carcinogenicity, *Am. J. Ind. Med.*, **21**, 253–273.
7. Brody, A.R. (1993) Asbestos-induced lung disease, *Environ. Health Perspect.*, **100**, 21–30.
8. Everitt, J.I. (1994) Mechanisms of fiber-induced diseases: implications for the safety evaluation of synthetic vitreous fibers, *Regul. Toxicol. Pharmacol.*, **20**, S68–S75.
9. Hei, T.K., Piao, C.Q., He, Z.Y., Vannais, D. and Waldren, C.A. (1992) Chrysotile fiber is a strong mutagen in mammalian cells, *Cancer Res.*, **52**, 6305–6309.

10. Both, K., Henderson, D.W. and Turner, D.R. (1994) Asbestos and erionite fibres can induce mutations in human lymphocytes that result in loss of heterozygocity. *Int. J. Cancer*, **59**, 538–542.
11. Denizeau, F., Marion, M., Chevalier,G. and Cote, M. (1985) Inability of chrysotile asbestos fibers to modulate the 2-acetylaminofluorene-induced UDS in primary cultures of hepatocytes, *Mutat. Res.*, **155**, 83–90.
12. Dong, H.Y., Buard, A., Renier, A., Levy, F., Saint-Etienne, L. and Jaurand, M.C. (1994) Role of oxygen derivatives in the cytotoxicity and DNA damage produced by asbestos on rat pleural mesothelial cells *in vivo*, *Carcinogenesis*, **15**, 1251–1255.
13. Dong, H.Y., Buard, A., Levy, F., Renier, A., Laval, F. and Jaurand, M.C. (1995) Synthesis of poly(ADP-ribose) in asbestos treated rat pleural mesothelial cells in culture, *Mutat. Res.*, (in press).
14. Renier, A., Levy, F., Pilliere, F. and Jaurand, M.C. (1990) Unscheduled DNA synthesis in rat pleural mesothelial cells treated with mineral fibres or benzo[*a*]pyrene, *Mutat. Res.*, **241**, 361–367.
15. Marczynski, B., Czuppon, A.B., Marek, W., Reichel, G. and Baur, X. (1994) Increased incidence of DNA double-strand breaks and anti-ds DNA antibodies in blood of workers occupationally exposed to asbestos, *Hum. Exp. Toxicol.*, **13**, 3–9.
16. Butterworth, B.E. (1990) Consideration of both genotoxic and nongenotoxic mechanisms in predicting carcinogenic potential, *Mutat. Res.*, **239**, 117–132.
17. Jackson, M.A., Stack, H.F. and Waters, M.D. (1993) The genetic toxicology of putative nongenotoxic carcinogens, *Mutat. Res.*, **296**, 241–277.
18. Emerit, I., Jaurand, M.C., Saint-Etienne, L. and Levy, F. (1991) Formation of a clastogenic factor by asbestos-treated rat pleural mesothelial cells, *Agents Actions*, **34**, 410–415.
19. Korkina, L.G., Durnev, A.D., Suslova, T.B., Cheremisina, Z.P., Daugel-Dauge, N.O. and Afanas'ev, I.B. (1992) Oxygen radical-mediated mutagenic effect of asbestos on human lymphocytes suppression by oxygen radical scavengers, *Mutat. Res.*, **265**, 245–253.
20. Oshimura, M., Hesterberg, T.W. and Barrett, J.C. (1986) An early, nonrandom Karyotypic Change in Immortal Syrian hamster cell lines transformed by asbestos: trisomy of chromosome 11, *Cancer Genet. Cytogenet.*, **22**, 225–237.
21. Osgood, C. and Sterling, D. (1991) Chrysotile and amosite asbestos induce germ-line aneuploidy in drosophila, *Mutat. Res.*, **261**, 9–13.
22. Ault, J.G., Cole, R.W., Jensen, C.G., Jensen, L.C.W., Bachert, L.A. and Rieder, C.L. (1995) Behavior of crocidolite asbestos during mitosis in living vertebrate lung epithelial cells, *Cancer Res.*, **55**, 792–798.
23. Hesterberg, T.W. and Barrett, J.C. (1985) Induction by asbestos fibers of anaphase abnormalities: mechanism for aneuploidy induction and possibly carcinogenesis. *Carcinogenesis*, **6**, 473–475.
24. Palekar, L.D., Eyre, J.F., Most, B.M. and Coffin, D.L. (1987) Metaphase and anaphase analysis of V79 cells exposed to erionite, UICC chrysotile and UICC crocidolite, *Carcinogenesis*, **8**, 553–560.
25. Yegles, M., Saint-Etienne, L., Renier, A., Janson, X. and Jaurand, M.C. (1993) Induction of metaphase and anaphase/telophase abnormalities by asbestos fibers in rat pleural mesothelial cells *in vitro*, *Am. J. Respir. Cell Mol. Biol.*, **9**, 186–191.
26. Gibas, Z., Li, F.P., Antman, K.H., Bernal, S., Stahel, R. and Sandberg, A.A. (1986) Chromosome changes in malignant mesothelioma, *Cancer Genet. Cytogenet.*, **20**, 191–201.
27. Tiainen, M., Tammilehto, L., Rautonen, J., Tuomi, T., Mattson, K. and Knuutila, S. (1989) Chromosomal abnormalities and their correlations with asbestos exposure and survival in patients with mesothelioma, *Br. J. Cancer*, **60**, 618–626.

28. Hagemeijer, A., Versnel, M.A., Van Drunen, E., Moret, M., Bouts, M.J., Van der Kwast, Th.H. and Hoogsteden, H.C. (1990) Cytogenetic analysis of malignant mesothelioma, *Cancer Genet. Cytogenet.*, **47**, 1–28.
29. Yegles, M., Janson, X., Dong, H.Y., Renier, A. and Jaurand, M.C. (1995) Role of fibre characteristics on cytotoxicity and induction of anaphase/telophase aberrations in rat pleural mesothelial cells *in vitro*. Correlations with *in vivo* animal findings, *Carcinogenesis*, (accepted for publication).
30. Hesterberg, T.W. and Barrett, J.C. (1984) Dependence of asbestos- and mineral dust-induced transformation of mammalian cells in culture on fiber dimension, *Cancer Res.*, **44**, 2170–2180.
31. Hesterberg, T.W., Butterick, C.J., Oshimura, M., Brody, A.R. and Barrett, J.C. (1986) Role of phagocytosis in syrian hamster cell transformation and cytogenetic effects induced by asbestos and short and long glass fibers, *Cancer Res.*, **46**, 5795–5802.
32. Hart, G.A., Kathman, L.M. and Hesterberg, T.W. (1994) *In vitro* cytotoxicity of asbestos and man-made vitreous fibers: Roles of fiber length, diameter and composition. *Carcinogenesis*, **15**, 971–977.
33. Jaurand, M.-C., Ellouk-Achard, S., Yegles, M., Dong, H., Levresse, V. and Renier, A. (20–24 February, Hannover, Germany) Relationship between DNA and chromosome damage and mesothelial cells transformation. Significance for *in vitro* assessment of carcinogenic potential, *5th Int. Inhalation Symposium* (in press).
34. Mossman, B.T. (1993) Mechanisms of asbestos carcinogenesis and toxicity – The amphibole hypothesis revisited, *Br. J. Ind. Med.*, **50**, 673–676.
35. Eastman, A. and Barry, M.A. (1992) The origins of DNA breaks – A consequence of DNA damage, DNA repair, or apoptosis? *Cancer Invest.*, **10**, 229–240.
36. Fishel, R. and Kolodner, R.D. (1995) Identification of mismatch repair genes and their role in the development of cancer, *Curr. Opin. Genet. Dev.*, **5**, 382–395.
37. Kuerbitz, S.J., Plunkett, B.S., Walsh, W.V. and Kastan, M.B. (1992) Wild-type p53 is a cell cycle checkpoint determinant following irradiation, *Proc. Natl. Acad. Sci. U.S.A.*, **89**, 7491–7495.
38. Kastan, M.B., Onyekwere, O., Sidransky, D., Vogelstein, B. and Craig, R.W. (1990) Participation of p53 protein in the cellular response to DNA damage, *Cancer Res.*, **51**, 6304–6311.
39. Bhatia, K., Pommier, Y., Fornace, A.H., Imaitzumi, M., Breitman, T.R., Cherney, B.W. and Smulson, M.E. (1990) Expression of the poly(ADP-ribose) polymerase gene following natural and induced DNA strand breakage and effect of hyperexpression on DNA repair, *Carcinogenesis*, **11**, 123–128.
40. Janssen, Y.M.W., Heintz, N.H., Marsh, J.P., Borm, P.J.A. and Mossman, B.T. (1994) Induction of *c-fos* and *c-jun* proto-oncogenes in target cells of the lung and pleura by carcinogenic fibers, *Am. J. Respir. Cell Mol. Biol.*, **11**, 522–530.
41. Levresse, V., Renier, A., Levy, F., Greffard, A., Pilatte, Y. and Jaurand, M.C. (1995) Induction de l'apoptose et expression de p53 dans les cellules mésothéliales pleurales de rat traitées, *in vitro*, par les fibres d'amiante. In, 'Eurocancer', Boiron, M., Marty, M. eds., (in press).
42. Becklake, M.R. and Case, B.W. (1994) Fiber burden and asbestos-related lung disease: determinants of dose- response relationships, *Am. J. Respir. Critical Care Med.*, **150**, 1488–1492.

Sequence of Events in Lung Carcinogenesis

C. H. KENNEDY AND J. F. LECHNER

INHALATION TOXICOLOGY RESEARCH INSTITUTE, ALBUQUERQUE, USA

1 Interindividual Variation in Cancer Susceptibility

Dose–response studies have shown that the magnitude of exposure to a chemical or physical carcinogen is a major determinant of risk for developing cancer.[1] However, other variables such as the nutritional status and the presence of co-carcinogens and/or tumour-promoting agents can markedly alter this risk.[2,3] In addition, for outbred animals and humans, it is now recognised that intrinsic tumour susceptibility genes play a major role in defining the risk of whether a cancer will develop after exposure to a carcinogen.[4-9] These genes operate as polymorphic variants of 'normal' genes or genetic loci that either positively or negatively change the probability that cancer will develop in each carcinogen-exposed individual.

Several enzyme polymorphisms that putatively affect the rate of metabolism and/or detoxification of chemical carcinogens are candidate tumour susceptibility genes. The archetype example is cytochrome P-450 CYP2D6. For reasons only partially explained by the metabolism of tobacco smoke constituents, individuals homozygous for the variant of this enzyme who extensively metabolize the antihypertensive drug debrisoquine are at a two- to five-fold greater risk for developing lung cancer than those individuals who are homozygous for the poor-metabolizer phenotype. Another example is the restriction fragment length polymorphism in the second intron of cytochrome P-450IIEl that is detected by the endonuclease *Dra* I.[10] Japanese, who are homozygous for this enzyme restriction site, have a disproportionately higher frequency of lung cancer than heterozygotic individuals and those without the enzyme restriction site. Similar examples, relating inherent cancer susceptibility to rare polymorphic forms of *N*-acetyltransferase[11] and absence of glutathione *S*-transferase I activity,[12] have been reported. Other examples of tumour susceptibility genes include rare polymorphic variants of proto-oncogenes, such as the *Eco* RI restriction site variant of human c-*mos*,[13] the variable number tandem repeat minisatellite that is downstream of human H-*ras*,[14] the *Eco* RI restriction site variant in the second intron of human L-*myc*[15] and deletion of the tandem repeat in the second intron of murine K-*ras*.[16] The molecular mechanisms whereby these variants influence susceptibility are obscure because, putatively, none of these genetic polymorphisms directly affects the nature of the protein products of these proto-oncogenes.

Rare cases of inherited heterozygocity of a dysfunctional tumour suppressor gene include *p53* (Li-Fraumeni Syndrome), *Rb-1* (inherited retinoblastoma), *WT1* (Wilms's tumour), *APC* (Familial adenomatous polyposis coli), and *NF1* (neurofibromatosis).[17] These represent a different mechanism of the tumour susceptibility genes. Afflicted individuals are born with only one functional allele of a particular tumour suppressor gene. Thus, these polymorphisms function as tumour susceptibility genes because of the increased probability that a single somatic mutation will result in the total loss of activity of the tumour suppressor gene. Analogous to these rare syndromes are several rare inherited clinical syndromes such as Fanconis anemia, Blooms syndrome, Xeroderma pigmentosum, Cockayne's syndrome and Ataxia-Telangiectasia. Individuals with Fanconis anemia suffer a high risk for cancer from exposure to alkylating agents, whereas patients having any of the other four diseases show enhanced susceptibility to cancer development if they are exposed to ultraviolet light or ionizing radiation.[17]

These examples represent only a small portion of the potential list of polymorphisms that affect interindividual differences in susceptibility.[8] Presumably, many tumour susceptibility genes remain to be discovered that modulate susceptibility to cancer following exposure to chemical and physical carcinogens.

2 Field Cancerization Theory

Molecular, cellular and epidemiological models of carcinogenesis indicate that cells undergo multiple genetic, epigenetic and morphological changes prior to the appearance of an invasive cancer.[18,3] The absolute number and specificity of gene changes leading to tumour formation are likely to be different for tumours that develop independently, suggesting that alternative patterns of gene dysfunction can result in the development of cancer.[3] Thus, in the case of lung cancer, cells with genetic alterations should populate the lung epithelium long before frank cancer is evident. In the early 1960s, the presence of progressive histolopathological changes throughout the bronchial epithelium of cigarette smokers was first reported.[19,20] This phenomenon, now referred to as field cancerization, is characterised by premalignant lesions, ranging from hyperplasia and metaplasia to severe dysplasia and carcinoma *in situ*, distributed diffusely throughout the bronchial mucosa of smokers' lungs.[21] The number and severity of these changes were directly related to cigarette smoke exposure; smokers diagnosed with lung cancer had more areas containing carcinoma *in situ* and atypical cells than heavy smokers (> 2 packs/day) without tumours. The high occurrence of second primary tumours in lung cancer patients supports the supposition that aberrant cells within the cancerization field can develop into invasive cancer.[22-24] Further, a study of the genetic alterations in these secondary tumours has confirmed that these cancers developed independently of the primary tumour.[25]

3 Somatic Genetic Changes Commonly Detected in Lung Cancer

Neoplasia arises because certain genetic alterations provide a clone of cells with a relative growth advantage. These genetic changes occur in two different categories of genes involved in the control of cellular proliferation: oncogenes and tumour suppressor genes. Proto-oncogenes encode proteins that stimulate cell division by functioning as growth factors, growth factor receptors, signal transducers and positive regulators of nuclear gene expression.[26] Tumour suppressor genes encode proteins that inhibit cell division by functioning as negative regulators of cell cycling.[27] Oncogenes are dominant because mutation of only one allele, *via* point mutation or gene amplification, is necessary to elicit a phenotypic effect. Conversely, tumour suppressor genes are usually recessive because both copies of the gene must be either deleted or mutated to have a phenotypic effect.

Detection of an extra copy of any chromosome is evidence of generalized aneuploidy, which is an early, ubiquitous characteristic of lung cancer.[28] Trisomy 7, an important chromosomal anomaly for lung cancer, is observed in 50% of non-small cell lung carcinomas (NSCLCs) and has been frequently detected in the far margins from resected lung tumours.[29] Thus, an extra copy of chromosome 7 may play a role in the early stages of neoplastic development in some tumours. The molecular mechanism by which the extra chromosome provides a growth advantage is unknown. However, this chromosome harbours the proto-oncogenes *EGFR* and *Met*, and the gene products of both of these loci are frequently overexpressed in human lung tumour cells.[30,31]

In addition to aneuploidy, specific gene mutations are also present in lung tumours. A major player in the development of NSCLC is the proto-oncogene K-*ras*, a member of the *ras* family that also includes the proto-oncogenes H-*ras* and N-*ras*. In its activated (K-ras-GTP) form, the K-ras protein interacts with the raf protein to stimulate cellular growth via a signal transduction pathway.[32] Thirty percent of adenocarcinomas, a histological subtype of NSCLCs, have K-*ras* point mutations at codons 12, 13 or 61;[33] however, these mutations are rarely detected in other histological subtypes of NSCLCs[34,33] and are seldom present in either small cell lung carcinomas (SCLCs) or SCLC-derived cell lines.[35] These mutations inactivate the intrinsic GTPase activity of the K-ras protein so that it cannot cycle back to its inactive (ras-GDP) form, resulting in a continuous signal for cellular growth. It is, therefore, not surprising that detection of mutant K-*ras* in early stage adenocarcinomas is indicative of a poor prognosis for lung cancer patients.[36] While H-*ras* and N-*ras* mutations are infrequently detected in lung tumours, these aberrations also correlate with a poor prognosis for the patients.[37] Likewise, expression of the protein product of H-*ras* in well- or moderately-differentiated adenocarcinomas is correlated with poor prognosis.[38]

The proto-oncogenes c-*erb*-B1, which encodes the epidermal growth factor receptor (EGFR), and c-*erb*-B2 (or HER2/*neu*) are members of the type 1 family of tyrosine kinase growth factor receptors;[39] these genes also play an important role in the genesis of NSCLC. The role of EGFR will be discussed in a later section on autocrine loops. Unlike the *ras* genes, c-*erb*-B2 is not activated via point mutations; instead, gene amplification (*i.e.* increased gene copy number) results in aberrantly high expression of its protein product p185^{erbB}.[40] This overexpression occurs predominantly in lung adenocarcinomas[41] and is correlated with decreased survival of NSCLC patients.[42] Co-expression of EGFR and c-erb-B2 proteins is correlated with an even poorer prognosis, and it has been suggested that amplification of the c-*erb*-B1 and c-*erb*-B2 proto-oncogenes may play a significant role in either cancer invasion or metastasis in patients with adenocarcinoma of the lung.[43]

The *myc* family of dominant oncogenes, which includes c-*myc*, L-*myc* and N-*myc*, encodes a group of closely related nuclear proteins that stimulate cellular proliferation by transcriptional regulation of gene expression.[44] Myc protein overexpression occurs as a result of either gene amplification or by mutations in a transcriptional regulatory sequence in the first intron.[45] This aberration is a frequent event in SCLCs but is rarely detected in NSLCs.[46-48] Although all three members of the *myc* family are amplified in lung cancer, only one member is amplified per tumour. Clinical studies have shown a correlation between *myc* amplification and decreased patient survival.[49,50] Further, c-*myc* amplification is proposed to play a role in the progression of SCLCs and the development of resistance to chemotherapy, based on the fact that amplification of this proto-oncogene is usually detected only after chemotherapy and clinical relapse.[49-51]

Although variations in expression of proto-oncogenes such as c-*raf*-1, *jun* and *src*-related tyrosine kinase genes have also been identified in lung tumours, their role in carcinogenesis is unclear because they are also expressed in normal lung tissue.[45]

General chromosomal losses are also important in the genesis of a lung tumour, particularly regions of chromosomes 3p, 5q, 9p, 11p, 13q and 17p.[28] However, loss of heterozygocity (LOH) is most frequently detected at 3p, 9p21 and 17p13. Deletion of these regions contributes to the development of cancer because LOH results in the loss of one copy of a tumour suppressor gene.

The first tumour suppressor gene to be characterised was the *rb* gene, which resides on chromosome 13 at q14. *Rb* encodes a nuclear phosphoprotein that regulates the cell cycle by binding to other regulatory proteins such as c-*myc* and the transcription factor E2F.[52,53] The binding ability of the Rb protein is activated *via* phosphorylation by cdc2 kinase.[54] Therefore, LOH at 13q14 followed by point mutation of the second copy of *rb* (so that the resulting mutant protein could not be phosphorylated) would result in a loss of E2F binding at this regulatory step. LOH on the 13q region containing the *rb* gene has been detected in SCLCs and NSCLCs.[55,56] In addition, > 95% of SCLC cells and 20% of NSCLC cells exhibit point mutations in the *rb* gene that result in expression of a defective Rb protein. Thus, *rb* appears to be a major player in SCLC.

Another tumour suppressor gene is *p53*, one of the most frequently mutated genes in cancer. The *p53* gene, which is located on chromosome 17 at p13.1, encodes a nuclear phosphoprotein that activates transcription via binding to the regulatory region of the *waf*1/*cip*1 gene.[57] The Cip1 protein, in turn, causes cell-cycle arrest at the G1-S checkpoint by inhibiting multiple cyclin D kinase (cdk)-cyclin complexes.[58] This arrest allows the cell to repair any DNA damage caused by either chemical or physical carcinogens and/or genotoxicants prior to

DNA replication. Mutation of the *p53* gene leads to overexpression of a mutant p53 protein that cannot bind to *waf*1/*cip*1 to initiate cell-cycle arrest.[59] In this case, the cell cannot repair the damaged DNA prior to the S phase of the cell cycle, resulting in 'fixed' mutations that are passed on to progeny of the exposed cell. The half life of the mutant p53 protein is significantly longer than that of the corresponding wild-type protein, facilitating detection of p53 mutations by immunohistochemical analysis.[60] In lung cancer, mutations in the *p53* gene have been detected in approximately 90% and 60% of SCLC and NSCLC cases, respectively.[61–64] To date, it has been difficult to correlate *p53* mutation with prognosis. This may be due to the fact that p53 protein can be inactivated by mechanisms other than mutation, such as binding to viral proteins, and conversion from a wild-type conformation to a mutant conformation *via* displacement of zinc ions by either mercury or cadmium ions.[65] However, loss of p53 protein function may result in cancer promotion by preventing the cell from maintaining the fidelity of genetic material, thereby contributing to the induction of genomic instability, which is characteristic of solid tumour cells.[66] In addition, it has been shown[67] that wild-type p53 protein induces apoptosis (*i.e.* programmed cell death); therefore, loss of p53 protein function may also inactivate this form of cellular protection.

Region 9p21, which is lost in up to 60% of SCLCs and 70% of NSCLCs,[28,68] contains the gene $p16^{INK4a}$. This gene encodes a protein that regulates cell cycling by governing the kinase activity of the complex consisting of cdk4, proliferating cell nuclear antigen, and cyclin D.[69] A related cell-cycle regulating gene, referred to as $p15^{INK4b}$, is also located in this region.[70] Recently, deletions in these genes have been identified as frequent mutations in up to approximately 40% of NSCLCs; however, no deletions have been detected in SCLCs.[71,72] An inverse relationship exists between the expression of the $p16^{INK4a}$ and Rb proteins, and it has been suggested that $p16^{INK4a}$ may act in a feedback loop to regulate the ability of Rb to inhibit cellular proliferation.[69] Because mutations in the $p16^{INK4a}$ gene have been detected in 27% of metastatic NSCLCs, but not in primary NSCLCs or in SCLCs (primary or metastatic), these mutations are proposed to play a role in the tumour progression of a subset of NSCLC.[71] Likewise, mutations in the $p15^{INK4b}$ gene have been detected in 12% of primary NSCLCs and 23% of metastatic NSCLCs, but not in SCLCs (primary or metastatic); therefore, these mutations, in addition to contributing to tumour progression of certain NSCLCs, are also proposed to be an early alteration in the development of a subset of NSCLCs.[71]

Deletion on the short arm of chromosome 3 was the first common genetic aberration detected in lung tumours.[73] LOH of 3p is observed in >90% of SCLCs and approximately 60% of NSCLCs.[55,74,75] Mapping studies have shown that the regions most frequently deleted are 3p14, 3p21 and 3p25.[76] Although several tumour suppressor genes have been hypothesized, none of them has been authenticated.[45]

In addition to deletion and point mutation, tumour suppressor genes can be inactivated by 5′ CpG island methylation. For example, the 5′ CpG island of *p16* was found to be methylated in 6/9 NSCLC cell lines and in 21/27 NSCLCs; conversely, methylation was only detected in 1/10 SCLC cell lines and in 0/5

SCLCs.[77] Further, no methylation was detected in normal lung tissue. Hemizygous deletion of the wild-type *p16* allele was also detected in the majority of the methylated NSCLC cell lines. In two cases where LOH was not detected, both alleles were found to be methylated. Thus, methylation of the 5' CpG island of *p16* appears to be an important alternative pathway for gene inactivation in NSCLCs; however, it is currently not understood whether *de novo* methylation constitutes a regulated mechanism of base modification or a random epigenetic change.

The oncogenes discussed so far all share the characteristic that overexpression of their protein products ultimately stimulates cell proliferation. The proto-oncogene *bcl-2* represents the first of a new category of oncogenes that contribute to the development of cancer by inhibiting programmed cell death. Thus, overexpression of the Bcl-2 protein results in enhanced cellular viability rather than enhanced proliferation.[78] Although a number of different proteins are involved in the cellular decision of whether to undergo apoptosis in response to DNA damage,[79] the Bcl-2 protein is the only member of this group that is clearly established as being oncogenic. The *bcl-2* gene, located on 18q21, was first shown to be oncogenic in follicular lymphoma where translocation [t(14;18)] results in inappropriate expression of high levels of Bcl-2 protein.[80] It has been shown that the Bcl-2 protein inhibits apoptosis by forming a heterodimer with the protein product of *bax*,[81] an inducer of apoptosis that is transcriptionally activated by binding wild-type p53 protein.[82] Therefore, the cellular ratio of these two proteins plays a role in determining whether a cell will undergo apoptosis.

In the first lung cancer study of *bcl-2*, expression of Bcl-2 protein was detected in 20% of NSCLCs.[83] Because neither translocation of the *bcl-2* gene nor other abnormalities of chromosome 18 have been detected in lung carcinomas, these authors have suggested that the expression of Bcl-2 protein may be controlled by post-transcriptional regulation. In SCLC cell lines, expression of Bcl-2 protein was detected in 5/6 cell lines examined; however, Bcl-2 protein expression was correlated with *bcl-2* mRNA expression, providing evidence against post-transcriptional modification in SCLC cells.[84] Recently, *bcl-2* was shown to co-operate with c-*myc* in carcinogenesis by specifically inhibiting the induction of apoptosis by c-*myc* without affecting its mitogenic function.[85,86] In SCLC cell lines, co-expression of Bcl-2 and c-myc proteins was detected in 4/6 cell lines examined,[84] suggesting that *bcl-2* expression may provide a growth advantage to preneoplastic cells which exhibit deregulated c-*myc* expression by inhibiting the induction of apoptosis by c-myc protein. Expression of *bcl-2* in NSCLC has been correlated with less aggressive tumour behavior; the relatively slow progression of these tumours is proposed to be a result of *bcl-2* acting as an initial oncogene.[83] It has been suggested that the rate of acquiring additional genetic mutations may be slower in clones in which a low mitotic rate is offset by *bcl-2* expression compared with clones with a high mitotic rate.[78] However, the effect of *bcl-2* expression on tumour progression in SCLC remains to be determined.

In addition to chromosomal alterations, there are now several reports that microsatellite instability (*i.e.* variations in the number of repetitive unit sequences in each microsatellite) is also a common feature of SCLC[87] and NSCLC.[88] Microsatellite instability, first identified in tumours from hereditary nonpolyposis colorectal carcinoma patients,[89] has been established as a sensitive marker of genomic instability, a common feature of tumour cells.[66] Genomic instability is caused by mutational or exogenous interference with pathways governing the accurate duplication and distribution of DNA to progeny cells and/or with pathways controlling regulatory modifications of DNA during normal development.[66] During the process of carcinogenesis, mutations in so-called 'stability' genes are believed to result in the development of a mutator phenotype, a hypothesis invoked to explain the fact that the spontaneous mutation rate in normal cells is insufficient to account for the high frequency of mutations detected in human cancer cells.[90] The role of genomic instability in carcinogenesis is currently an area of active investigation in which this phenotypic change is being used as a marker *in vitro* to predict inter-individual differences in susceptibility to tumour development following carcinogen exposure.[91]

4 Putative Role For Autocrine Loops in the Development of Lung Cancer

With the advent of serum-free culture conditions for normal human bronchial epithelial (NHBE) cells, it has been shown that there are marked differences in the growth factor requirements of normal and tumour cells. On average, normal human epithelial cells inoculated in nutritional- and growth factor-optimised media will divide once a day. In contrast, tumour cells often exhibit significantly slower clonal growth rates in this medium. However, tumour cells will grow at rates approaching those of normal cells if the medium is supplemented with serum.[92] Thus, two seminal characteristics distinguish tumour cells from their normal counterparts. First, the tumour cells do not undergo terminal squamous differentiation in response to transforming growth factor-β (TGF-β), a serum constituent. Second, tumour cells have requirements for mitogens that are not necessary for optimal growth of normal cells. However, it is not clear at what phase of carcinogenesis (*i.e.* initiation, promotion or progression) the cells acquire requirements for serum-borne mitogens and lose the ability to recognise TGF-β as an inducer of terminal squamous differentiation. This difference in growth factor requirements between NHBE and lung tumour cells has generated a great deal of interest in the role of growth factors in the development of lung cancer.

Several studies have demonstrated that transforming growth factor-α (TGF-α) is an autocrine growth factor for the proliferation of certain lung carcinoma cell lines.[93,94] Although TGF-α is expressed at high levels in three subtypes of NSCLC (adenocarcinoma, squamous cell carcinoma and large cell carcinoma), the receptor for TGF-α, EGFR, is only overexpressed in squamous cell carcinomas.[30] Recently, it has been shown that overexpression of EGFR is a frequent event in the early stages of bronchial neoplasia.[95] Conversely, expression of TGF-α appears to be a late event in carcinogenesis, as positive staining has been consistently detected only in invasive cancers and not in normal epithelium or dysplastic, atypic or metaplastic lesions.[95] These results suggest that the autocrine loop of TGF-α and EGFR may be important in the late stages of development of squamous cell carcinoma.

The c-*kit* proto-oncogene encodes a protein that has been identified as the receptor for stem cell factor (SCF).[96,97] Co-expression of c-*kit* and SCF occurs frequently (18/22) in SCLC cell lines as well as in normal lung tissue.[98] It has been suggested that expression of SCF by both normal lung cells and SCLC cells may provide a growth advantage to SCLC cells expressing the c-*kit* receptor if binding of SCF does, in fact, stimulate cellular proliferation.[98] If this hypothesis is correct, then SCF functions as autocrine and paracrine factors in this system. These results suggest that co-expression of c-*kit* and SCF may be important in the development of human SCLC.

The human c-*met* oncogene encodes a transmembrane tyrosine kinase termed $p190^{c\text{-}met}$ (Ref. 99) that functions as a receptor for hepatocyte growth factor (HGF).[100,101] Expression of immunoreactive $p190^{c\text{-}met}$ has been detected in hepatomas, carcinomas of the colon and rectum, and carcinomas of the stomach, kidney, ovary, skin, lung, thyroid and pancreas.[102] With regard to lung cancer, expression of $p190^{c\text{-}met}$ in solid tumours appears to be tumour-type specific; immunoreactivity has been detected in 11/13 NSCLCs but not in SCLCs (0/3) or in normal lung samples (0/3).[102] A more recent study has shown that, between the histological subtypes of NSCLC, adenocarcinomas most frequently overexpress c-*met*; however, overexpression is rarely the result of gene amplification.[30] In addition to solid tumours, $p190^{c\text{-}met}$ is also expressed in NSLC cell lines; the ability of primary lung tumour cells to form cell lines *in vitro* has been correlated with high mRNA levels for TGF-α and c-*met*, suggesting that autocrine growth loops play a major role in the continuous growth of NSCLC cells in monolayer culture.[31]

Studies on the induction of tumourigenicity in murine NIH 3T3 cells by co-expression of HGF and c-*met* have provided the first indication for the role of this autocrine loop in carcinogenesis.[103,104] In human lung carcinomas, co-expression of HGF and c-*met* appears to be tumour-type specific, occurring at a low frequency (1/22) in SCLC cell lines[98] and at a high frequency (7/7) in NSCLC cell lines.[30,31] Although co-expression has also been demonstrated in NHBE cells,[30,31] this study has shown that there is a difference in expression of HGF mRNA between NHBE and NSCLC cells; the normal cells express predominantly a 1.3 kB transcript, whereas the cancer cells also exhibit expression of 3 kB and 6 kB transcripts. The addition of anti-recombinant human HGF antiserum inhibits the proliferation of 4/9 NSCLC cell lines studied and stimulates the growth of NHBE cells. This difference in response can be attributed to the fact that the 3 kB and 6 kB transcripts encode full-length HGF protein, while the alternatively spliced 1.3 kB transcript encodes a protein (HGF/NK2) that competitively binds to $p190^{c\text{-}met}$ but is not mitogenic. The ratio of HGF/NK2 mRNA to the 6 kB HGF mRNA is 10 in NHBE cells; this molar ratio is sufficient to observe a 50% inhibition of HGF mitogenic activity.[105] It has been suggested that anti-HGF may disrupt the low affinity binding of HGF/NK2 to $p190^{c\text{-}met}$ in NHBE cells, thereby displaying the mitogenic activity of the high affinity HGF protein.[30,31] On the basis of these studies, it appears that upregulation of expression of the full-length HGF mRNA transcripts may represent a critical event in the activation of the HGF/$p190^{c\text{-}met}$ autocrine loop. An alternative explanation has been proposed: another HGF-like protein encoded by a gene on a human chromosome 3p21, a region that is frequently deleted in human lung carcinomas,[106–108] may be an autocrine inhibitor of growth in NHBE cells.[109,110] Therefore, this HGF-like protein may function as a tumour suppressor gene, and loss of expression may also be a critical event in the activation of the HGF/p190$^{c\text{-met}}$ autocrine loop and subsequent malignant transformation of NHBE cells.

Together, the data to date suggest that three different autocrine loops may be important in the development of human lung cancer. Furthermore, each loop appears to be tumour-type specific. Co-expression of TGF-α and EGFR may play a role in squamous cell carcinoma, whereas SCF and c-*kit* may be involved in small cell carcinoma. Finally, HGF and c-*met* appear to be important players in the development of other subtypes of NSCLCs such as adenocarcinoma. Further elucidation of the timing of overexpression of these growth factors and their receptors in the development and progression of lung tumours should provide additional insights into the processes involved in lung carcinogenesis.

5 Sequence of Events in Lung Carcinogenesis

Because of the relative ease of obtaining biopsy samples during colonoscopy, it has been possible for cancer researchers to assemble an elegant model to describe the nature and timing of genetic alterations in hereditary nonpolyposis colorectal carcinoma.[111] Unfortunately, to date, a comparable model has not been developed to describe the events involved in the development of lung cancer. Although several genetic alterations have been identified in early bronchial neoplasia, it has not been feasible to examine earlier changes *in vivo* except by analysis of exfoliated cells present in sputum samples.[112–114] However, lung biopsy samples have recently become available from at-risk individuals (*e.g.* smokers and uranium miners) who undergo diagnostic bronchoscopy. These samples are particularly useful because they can be used to establish cell cultures *in vitro*, providing a valuable resource of viable normal, premalignant and malignant cells from the same individual. Further, because of field cancerization effects, samples collected from morphologically normal tissue in a tumour-containing lung must be rigorously scrutinized for mutations to rule out premalignant changes in phenotype. One molecular marker is insufficient to identify all premalignant cells because there is no evidence to suggest that all lung cancers develop along a single temporal pathway of gene/chromosome alterations. Thus, assaying for several alterations is necessary to be reasonably assured that morphologically normal tissue does not harbour mutated cells.

Molecular studies of premalignant lung epithelial cells have revealed some clues as to the nature and timing of early genetic alterations in lung cancer. For example, mutations in the *p53* gene have been observed in dysplastic bronchial epithelium microdissected from areas adjacent to some, but not all, lung tumours.[115] In addition, a study of lung cancers recovered from uranium miners reported *p53* mutations in preinvasive cells from one patient.[116] Further, LOH of chromosome 3p has been reported in bronchial epithelium exhibiting mild and severe dysplasia recovered from areas near the tumour mass.[117,118] Finally, simple gains, losses and rearrangements of chromosomes 3, 7, 9, 11 and 17 have been found in the proximal airway cells of 46% of patients with lung cancer.[119] Together, these studies have identified several frequent chromosome alterations that may be markers for dysplastic disease that precedes clinical lung cancer. Molecular and immunohistochemistry studies on sputum samples and lung bronchoscopy samples collected from at-risk individuals prior to clinical cancer should ultimately provide the information necessary to understand fully the nature and timing of genetic alterations in the development of SCLCs and NSCLCs. Once this has been accomplished, chemoprevention strategies can be developed and applied to protect individuals who, by molecular analyses, are determined to be susceptible for developing lung cancer.

(Research sponsored by the Office of Health and Environmental Research, US Department of Energy, under Contract No. DE-AC04-76EV01013.)

References

1. Farber, E. (1984) The multistep nature of cancer development, *Cancer Res.*, **44**, 4217–4223.
2. Harris, C.C. (1991) Chemical and physical carcinogenesis: advances and perspectives for the 1990s, *Cancer Res.*, **51**, 5023s–5044s.
3. Harris, C.C. (1992) Tumour suppressor genes, multistage carcinogenesis and molecular epidemiology, IARC.-Sci.-Publ. **118**, 67–85.
4. Friend, S.H. (1993) Genetic models for studying cancer susceptibility, *Science*, **259**, 774–775.
5. Demant, P. (1992) Genetic resolution of susceptibility to cancer – new perspectives, *Semin. Cancer Biol.*, **3**, 159–166.
6. Ponder, B.A. (1991) Genetic predisposition to cancer, *Br. J. Cancer*, **64**, 203–204.
7. Malkin, D. and Friend, S.H. (1992) The role of tumour suppressor genes in familial cancer, *Semin. Cancer Biol.*, **3**, 121–130.
8. Easton, D. and Peto, J. (1990) The contribution of inherited predisposition to cancer incidence, *Cancer Surv.*, **9**, 395–416.
9. Caporaso, N. (1991) Study design and genetic susceptibility factors in the risk assessment of chemical carcinogens, *Ann. Ist. Super. Sanita.*, **27**, 621–630.
10. Uematsu, F., Kikuchi, H., Motomiya, M., Abe, T., Sagami, I., Ohmachi, T., Wakui, A., Kanamaru, R. and Watanabe, M. (1991) Association between restriction fragment length polymorphism of the human cytochrome P450IIE1 gene and susceptibility to lung cancer, *Jpn. J. Cancer Res.*, **82**, 254–256.
11. Grant, D.M. (1993) Molecular genetics of the N-acetyltransferases, *Pharmacogenetics*, **3**, 45–50.
12. Hayashi, S., Watanabe, J. and Kawajiri, K. (1992) High susceptibility to lung cancer analyzed in terms of combined genotypes of P450IA1 and Mu-class glutathione S-transferase genes, *Jpn. J. Cancer Res.*, **83**, 866–870.
13. Lidereau, R., Mathieu-Mahul, D., Theillet, C., Renaud, M., Mauchauffe, M., Gest, J. and Larsen, C.J. (1985) Presence of an allelic EcoRI restriction fragment of the c-mos locus in leukocyte and tumour cell DNAs of breast cancer patients, *Proc. Natl. Acad. Sci. U.S.A.*, **82**, 7068–7070.
14. Krontiris, T.G. (1990) Detection of cancer predisposition by hypervariable region analysis, *Birth Defects*, **26**, 129–140.

15. Kawashima, K., Shikama, H., Imoto, K., Izawa, M., Naruke, T., Okabayashi, K. and Nishimura, S. (1988) Close correlation between restriction fragment length polymorphism of the L-*myc* gene and metastasis of human lung cancer to the lymph nodes and other organs, *Proc. Natl. Acad. Sci. U.S.A.*, **85**, 2353–2356.
16. You, M., Wang, Y., Stoner, G., You, L., Maronpot, R., Reynolds, S.H. and Anderson, M. (1992) Parental bias of Ki-*ras* oncogenes detected in lung tumours from mouse hybrids, *Proc. Natl. Acad. Sci. U.S.A.*, **89**, 5804–5808.
17. Digweed, M. (1993) Human genetic instability syndromes: single gene defects with increased risk of cancer, *Toxicol. Lett.*, **67**, 259–281.
18. Fearon, E.R. and Jones, P.A. (1992) Progressing toward a molecular description of colorectal cancer development, *FASEB J.*, **6**, 2783–2790.
19. Auerbach, O., Stout, A.P., Hammond, E.C. and Garfinkel, L. (1961) Changes in bronchial epithelium in relation to cigarette smoking and in relation to lung cancer, *New Eng. J. Med.*, **265**, 253–267.
20. Auerbach, O., Stout, A.P., Hammond, E.C. and Garfinkel, L. (1962) Changes in bronchial epithelium in relation to sex, age, residence, smoking and pneumonia, *New Engl. J. Med.*, **267**, 111–119.
21. Strong, M.S., Incze, J. and Vaughan, C.W. (1984) Field cancerization in the aerodigestive tract – its etiology, manifestation, and significance, *J. Otolaryngol.*, **13**, 1–6.
22. Shields, T.W., Humphrey, E.W., Higgins, Jr., G.A. and Keehn R.J. (1978) Long-term survivors after resection of lung carcinoma, *J. Thorac. Cardiovasc. Surg.*, **76**, 439–445.
23. van Bodegom, P.C., Wagenaar, S.S., Corrin, B., Baak, J.P., Berkel, J. and Vanderschueren, R.G. (1989) Second primary lung cancer: importance of long term follow up, *Thorax*, **44**, 788–793.
24. Pairolero, P.C., Williams, D.E., Bergstralh, E.J., Piehler, J.M., Bernatz, P.E. and Payne, W.S. (1984) Postsurgical stage I bronchogenic carcinoma: morbid implications of recurrent disease, *Ann. Thorac. Surg.*, **38**, 331–338.
25. Sozzi, G., Miozzo, M., Pastorino, U., Pilotti, S., Donghi, R., Giarola, M., De Gregorio, L., Manenti, G., Radice, P. and Minoletti, F. (1995) Genetic evidence for an independent origin of multiple preneoplastic and neoplastic lung lesions, *Cancer Res.*, **55**, 135–140.
26. Varmus, H. (1989) An historical overview of oncogenes. In 'Oncogenes and the molecular origins of cancer', Weinberg, R., ed., Cold Spring Harbor Lab Press, New York, pp. 3–44.
27. Friend, S.H., Dryja, T.P. and Weinberg, R.A. (1988) Oncogenes and tumor-suppressing genes, *New Engl. J. Med.*, **318**, 618–622.
28. Testa, J.R. and Siegfried, J.M. (1992) Chromosome abnormalities in human non-small cell lung cancer, *Cancer Res.*, **52**, 2702s–2706s.
29. Lee, J.S., Pathak, S., Hopwood, V., Tomasovic, B., Mullins, T.D., Baker, F.L., Spitzer, G. and Neidhart, J.A. (1987) Involvement of chromosome 7 in primary lung tumor and nonmalignant normal lung tissue, *Cancer Res.*, **47**, 6349–6352.
30. Liu, C. and Tsao, M.S. (1993a) *In vitro* and *in vivo* expressions of transforming growth factor-alpha and tyrosine kinase receptors in human non-small-cell lung carcinomas, *Am. J. Pathol.*, **142**, 1155–1162.
31. Liu, C. and Tsao, M.S. (1993b) Proto-oncogene and growth factor/receptor expression in the establishment of primary human non-small cell lung carcinoma cell lines, *Am. J. Pathol.*, **142**, 413–423.
32. Moodie, S.A., Willumsen, B.M., Weber, M.J. and Wolfman, A. (1993) Complexes of Ras-GTP with Raf-1 and mitogen-activated protein kinase kinase, *Science*, **260**, 1658–1661.

33. Rodenhuis, S., Slebos, R.J., Boot, A.J., Evers, S.G., Mooi, W.J., Wagenaar, S.S., van Bodegom, P.C. and Bos, J.L. (1988) Incidence and possible clinical significance of K-*ras* oncogene activation in adenocarcinoma of the human lung, *Cancer Res.*, **48**, 5738–5741.
34. Rodenhuis, S., van de Wetering, M.L., Mooi, W.J., Evers, S.G., van Zandwijk, N. and Bos, J.L. (1987) Mutational activation of the K-*ras* oncogene. A possible pathogenetic factor in adenocarcinoma of the lung, *New Engl. J. Med.*, **317**, 929–935.
35. Mitsudomi, T., Viallet, J., Mulshine, J.L., Linnoila, R.I., Minna, J.D. and Gazdar, A.F. (1991b) Mutations of *ras* genes distinguish a subset of non-small-cell lung cancer cell lines from small-cell lung cancer cell lines, *Oncogene*, **6**, 1353–1362.
36. Slebos, R.J., Kibbelaar, R.E., Dalesio, O., Kooistra, A., Stam, J., Meijer, C.J., Wagenaar, S.S., Vanderschueren, R.G., van Zandwijk, N. and Mooi, W.J. (1990) K-*ras* oncogene activation as a prognostic marker in adenocarcinoma of the lung, *New Engl. J. Med.*, **323**, 561–565.
37. Mitsudomi, T., Steinberg, S.M., Oie, H.K., Mulshine, J.L., Phelps, R., Viallet, J., Pass, H., Minna, J.D. and Gazdar, A.F. (1991a) *ras* gene mutations in non-small cell lung cancers are associated with shortened survival irrespective of treatment intent, *Cancer Res.*, **51**, 4999–5002.
38. Nishio, H., Nakamura, S., Horai, T., Ikegami, H. and Matsuda, M. (1992) Clinical and histopathologic evaluation of the expression of Ha-*ras* and *fes* oncogene products in lung cancer, *Cancer*, **69**, 1130–1136.
39. Yarden, Y. and Ullrich, A. (1988) Growth factor receptor tyrosine kinases, *Annu. Rev. Biochem.*, **57**, 443–478.
40. Slamon, D.J., Godolphin, W., Jones, L.A., Holt, J.A., Wong, S.G., Keith, D.E., Levin, W.J., Stuart, S.G., Udove, J. and Ullrich, A. (1989) Studies of the HER-2/neu proto-oncogene in human breast and ovarian cancer, *Science*, **244**, 707–712.
41. Schneider, P.M., Hung, M.C., Chiocca, S.M., Manning, J., Zhao, X.Y., Fang, K. and Roth, J.A. (1989) Differential expression of the c-*erb*-B2 gene in human small cell and non-small cell lung cancer, *Cancer Res.*, **49**, 4968–4971.
42. Kern, J.A., Schwartz, D.A., Nordberg, J.E., Weiner, D.B., Greene, M.I., Torney, L. and Robinson, R.A. (1990) p185neu expression in human lung adenocarcinomas predicts shortened survival, *Cancer Res.*, **50**, 5184–5187.
43. Tateishi, M., Ishida, T., Kohdono, S., Hamatake, M., Fukuyama, Y. and Sugimachi, K. (1994) Prognostic influence of the co-expression of epidermal growth factor receptor and c-*erb*-B2 protein in human lung adenocarcinoma, *Surg. Oncol.*, **3**, 109–113.
44. Evan, G.I. and Littlewood, T.D. (1993) The role of c-myc in cell growth, *Curr. Opin. Genet. Dev.*, **3**, 44–49.
45. Gazdar, A.F. and Carbone, D.P. (1994) 'The biology and molecular genetics of lung cancer', R.G. Landes Company, Austin, TX.
46. Little, C.D., Nau, M.M., Carney, D.N., Gazdar, A.F. and Minna, J.D. (1983) Amplification and expression of the c-*myc* oncogene in human lung cancer cell lines, *Nature*, **306**, 194–196.
47. Nau, M.M., Brooks, B.J., Battey, J., Sausville, E., Gazdar, A.F., Kirsch, I.R., McBride, O.W., Bertness, V., Hollis, G.F. and Minna, J.D. (1985) L-*myc*, a new *myc*-related gene amplified and expressed in human small cell lung cancer, *Nature*, **318**, 69–73.
48. Slebos, R.J., Evers, S.G., Wagenaar, S.S. and Rodenhuis, S. (1989) Cellular protoonocogenes are infrequently amplified in untreated non-small cell lung cancer, *Br. J. Cancer*, **59**, 76–80.
49. Johnson, B.E., Ihde, D.C., Makuch, R.W., Gazdar, A.F., Carney, D.N., Oie, H., Russell, E., Nau, M.M. and Minna, J.D. (1987) *myc* family oncogene amplification in

tumor cell lines established from small cell lung cancer patients and its relationship to clinical status and course, *J. Clin. Invest.*, **79**, 1629–1634.
50. Funa, K., Steinholtz, L., Nou, E. and Bergh, J. (1987) Increased expression of N-myc in human small cell lung cancer biopsies predicts lack of response to chemotherapy and poor prognosis, *Am. J. Clin. Pathol.*, **88**, 216–220.
51. Brennan, J., O'Connor, T., Makuch, R.W., Simmons, A.M., Russell, E., Linnoila, R.I., Phelps, R.M., Gazdar, A.F., Ihde, D.C. and Johnson, B.E. (1991) *myc* family DNA amplification in 107 tumours and tumour cell lines from patients with small cell lung cancer treated with different combination chemotherapy regimens, *Cancer Res.*, **51**, 1708–1712.
52. Rustgi, A.K., Dyson, N., Hill, D. and Bernards, R. (1991) The c-myc oncoprotein forms a specific complex with the product of the retinoblastoma gene, *Cold Spring Harb. Symp. Quant. Biol.*, **56**, 163–167.
53. Hiebert, S.W., Chellappan, S.P., Horowitz, J.M. and Nevins, J.R. (1992) The interaction of RB with E2F coincides with an inhibition of the transcriptional activity of E2F, *Genes Dev.*, **6**, 177–185.
54. Lin, B.T., Gruenwald, S., Morla, A.O., Lee and Wang, J.Y. (1991) Retinoblastoma cancer suppressor gene product is a substrate of the cell cycle regulator cdc2 kinase, *EMBO J.*, **10**, 857–864.
55. Yokota, J., Wada, M., Shimosato, Y., Terada, M. and Sugimura, T. (1987) Loss of heterozygosity on chromosomes 3, 13, and 17 in small-cell carcinoma and on chromosome 3 in adenocarcinoma of the lung, *Proc. Natl. Acad. Sci. U.S.A.*, **84**, 9252–9256.
56. Harbour, J.W., Lai, S.L., Whang-Peng, J., Gazdar, A.F., Minna, J.D. and Kaye, F.J. (1988) Abnormalities in structure and expression of the human retinoblastoma gene in SCLC, *Science*, **241**, 353–357.
57. el-Deiry, W.S., Tokino, T., Velculescu, V.E., Levy, D.B., Parsons, R., Trent, J.M., Lin, D., Mercer, W.E., Kinzler, K.W. and Vogelstein, B. (1993) WAF1, a potential mediator of p53 tumour suppression, *Cell*, **75**, 817–825.
58. Xiong, Y., Hannon, G.J., Zhang, H., Casso, D., Kobayashi, R. and Beach, D. (1993) p21 is a universal inhibitor of cyclin kinases, *Nature*, **366**, 701–704.
59. el-Deiry, W.S., Harper, J.W., O'Connor, P.M., Velculescu, V.E., Canman, C.E., Jackman, J., Pietenpol, J.A., Burrell, M., Hill, D.E. and Wang, Y. (1994) WAF1/CIP1 is induced in p53-mediated G1 arrest and apoptosis, *Cancer Res.*, **54**, 1169–1174.
60. Bodner, S.M., Minna, J.D., Jensen, S.M., D'Amico, D., Carbone, D., Mitsudomi, T., Fedorko, J., Buchhagen, D.L., Nau, M.M. and Gazdar, A.F. (1992) Expression of mutant p53 proteins in lung cancer correlates with the class of *p53* gene mutation, *Oncogene*, **7**, 743–749.
61. Takahashi, T., Nau, M.M., Chiba, I., Birrer, M.J., Rosenberg, R.K., Vinocour, M., Levitt, M., Pass, H., Gazdar, A.F. and Minna, J.D. (1989) p53: a frequent target for genetic abnormalities in lung cancer, *Science*, **246**, 491–494.
62. Takahashi, T., D'Amico, D., Chiba, I., Buchhagen, D.L. and Minna, J.D. (1990) Identification of intronic point mutations as an alternative mechanism for p53 inactivation in lung cancer, *J. Clin. Invest.*, **86**, 363–369.
63. D'Amico, D., Carbone, D., Mitsudomi, T., Nau, M,. Fedorko, J., Russell, E., Johnson, B., Buchhagen, D., Bodner, S. and Phelps, R. (1992) High frequency of somatically acquired p53 mutations in small-cell lung cancer cell lines and tumors, *Oncogene*, **7**, 339–346.
64. Mitsudomi, T., Steinberg, S.M., Nau, M.M., Carbone, D., D'Amico, D., Bodner, S., Oie, H.K., Linnoila, R.I., Mulshine, J.L. and Minna, J.D. (1992) *p53* gene mutations in

non-small-cell lung cancer cell lines and their correlation with the presence of *ras* mutations and clinical features, *Oncogene*, **7**, 171–180.
65. Hainaut, P. and Milner, J. (1993) A structural role for metal ions in the wild-type conformation of the tumor suppressor protein p53, *Cancer Res.*, **53**, 1739–1742.
66. Cheng, K.C. and Loeb, L.A. (1993) Genomic instability and tumor progression: mechanistic considerations, *Adv. Cancer Res.*, **60**, 121–156.
67. Ramqvist, T., Magnusson, K.P., Wang, Y., Szekely, L., Klein, G. and Wiman, K.G. (1993) Wild-type p53 induces apoptosis in a Burkitt lymphoma (BL) line that carries mutant p53, *Oncogene*, **8**, 1495–1500.
68. Merlo, A., Gabrielson, E., Mabry, M., Vollmer, R., Baylin, S.B. and Sidransky, D. (1994a) Homozygous deletion on chromosome 9p and loss of heterozygosity on 9q, 6p, and 6q in primary human small cell lung cancer, *Cancer Res.*, **54**, 2322–2326.
69. Serrano, M., Hannon, G.J. and Beach, D. (1993) A new regulatory motif in cell-cycle control causing specific inhibition of cyclin D/CDK4, *Nature*, **366**, 704–707.
70. Hannon, G.J. and Beach, D. (1994) p15INK4B is a potential effector of TGF-beta-induced cell cycle arrest, *Nature*, **371**, 257–261.
71. Okamoto, A., Hussain, S.P., Hagiwara, K, Spillare, E.A., Rusin, M.R., Demetrick, D.J., Serrano, M., Hannon, G.J., Shiseki, M. and Zariwala, M. (1995) Mutations in the *p16*INK4/MTS1/CDKN2, *p15*INK4B/MTS2, and *p18* genes in primary and metastatic lung cancer, *Cancer Res.*, **55**, 1448–1451.
72. Washimi, O., Nagatake, M., Osada, H., Ueda, R., Koshikawa, T., Seki, T. and Takahashi, T. (1995) *In vivo* occurrence of *p16* (MTS1) and *p15* (MTS2) alterations preferentially in non-small cell lung cancers, *Cancer Res.*, **55**, 514–517.
73. Whang-Peng, J. (1989) 3p deletion and small cell lung carcinoma, *Mayo Clin. Proc.*, **64**, 256–260.
74. Naylor, S.L., Johnson, B.E., Minna, J.D. and Sakaguchi, A.Y. (1987) Loss of heterozygosity of chromosome 3p markers in small-cell lung cancer, *Nature*, **329**, 451–454.
75. Rabbitts, P., Douglas, J., Daly, M., Sundaresan, V., Fox, B., Haselton, P., Wells, F., Albertson, D., Waters, J. and Bergh, J. (1989) Frequency and extent of allelic loss in the short arm of chromosome 3 in nonsmall-cell lung cancer, *Genes Chromosom. Cancer*, **1**, 95–105.
76. Hibi, K., Takahashi, T., Yamakawa, K., Ueda, R., Sekido, Y., Ariyoshi, Y., Suyama, M., Takagi, H. and Nakamura, Y. (1992) Three distinct regions involved in 3p deletion in human lung cancer, *Oncogene*, **7**, 445–449.
77. Merlo, A., Herman, J.G., Mao, L., Lee, D.J., Gabrielson, E., Burger, P.C., Baylin, S.B. and Sidransky, D. (1995) 5' CpG island methylation is associated with transcriptional silencing of the tumour suppressor *p16/CDKN2/MTS1* in human cancers, *Nat. Med.*, **1**, 686–692.
78. McDonnell, T.J., Deane, N., Platt, F.M., Nunez, G., Jaeger, U., McKearn, J.P. and Korsmeyer, S.J. (1989) bcl-2-immunoglobulin transgenic mice demonstrate extended B cell survival and follicular lymphoproliferation, *Cell*, **57**, 79–88.
79. Korsmeyer, S.J. (1995) Regulators of cell death, *Trends Genet.*, **11**, 101–105.
80. Seto, M., Jaeger, U., Hockett, R.D., Graninger, W., Bennett, S., Goldman, P. and Korsmeyer, S.J. (1988) Alternative promoters and exons, somatic mutation and deregulation of the *Bcl*-2-Ig fusion gene in lymphoma, *EMBO J.*, **7**, 123–131.
81. Korsmeyer, S.J., Shutter, J.R., Veis, D.J., Merry, D.E. and Oltvai, Z.N. (1993) Bcl-2/Bax: a rheostat that regulates an anti-oxidant pathway and cell death, *Semin. Cancer Biol.*, **4**, 327–332.
82. Miyashita, T., Krajewski, S., Krajewska, M., Wang, H.G., Lin, H.K., Liebermann, D.A., Hoffman, B. and Reed, J.C. (1994) Tumor suppressor p53 is a regulator of *bcl*-2

and *bax* gene expression *in vitro* and *in vivo*, *Oncogene*, **9**, 1799–1805.
83. Pezzella, F., Turley, H., Kuzu, I. Tungekar, M.F., Dunnill, M.S., Pierce, C.B., Harris, A., Gatter, K.C. and Mason, D.Y. (1993) bcl-2 protein in non-small-cell lung carcinoma, *New Engl. J. Med.*, **329**, 690–694.
84. Ikegaki, N., Katsumata, M., Minna, J. and Tsujimoto, Y. (1994) Expression of bcl-2 in small cell lung carcinoma cells, *Cancer Res.*, **54**, 6–8.
85. Evan, G.I., Wyllie, A.H., Gilbert, C.S., Littlewood, T.D., Land, H., Brooks, M., Waters, C.M., Penn, L.Z. and Hancock, D.C. (1992) Induction of apoptosis in fibroblasts by c-myc protein, *Cell*, **69**, 119–128.
86. Fanidi, A., Harrington, E.A. and Evan, G.I. (1992) Cooperative interaction between *c-myc* and *bcl*-2 proto-oncogenes, *Nature*, **359**, 554–556.
87. Merlo, A., Mabry, M., Gabrielson, E., Vollmer, R., Baylin, S.B. and Sidransky, D. (1994b) Frequent microsatellite instability in primary small cell lung cancer, *Cancer Res.*, **54**, 2098–2101.
88. Shridhar, V., Siegfried, J., Hunt, J., del Mar Alonso, M. and Smith, D.I. (1994) Genetic instability of microsatellite sequences in many non-small cell lung carcinomas, *Cancer Res.*, **54**, 2084–2087.
89. Peltomaki, P., Lothe, R.A., Aaltonen, L.A., Pylkkanen, L., Nystrom-Lahti, M., Seruca, R., David, L., Holm, R., Ryberg, D. and Haugen, A. (1993) Microsatellite instability is associated with tumors that characterize the hereditary non-polyposis colorectal carcinoma syndrome, *Cancer Res.*, **53**, 5853–5855.
90. Loeb, L.A. (1991) Mutator phenotype may be required for multistage carcinogenesis, *Cancer Res.*, **51**, 3075–3079.
91. Kadhim, M.A., Lorimore, S.A., Hepburn, M.D., Goodhead, D.T., Buckle, V.J. and Wright, E.G. (1994) Alpha-particle-induced chromosomal instability in human bone marrow cells, *Lancet*, **344**, 987–988.
92. Lechner, J.F., McClendon, I.A,. LaVeck, M.A., Shamsuddin, A.M. and Harris, C.C. (1983) Differential control by platelet factors of squamous differentiation in normal and malignant human bronchial epithelial cells, *Cancer Res.*, **43**, 5915–5921.
93. Imanishi, K., Yamaguchi, K., Kuranami, M., Kyo, E. Hozumi, T. and Abe, K. (1989) Inhibition of growth of human lung adenocarcinoma cell lines by anti-transforming growth factor-alpha monoclonal antibody, *J. Natl. Cancer Inst.*, **81**, 220–223.
94. Siegfried, J.M. (1987) Detection of human lung epithelial cell growth factors produced by a lung carcinoma cell line: use in culture of primary solid lung tumors, *Cancer Res.*, **47**, 2903–2910.
95. Rusch, V., Klimstra, D., Linkov, I. and Dmitrovsky, E. (1995) Aberrant expression of p53 or the epidermal growth factor receptor is frequent in early bronchial neoplasia and coexpression precedes squamous cell carcinoma development, *Cancer Res.*, **55**, 1365–1372.
96. Williams, D.E., Eisenman, J., Baird, A., Rauch, C., Van Ness, K., March, C.J., Park, L.S., Martin, U., Mochizuki, D.Y. and Boswell, H.S. (1990) Identification of a ligand for the c-*kit* proto-oncogene, *Cell*, **63**, 167–174.
97. Zsebo, K.M., Williams, D.A., Geissler, E.N., Broudy, V.C., Martin, F.H., Atkins, H.L., Hsu, R.Y., Birkett, N.C., Okino, K.H. and Murdock, D.C. (1990) Stem cell factor is encoded at the Sl locus of the mouse and is the ligand for the c-kit tyrosine kinase receptor, *Cell*, **63**, 213–224.
98. Rygaard, K., Nakamura, T. and Spang-Thomsen, M. (1993) Expression of the proto-oncogenes c-*met* and c-*kit* and their ligands, hepatocyte growth factor/scatter factor and stem cell factor, in SCLC cell lines and xenografts, *Br. J. Cancer*, **67**, 37–46.

99. Giordano, S., Ponzetto, C., Di Renzo, M.F., Cooper, C.S. and Comoglio, P.M. (1989) Tyrosine kinase receptor indistinguishable from the c-met protein, *Nature*, **339**, 155–156.
100. Naldini, L., Vigna, E., Ferracini, R., Longati, P., Gandino, L., Prat, M. and Comoglio, P.M. (1991) The tyrosine kinase encoded by the *MET* proto-oncogene is activated by autophosphorylation, *Mol. Cell Biol.*, **11**, 1793–1803.
101. Vigna, E., Naldini, L., Tamagnone, L., Longati, P., Bardelli, A., Maina, F., Ponzetto, C. and Comoglio, P.M. (1994) Hepatocyte growth factor and its receptor, the tyrosine kinase encoded by the c-*MET* proto-oncogene, *Cell Mol. Biol.*, **40**, 597–604.
102. Prat, M., Narsimhan, R.P., Crepaldi, T., Nicotra, M.R., Natali, P.G. and Comoglio, P.M. (1991) The receptor encoded by the human c-*MET* oncogene is expressed in hepatocytes, epithelial cells and solid tumors, *Int. J. Cancer*, **49**, 323–328.
103. Rong, S., Bodescot, M., Blair, D., Dunn, J., Nakamura, T., Mizuno, K., Park, M., Chan, A., Aaronson, S. and Vande Woude, G.F. (1992) Tumorigenicity of the *met* proto-oncogene and the gene for hepatocyte growth factor, *Mol. Cell Biol.*, **12**, 5152–5158.
104. Iyer, A., Kmiecik, T.E., Park, M., Daar, I., Blair, D., Dunn, K.J., Sutrave, P., Ihle, J.N., Bodescot, M. and Vande Woude, G.F. (1990) Structure, tissue-specific expression, and transforming activity of the mouse *met* protooncogene, *Cell Growth Differ.*, **1**, 87–95.
105. Chan, A.M., Rubin, J.S., Bottaro, D.P., Hirschfield, D.W., Chedid, M. and Aaronson, S.A. (1991) Identification of a competitive HGF antagonist encoded by an alternative transcript, *Science*, **254**, 1382–1385.
106. Weston, A., Willey, J.C., Modali, R., Sugimura, H., McDowell, E.M., Resau, J., Light, B., Haugen, A., Mann, D.L. and Trump, B.F. (1989) Differential DNA sequence deletions from chromosomes 3, 11, 13, and 17 in squamous-cell carcinoma, large-cell carcinoma, and adenocarcinoma of the human lung, *Proc. Natl. Acad. Sci. U.S.A.*, **86**, 5099–5103.
107. Whang-Peng, J., Kao-Shan, C.S., Lee, E.C., Bunn, P.A., Carney, D.N., Gazdar, A.F. and Minna, J.D. (1982) Specific chromosome defect associated with human small-cell lung cancer; deletion 3p(14–23), *Science*, **215**, 181–182.
108. Brauch, H., Johnson, B., Hovis, J., Yano, T. Gazdar, A., Pettengill, O.S., Graziano, S., Sorenson, G.D., Poiesz, B.J. and Minna, J. (1987) Molecular analysis of the short arm of chromosome 3 in small-cell and non-small-cell carcinoma of the lung, *New Engl. J. Med.*, **317**, 1109–1113.
109. Han, S., Stuart, L.A. and Degen, S.J. (1991) Characterization of the DNF15S2 locus on human chromosome 3: identification of a gene coding for four kringle domains with homology to hepatocyte growth factor, *Biochemistry*, **30**, 9768–9780.
110. Degen, S.J., Stuart, L.A., Han, S. and Jamison, C.S. (1991) Characterization of the mouse cDNA and gene coding for a hepatocyte growth factor-like protein: expression during development, *Biochemistry*, **30**, 9781–9791.
111. Thomas, G. (1994) Advances in the genetics and molecular biology of colorectal tumors, *Curr. Opin. Oncol.*, **6**, 406–412.
112. Takeda, S., Ichii, S. and Nakamura, Y. (1993) Detection of K-ras mutation in sputum by mutant-allele-specific amplification (MASA), *Hum. Mutat.*, **2**, 112–117.
113. Mao, L., Hruban, R.H., Boyle, J.O., Tockman, M. and Sidransky, D. (1994a) Detection of oncogene mutations in sputum precedes diagnosis of lung cancer, *Cancer Res.*, **54**, 1634–1637.
114. Mao, L., Lee, D.J., Tockman, M.S., Erozan, Y.S., Askin, F. and Sidransky, D. (1994b) Microsatellite alterations as clonal markers for the detection of human cancer, *Proc. Natl. Acad. Sci. U.S.A.*, **91**, 9871–9875.

115. Bennett, W.P., Colby, T.V., Travis, W.D., Borkowski, A., Jones, R.T., Lane, D.P., Metcalf, R.A., Samet, J.M., Takeshima, Y. and Gu, J.R. (1993) p53 protein accumulates frequently in early bronchial neoplasia, *Cancer Res.*, **53**, 4817–4822.
116. Nuorva, K., Soini, Y., Kamel, D., Autio-Harmainen, H., Risteli, L., Risteli, J., Vähäkangas, K. and Pääkkö, P. (1993) Concurrent p53 expression in bronchial dysplasias and squamous cell lung carcinomas, *Am. J. Pathol.*, **142**, 725–732.
117. Sozzi, G., Miozzo, M., Donghi, R., Pilotti, S., Cariani, C.T., Pastorino, U., Della Porta, G. and Pierotti, M.A. (1992) Deletions of 17p and p53 mutations in preneoplastic lesions of the lung, *Cancer Res.*, **52**, 6079–6082.
118. Sundaresan, V., Ganly, P., Hasleton, P., Rudd, R., Sinha, G., Bleehen, N.M. and Rabbitts, P. (1992) p53 and chromosome 3 abnormalities, characteristic of malignant lung tumours, are detectable in preinvasive lesions of the bronchus, *Oncogene*, **7**, 1989–1997.
119. Pastorino, U., Sozzi, G., Miozzo, M., Tagliabue, E., Pilotti, S. and Pierotti, M.A. (1993) Genetic changes in lung cancer, *J. Cell Biochem.*, **17F** (Suppl.), 237–248.

The Significance of Toxicokinetics of Solid Particles in the Rat Lung

H. MUHLE

FRAUNHOFER INSTITUT FÜR TOXIKOLOGIE UND
AEROSOLFORSCHUNG, HANNOVER, GERMANY

1 Introduction

This paper deals primarily with poorly-soluble particles. The reason for creating a joint group for these materials is two-fold. Firstly, these materials have the potential to accumulate in lungs after chronic exposure. Secondly, in toxicology it is usually the dissolved, bioavailable molecules or ions that cause toxic effects; poorly-soluble materials act in a different way on the body; it is the entire particle that interacts with cells.

Lung clearance of solid particles with low solubility is dependent on the location of the deposition in the respiratory tract. The mucociliary escalator removes particles from the tracheobronchial region in the range of one day.[1,2] However, Stahlhofen et al. (1994)[3] have reported a slow phase clearance in the ciliated airways of humans. The authors have used an aerosol bolus of radioactively-labelled particles of defined volume injected at a selected moment during inhalation and investigated the lung clearance from the tracheobronchial tract. Because of the small dimensions of the airways of rats, this technique was not used in this species. The significance of a slow tracheobronchial clearance phase for the toxicity of solid particles is not fully understood.

In the alveolar region, the macrophage-mediated clearance is the predominant mechanism of removing particles. This process is in general much slower than the ciliated clearance; therefore, an impairment of this defence mechanism is of particular relevance.

The specific principles of the toxicokinetics of poorly-soluble particles are discussed because long-term retention in the respiratory tract is of major concern in occupational dust exposure; e.g. there are reports on an accumulation of up to 20 g of dust in the lungs of miners.[4] While clearance of particles from the lungs is important because it represents an essential defence mechanism for the respiratory system, failure to clear particles leads to their accumulation, resulting in histopathological alterations.

Several papers deal with the significance of these results to dust limit values.[5–8]

2 Design of Retention Measurements

In toxicity studies, relatively high concentrations are usually used compared with the exposure that humans experience in the occupational setting. One of the reasons for this approach is to detect significant effects with a minimum number of animals. The mechanism of damage at high exposure levels may be different compared with low levels.

For optimal information on the retained dose in rodent inhalation experiments, the pattern of the time course during exposure and after cessation of exposure should be determined. This would require frequent serial sacrifices. To minimize the number of animals in chronic studies, the retained mass is determined at three and 24 months. This information is essential for establishing dose–response curves.

The ability of alveolar macrophages to clear particles can be determined by using γ-labelled particles with low solubility.[9] This non-invasive method has been used in many studies. The principle is a short-term inhalation of, for example, ^{85}Sr-polystyrene particles, which serve as surrogate materials. The thoracic g-activity in the test animals is measured about twice a week for several months. These measurements can be done in parallel to an ongoing exposure to test materials. A comparison of the clearance of the surrogate particles in the exposed group *versus* a clean air-exposed group gives a good indication of effects on macrophage-mediated clearance.

3 Generic Response in Particle Inhalation Studies

Various studies have been performed with solid particles in the last 20 years, among them studies with coal mine dust,[10] volcanic ash,[11] fly ash from coal power plants[12,13] and petroleum coke.[14] In some of the studies, the documentation of the retained materials was incomplete. Among the described effects were an increase in lung weight, chronic inflammatory processes, septal thickening, lipoproteinosis, fibrosis and in some cases the induction of lung tumours. A systematic review of these effects is given by Morrow *et al.*, 1991.[6]

Unexpected pulmonary effects were detected in a study by Lee *et al.* (1985)[15] on titanium dioxide. The results of this long-term study showed that high titanium dioxide levels in the lung produced severe pathological changes, including carcinomas. The extraordinary levels of exposure (250 mg m^{-3}) were probably responsible for these pulmonary effects. At an exposure concentration of 10 or 50 mg m^{-3}, no increase of the tumour incidence was found.

4 Dust Overload in Lungs

Inhalation studies with carbon black were undertaken by Strom *et al.* (1989).[16] These rat studies, at an aerosol concentration of approx. 7 mg m^{-3}, 20 h day^{-1}, 7 days per week for up to 6 weeks, demonstrated a severe retardation in pulmonary particle clearance and the absence of first order kinetics at high pulmonary dust loads. An accumulation of dust-laden alveolar macrophages and the existence of alveolar lipoproteinosis were reported.

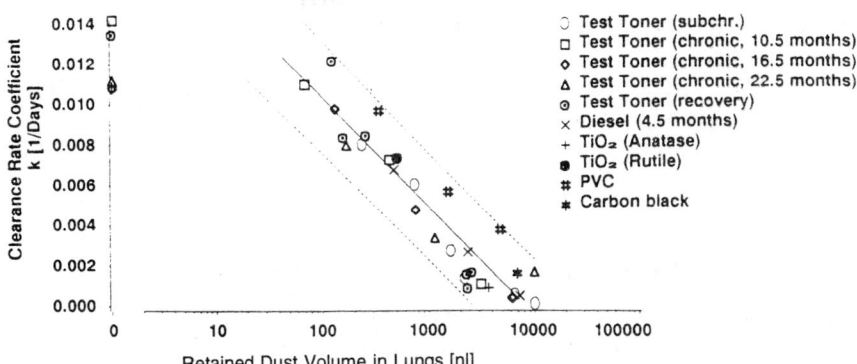

Figure 1 *Clearance rate coefficient of labelled particles (^{85}Sr-polystyrene) or toner particles as a function of the retained dust volume of various test materials. Compilation of various studies performed in rats. Dotted lines indicate the 95% confidence of the regression line (Muhle et al., 1990).* [20]

From the studies of Muhle et al. (1991)[17] and Bellmann et al. (1991)[18] a systematic investigation was made on the relationships between the lung burden of solid particles and ability of alveolar macrophages to clear radioactively-labelled particles. In the F-344 rat, dust overloading of the lung is generally reached in the range of 0.5–1.5 mg or approximately 1 mg of material per gram of lung tissue for insoluble particles with low intrinsic toxicity.

Comparing the effects of particles differing in density, the retained volume of particles is a more useful parameter than the retained mass. The reason could be that alveolar macrophages show an upper volumetric uptake limit.[5,19,20]

In Figure 1, a summary of the particle clearance constants measured in studies on polymers containing carbon black (toner for copy machines), titanium dioxide particles (TiO_2), diesel soot, carbon black, poly(vinyl chloride) particles (PVC) and on a radioactively-labelled surrogate dust administered at various times throughout the study to exposed F-344 or Wistar rats and age-matched unexposed controls is depicted as a function of the retained dust volume.[20]

The clearance rate is defined by $k = \ln(2/t_{1/2})$. A value of $k = 0.012$ thus corresponds to a half life of 58 days; $k = 0.001$ corresponds to a half life of 693 days. The results illustrate a progressive decrease in alveolar clearance rates once an excessive pulmonary dust burden is attained.

It should be noted that most of the chronic studies discussed above related to particles of relatively low intrinsic acute toxicity which showed first order retention in the rat lung at low dust burdens, namely, having half-lives of approximately 60–80 days. With excessive dust levels in the lungs, clearance rates for all of the test dusts used in the studies cited became progressively reduced until they were immeasurably small. Thus, the condition of dust overloading represents a serious, confounding complication to the toxicological assessment, one in which the intrinsic toxicity of the test material is either masked or modified by the non-specific effects of dusts on macrophage transport.

Studies of lung overloading provided evidence of macrophage accumulation, epithelial cell proliferation, inflammatory reactions and increased dust presence in the interstitium.[21] Dust overloading of lungs is characterised by an immobility and dysfunction of macrophages resulting from an excessive uptake of materials.

For human risk assessment, it must be taken into account that the clearance retardation induced by a high burden of particles with low toxicity in the lung will also affect the clearance of more toxic particles which could be inhaled at very low concentrations.

5 Irreversibility of Clearance Retardation

The reversibility of dust overloadings was investigated in a study of Bellmann *et al.* (1990).[22] Female SPF F-344 rats were exposed 6 h day^{-1}, 5 days per week for 3 months to test toner at 0, 10 or 40 mg m^{-3}. The quantity of test toner retained in the lungs at the end of exposure was 0.40 and 3.01 mg for the low and high exposure groups, respectively (see Figure 2). Fifteen months later, the corresponding values were 0.12 and 2.65 mg in the lungs. The calculated half lives of the alveolar clearance were 277 and 2845 days, respectively. Alveolar clearance of a polystyrene tracer aerosol with MMAD of 3.8 µm was investigated at the end of the toner exposure and subsequently up to 15 months later (see Table 1). A slightly retarded clearance at the low, and a substantially impaired clearance at the high, exposure level were observed. At the low exposure level, there was some recovery in the clearance behaviour up to six months post exposure. In contrast, in the high exposure group there was no indication of a reversal of the impaired clearance.

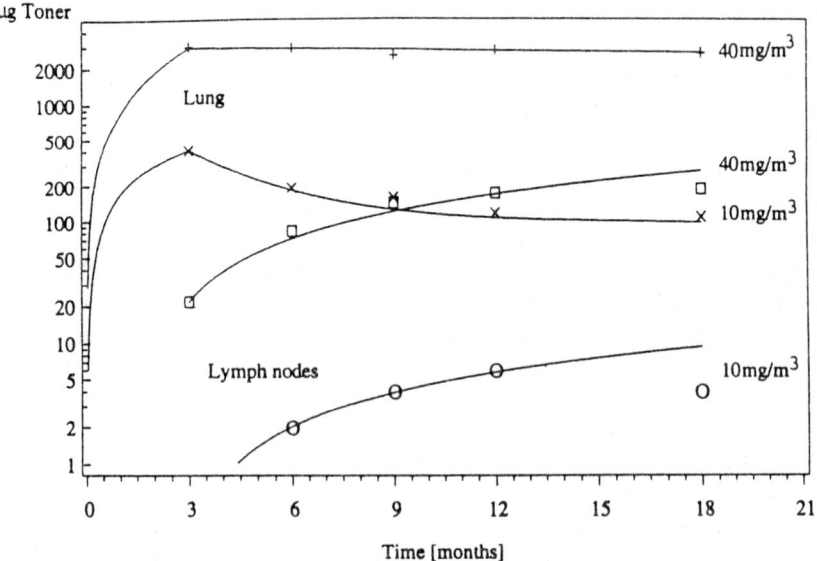

Figure 2 *Retention of toner in the lungs and lung-associated lymph nodes after 3 months of exposure and 15 months post treatment recovery period. (Lines: Model calculation, Bellmann et al., 1990).*[22]

Table 1 *Alveolar clearance half life of labelled particles [means and 95% confidence limit (95% C.L.) for 8 animals each], after cessation of a 3 month exposure period*

		Alveolar clearance half life [days]			
		Time of recovery [months]			
		0–3	3–6	6–9	12–15
		Mean (95% C.L.)	Mean (95% C.L.)	Mean (95% C.L.)	Mean (95% C.L.)
^{85}Sr-PS (3.8 μm)	Control	45 (39–54)	45 (42–49)	39 (35–45)	75 (58–105)
	Toner 10 mg m^{-3}	66 (55–84)	81 (71–96)	55 (49–61)	86 (68–117)
	Toner 40 mg m^{-3}	229 (170–352)	635 (378–1991)	329 (217–678)	308 (209–579)

The observation of persistent retardation of the alveolar clearance after reaching dust overload even after a 15 month clean air period was unexpected. The half life of particle clearance is at a value of a few hundred days, much longer than the turnover time of an alveolar macrophage, which was estimated to be about 4 days in mice under conditions of no particle load.[23] In rats, a daily loss of alveolar macrophages from the lung of 8–9% was measured;[24] 5–6% per day were estimated to be due to loss by cell death, which corresponds to a half-life of about 13 days. This means that there is a release of particles by dying macrophages and an immediate rephagocytosis by intact or newly arrived macrophages. The finding of this study that, even after a 12 month post exposure period, the alveolar clearance is retarded can be interpreted as follows: it is presumably not relevant for the alveolar macrophages whether the particles are recently deposited or released by dying macrophages. If a critical lung burden of insoluble particles is reached, the clearance of a small amount of recently inhaled g-labelled particles (up to 12 months post-exposure to test particles) is still influenced by the previous lung burden.

Another possible explanation for the persistent retardation of lung clearance was given by Lehnert *et al.* (1994).[25] After subchronic exposure to ultrafine titanium dioxide, the authors detected an occlusion of the pores of Kohn by Type II cells, which may conceivably effectively eliminate a pathway through which alveolar macrophages may gain access to neighbouring alveoli and perhaps travel on their way up the mucociliary apparatus.

Brown *et al.* (1992)[26] reported that the mobility of alveolar macrophages was impaired after cessation of exposure to coal mine dust and quartz.

6 Migration of Particles to Lung-associated Lymph Nodes

Ferin and Feldstein (1978)[27] reported that the transport to the lung-associated lymph nodes (LALN) was accelerated parallel to the decrease of lung clearance at high lung burden. Therefore, to evaluate the clearance processes under these conditions, an analysis of the particle content in lungs and LALN is necessary.

In the recovery study of Bellmann et al. (1990),[22] the transfer of particles to the lymph nodes was enhanced with increasing lung burden. Vincent et al. (1987)[28] found a threshold lung burden of 1.8 mg titanium dioxide; transportation of particles to the LALN was found to be proportional to the lung burden above this threshold value. In another study, Ferin and Feldstein (1978)[27] showed that, after intratracheal instillation of titanium dioxide, the transport rate to the lymph nodes increases by a factor of eight if the lung burden is above 0.7 mg. Strom et al. (1987)[29] concluded from their kinetic model for diesel particles that the predominant pathway of particles from the lung to the LALN is from a particle-sequestering compartment of the lung. The transport of particles from the normal macrophage compartment into this sequestering compartment is dependent on the particle burden of the macrophage compartment. They reported a clearance rate of 0.00082 per day from the sequestered compartment to the lymph nodes for diesel particles (MMD = 0.2 μm).

7 Model Calculation of Retention Kinetics of Solid Particles

In the last decade, various models have been published to analyse the relationship between aerosol concentration and lung burden of experimental animals. These models can be used in the design of chronic studies with poorly-soluble materials. The application of first order kinetics, which do not consider retardation of particle clearance at high dust loads, could lead to too high aerosol concentrations.

Vostal et al. (1982)[30] and Strom et al. (1987)[29] reported retention data of the lungs and LALN in rats during and after diesel soot inhalation. They developed a compartmental model to calculate the transport of diesel particles from the lung to the tracheobronchial tree and to the LALN. In this model a particle-sequestering compartment was introduced to explain the decreasing particle clearance with increasing lung burden. The transport of particles to this sequestering compartment is dependent on the particle burden in the macrophage compartment.

The Toxicology Design Committee (TDC) of the National Toxicology Program (NTP) published guidelines for use in inhalation toxicity studies.[31] This approach represents a step forward for the establishment of suitable aerosol exposure concentrations and the conduct of inhalation toxicity investigations. However, this recommendation does not sufficiently respect the increasing retardation of the alveolar clearance during a chronic study. A comparison of the predicted lung burdens for chronic studies using the method of Lewis et al. (1989)[31] and the actual retained masses of studies by Muhle et al. (1990)[20] shows considerable deviations.

The relationship between the clearance rate of labelled particles and the pulmonary burden of toner, TiO_2, diesel soot and carbon black (the two latter only to an exposure period of up to six months) was published previously as a 'semiempirical model'.[18,20] It was shown that, after the onset of overload, the alveolar clearance rate decreases as the lung burden increases. Simple first order modelling substantially underestimated pulmonary retention. The principal step in the semiempirical model is using the first order clearance rate coefficient as an explicit function of lung burden.

A mathematical model was developed to account for the retention and clearance behaviour of the toner and surrogate tracer aerosol particles.[22] The model is based upon the histopathological observation that particle-laden macrophages aggregate in focal areas, whereas in other areas of the lung, particle-laden macrophages were only seldom found. The following assumptions were used in the model:

- The lung is divided into two compartments, M and I, which contain mobile and immobile macrophages, respectively (Figure 3).
- Macrophages, and particles contained within them, are transferred from compartment M with a clearance rate k_{MT} to the GI-tract via the trachea.
- The particles in compartment I can only be cleared from I to the LALN at a clearance rate k_{IL}.
- The deposited dust, D, is distributed into compartments M and I in a ratio corresponding to their lung volume.
- Macrophages in M with a particle burden above a critical volume V_{crit} (corresponding to a critical number of particles N_{crit}) are immobilized. The nearest macrophages, the number of which is called R, can migrate by chemotaxis into this area. This volume of lung is defined as belonging to the I-compartment.

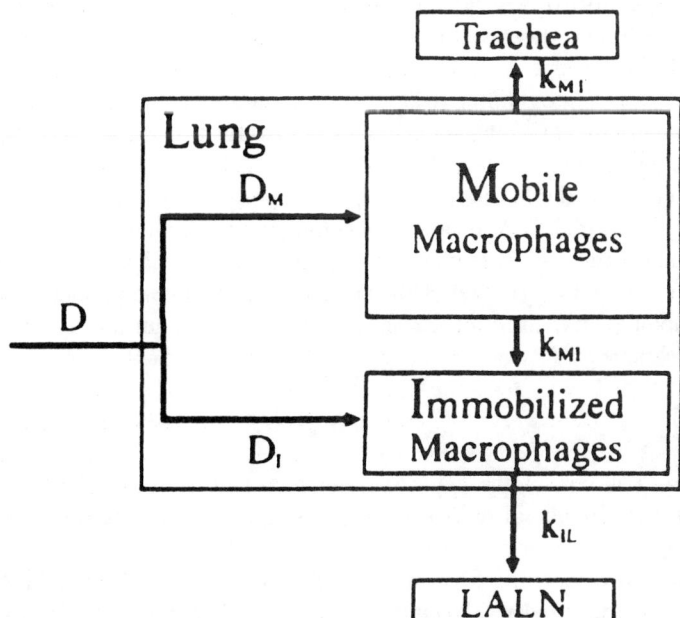

Figure 3 *Model of particle flow between the compartment of the lung, lung-associated lymph nodes (LALN) and trachea (according to Bellmann et al., 1990[22]).*

For each combination of parameters, the calculated values were compared with the experimental values. The least sum of the squares of differences between experimental and calculated values was used as an indicator for the best fit, similar to the method used by Strom et al. (1987).[29] For this procedure, the mass of toner in the lungs and the clearance rate k for ^{85}Sr-labelled particles, the mass of toner and ^{85}Sr particles in the lymph nodes were used. The lines drawn in Figure 2 are calculated by this model. The calculated values fit the experimental data in this relatively simple model quite well.

A more complex model was published by Stöber et al., 1990,[32] 1993.[33] The authors used a model of different compartments, which were linked together by distinct transfer pathways involving specific particle transport mechanisms. The development of this sophisticated model was possible on the basis of detailed experimental data on the particulate burden of the lung and the lung associated lymph nodes and the use of labelled tracer particles to analyse the impact of the lung burden on the macrophage-mediated clearance.

8 The Concept of Maximum Tolerated Dose (MTD) in Chronic Inhalation Studies

The study of Lee et al. (1985)[15] leads to the question whether the observed tumours at an exposure concentration of 250 mg m^{-3} could be an artefact. At that time, the author followed the concept of the Maximum Tolerated Dose (MTD). This concept has been widely used in long-term bio-assays. The MTD, as defined by Sontag et al. (1976),[34] is the highest dose of test agent that can be predicted not to alter the animals' normal longevity from effects other than carcinogenicity during a chronic study. A provisional guideline of the US Environmental Protection Agency indicated that a 10% increase in target organ weight also satisfies the MTD criterion.[35] For inhalation studies, a modification of the MTD definition to include a maximum functionally tolerated dose (MFTD) was suggested. This is the maximum lung burden above which macrophage-mediated lung clearance is significantly impaired.[36] A two- to four-fold increase in macrophage-mediated clearance half life, as measured by test material or suitable surrogate, was discussed as a suitable criterion. This concept has the advantage that clearance is related to pulmonary performance and is generally recognised as a significant functional parameter.

9 Fibres

Toxicokinetics is one of the key issues in investigating the toxicity of mineral fibres.[37] The persistence of man-made vitreous fibres in the lung tissue is thought to be related to their potency in inducing tumours.[38,39] It is proposed to distinguish between the persistence of a fibre, which can be defined as its long-term residence at the same location, and its biodurability, which is determined by the dissolution or disintegration of the fibre. A high biodurability is a precondition for the persistence.[40]

Fibre biopersistence can be defined as the retention of fibres at a defined location in the lung, over time, with regard to their number, dimensions, surface chemistry, chemical composition, surface area and similar physical characteristics.[41]

The elimination of durable fibres from the lung is dependent on fibre length, fibre diameter and fibre mass retained in the lungs.[42] Alveolar macrophages can only completely phagocytize fibres up to about 10 μm in length, prior to their removal by the ciliated airways.[43,44] In addition to the clearance mechanisms mentioned previously, fibres may break or disintegrate in the lungs.

To obtain an exact characterisation of the fibre elimination process, the total count of fibres retained in the lung and their length and diameter distribution for each date of serial sacrifice is required. Fibre length and diameter are usually log–normal distributions. An overview in the field of toxicokinetics of fibres is given in the proceedings of the workshop 'Biopersistence of respirable synthetic fibres and minerals' [*Environ. Health Perspect.* **102** (Suppl. 5), 1994].

Recent results have shown that fibres more than 20 μm long show a shorter retention time than shorter fibres.[45–48] This could be due to preferential breaking of the long fibres, or to faster dissolution of fibres outside macrophages. Half lives for different length fractions of fibres, *e.g.* < 5, 5–10, 10–20 and 20 μm can be calculated; they can give important information as the long fibres (> 10 μm) are supposed to be of special relevance for tumour induction.[49]

Yu *et al.* (1994)[50] proposed by model calculations that macrophage-mediated fibre clearance is impaired at a lower mass burden of the lungs compared with isometric, non-cytotoxic dusts.

10 Quartz

Solid particles with an intrinsic toxicity on alveolar macrophages show an impairment of alveolar clearance at much lower retained masses than those particles with a low inherent cytotoxicity.

Figure 4 shows an example of crystalline silica, which has an impact on the clearance of labelled particles. Fischer rats were exposed to 1 mg m^{-3} DQ12 Quartz 6 h day^{-1}, 5 days per week for up to 24 months.[51] During the ongoing exposure, the clearance of ^{85}Sr-polystyrene particles was measured at 9, 15 and 21 months for a period of 3 months at a time.

Polystyrene tracer clearance was impaired after SiO_2 exposure; however, there was a substantial reduction in the degree of impairment with elapsed time during the experiment. There appeared to be some reversal of an impaired clearance, from 9 to 21 months. The fraction of material retained in the lung-associated lymph nodes of the SiO_2-exposed rats appeared to be much higher than that in the TiO_2-exposed group and the toner groups at all exposure levels, which were done in parallel.[17,18] At the terminal sacrifice, the percentages of mass retained in the LALN compared with that found in the lungs of the silica-exposed group were 67 and 51% for the male and female rats, respectively. The corresponding values for the toner high-exposure group were only 8 and 6%, respectively. These results indicate massive movement of the crystalline silica from the lungs to the LALN. The high fraction of debris of macrophages found in the bronchoalveolar lavagate

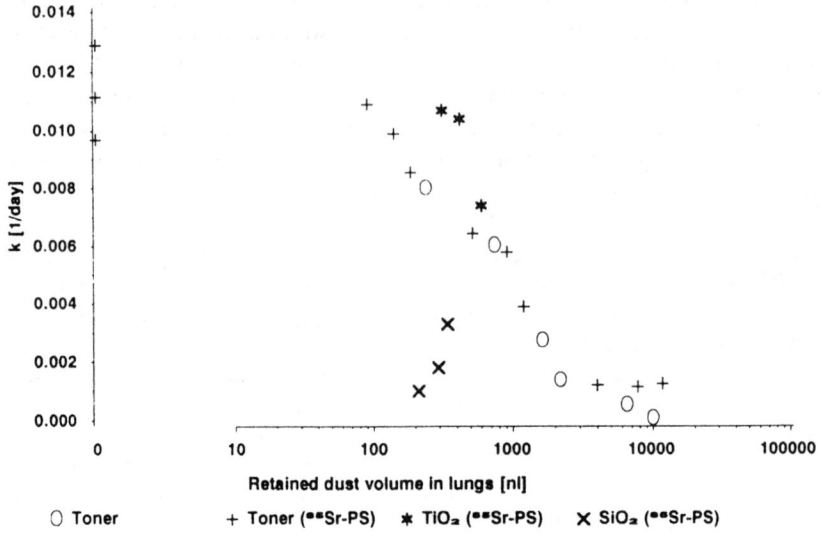

Figure 4 *Clearance rate coefficient of toner (Muhle et al., 1990)[20] and labelled particles in relation to the retained volume of the various test materials. Quartz particles show an impairment of lung clearance at much lower retained dust volumes compared with toner or TiO_2. Pooled data of male and female rats.*

of SiO_2-exposed animals[17] suggested that the cytotoxic effect of the quartz induced a shorter lifetime of the macrophages after phagocytosis of these particles. This may be the reason why quartz particles are more frequently observed outside the macrophages and have a higher chance of migrating to the interstitium and then to the lymph nodes.

11 Solid Ultrafine Particles

Recent studies with ultrafine particles (diameter ~20 nm) have shown that these particles cause greater inflammatory response in the lungs of rats than particles in the diameter range of 1 μm.[52–55] The translocation into the pulmonary interstitium into Type I and Type II cells is faster compared with particles with a diameter in the range of 1 μm.

In a recently completed study, Wistar rats were exposed for two years to diesel engine exhaust, carbon black (Printex 90, Degussa, FRG) and ultrafine TiO_2 (P25, Degussa, FRG) and were subsequently kept in clean air for six months.[56] The average particle exposure concentrations for diesel soot, carbon black and TiO_2 were 7, 11.6 and 10 mg m^{-3}, respectively. Two further diesel exhaust-exposed groups received particle concentrations of 2.5 and 0.8 mg m^{-3}. The half lives of the retardation of the alveolar clearance were about 80 days for controls and more than 350 days after exposure to titanium dioxide or carbon black.[57]

Effects on alveolar macrophages were seen after an exposure period of three months at a lung burden between 0.6 and 1.2 mg for the rat lung. Agglomerates of ultrafine carbon and TiO_2 particles seem particularly suited to exert toxic effects, primarily on alveolar macrophages and alveolar lung particle clearance. Although such lung toxic effects were also seen with the lowest diesel soot exposure concentration (0.8 mg m^{-3}) used, no increased lung tumour rate was detected in this group of rats. For the development of carcinogenic effects see Heinrich et al. (1995).[56]

These results show that the toxicokinetics of solid particles depend not only on the density of the particles but also on the diameter. The impact of these results for the regulation of these findings is under discussion.

References

1. Oberdörster, G. (1988) Lung clearance of inhaled insoluble and soluble particles, *J. Aerosol Med.*, **1**, 289–320.
2. Schlesinger, R.B. (1989) Deposition and clearance of inhaled particles. In, 'Concepts in inhalation toxicology', McClellan, R.O., Henderson, R.F., eds., Hemisphere Publ., New York, pp. 163–192.
3. Stahlhofen, W., Scheuch, G. and Bailey, M.R. (1994) Measurement of the tracheobronchial clearance of particles after aerosol bolus inhalation, *Ann. Occup. Hyg.*, **38** (Suppl. 1), 189–196.
4. Bergmann, I. and Casswell, C. (1972) Lung dust and lung iron contents of coal workers in different coalfields in Great Britain, *Br. J. Ind. Med.*, **29**, 160.
5. Morrow, P.E. (1988) Possible mechanisms to explain dust overloading of the lungs, *Fundam. Appl. Toxicol.*, **10**, 369–384.
6. Morrow, P.E., Muhle, H. and Mermelstein, R. (1991) Chronic inhalation study findings as a basis for proposing a new occupational dust exposure limit, *J. Am. Coll. Toxicol.*, **10**, 279–290.
7. Morrow, P.E. (1992) Contemporary issues in toxicology. Dust overloading in the lungs. Update and appraisal, *Toxicol. Appl. Pharmacol.*, **113**, 1–12.
8. Oberdörster, G. (1995) Lung particle overload: implications for occupational exposure to particles, *Regul. Toxicol. Pharmacol.*, **27**, 123–135.
9. Morrow, P.E., Gibb, F.R. and Gazioglu, K. (1967) The clearance of dust from the lower respiratory tract of man. An experimental study. In, 'Inhaled Particles and Vapours II', Davies, C.N., ed., Pergamon Press, Oxford, pp. 351–359.
10. Martin, J.C., Daniel, H. and LeBouffant, L. (1977) Short- and long-term experimental study of the toxicity of coal mine dust and some of its constituents. In, 'Inhaled Particles IV, Vol. 1', Walton, W.Hd., ed., Pergamon Press, Oxford, 361–370.
11. Wehner, A.P., Dagle, G.D., Clark, M.L. and Buschbom, R.L. (1986) Lung changes in rats following inhalation exposure to volcanic ash for two years. *Environ. Res.*, **40**, 499–517.
12. Muhle, H., Bellmann, B. and Heinrich, U. (1988) Overloading of lung clearance of experimental animals to particles, *Ann. Occup. Hyg.*, **32** (Suppl.), 141–148.
13. Raabe, O.G., Tyler, W.S., Last, J.A., Schwartz, L.W., Lollini, L.O., Fisher, G.L., Wilson, F.D. and Dungworth, D.L. (1982) Studies of the chronic inhalation of coal fly ash by rats, *Ann. Occup. Hyg.*, **29** (Suppl.), 189–211.
14. Klonne, D.R., Burns, J.M., Halder, C.A., Holdsworth, C.E. and Ulrich, C.E. (1987) Two-year inhalation toxicity study of petroleum coke in rats and monkeys, *Am. J. Ind. Med.*, **11**, 375–389.
15. Lee, K.P., Trochimowicz, H.J. and Reinhardt, C.F. (1985) Pulmonary response of rats exposed to titanium dioxide (TiO$_2$) by inhalation for two years, *Toxicol. Appl. Pharmacol.*, **79**, 179–182.

16. Strom, K.A., Johnson, J.T. and Chan, T.C. (1989) Retention and clearance of inhaled submicron carbon black particles, *J. Toxicol. Environ. Health*, **26**, 183–202.
17. Muhle, H., Bellmann, B., Creutzenberg, O., Dasenbrock, C., Ernst, H., Kilpper, R., MacKenzie, J., Morrow, P., Mohr, U., Takenaka, S. and Mermelstein, R. (1991) Pulmonary response to toner upon chronic inhalation exposure in rats, *Fundam. Appl. Toxicol.*, **17**, 280–299.
18. Bellmann, B., Muhle, H., Creutzenberg, O. *et al.* (1991) Lung clearance and retention of toner, utilizing a tracer technique during chronic inhalation exposure in rats, *Fundam. Appl. Toxicol.*, **17**, 300–313.
19. Bowden, D.H. (1987) Macrophages, dust and pulmonary disease, *Exp. Lung Res.*, 12, 89–107.
20. Muhle, H., Creutzenberg, O., Bellmann, B., Heinrich, U. and Mermelstein, R. (1990) Dust overloading of the lungs: Investigations of various material, species differences and irreversibility of effects, *J. Aerosol Med.*, 3 (Suppl. 1), 111–128.
21. McClellan, R.O. (1990) Particle overload in the lung: Approaches to improve our knowledge, *J. Aerosol Med.*, 3 (Suppl. 1), 197–207.
22. Bellmann, B., Muhle, H., Creutzenberg, O. and Mermelstein, R. (1990) Recovery behaviour after dust overloading of lungs, *J. Aerosol Sci.*, **21**, 377–380.
23. Bowden, D.H. and Adamson, Y.R. (1980) Role of monocyte and interstitial cells in the generation of alveolar macrophages: I. Kinetic studies in mice, *Lab. Invest.*, **42**, 511–517.
24. Fritsch, P. and Masse, R. (1992) Overview of pulmonary alveolar macrophages renewal in normal rats and during pathological processes, *Environ. Health Perspect.*, **97**, 95–67.
25. Lehnert, B.E., Sebring, R.J. and Oberdörster, G. (1994) Pulmonary macrophages: Phenomena associated with the particle "overload" condition. In, 'Toxic and carcinogenic effects of solid particles in the respiratory tract', Mohr, U., ed., ILSI Press, Washington, pp. 159–176.
26. Brown, G.M., Brown, D.M. and Donaldson, K. (1992) Persistent inflammation and impaired chemotaxis of alveolar macrophages on cessation of dust exposure, *Environ. Health Perspect.*, **97**, 91–94.
27. Ferin, J. and Feldstein, M.L. (1978) Pulmonary clearance and hilar lymph node content in rats after particle exposure, *Environ. Res.*, **16**, 342–352.
28. Vincent, J.H., Jones, A.D., Johnston, A.M., McMillan, C., Bolton, R.E. and Cowte, H. (1987) Accumulation of inhaled mineral dust in the lung and associated lymph nodes: Implications for exposure and dose in occupational lung disease, *Ann. Occup. Hyg.*, **31**, 375–393.
29. Strom, K.A., Chan, T.L. and Johnson, J.T. (1987) Pulmonary retention of inhaled submicron particles in rats: diesel exhaust exposure and lung retention model, *Research Publication GMR-5718*, General Motors Research Laboratories, Warren MI.
30. Vostal, J.J., Schreck, R.W., Lee, P.S. Chan, T.L. and Soderholm, S.C. (1982) Deposition and clearance of diesel particles from the lung. In, 'Toxicological effects of emission from diesel engines', Lewtas *et al.* eds., Elsevier Science, New York, pp. 143–159.
31. Lewis, T.R., Morrow, P.E., McClellan, R.O., Raabe, O.G., Kennedy, G.L., Schwetz, B.A., Goehl, T.J., Roycroft, J.H. and Chhabra, R.S. (1989) Contemporary Issues in Toxicology: Establishing Aerosol Exposure Concentrations for Inhalation Toxicology Studies, *Toxicol. Appl. Pharmacol.*, **99**, 377–383.
32. Stöber, W., Morrow, P.E., Morawietz, G., Koch, W. and Hoover, M.D. (1990) Developments in modeling alveolar retention of inhaled insoluble particles in rats, *J. Aerosol Med.*, **3** (Suppl. 1), 129–154.

33. Stöber, W., McClellan, R.O. and Morrow, P.E. (1993) Approaches to modeling disposition of inhaled particles and fibres in the lung. In, 'Toxicology of the Lung', 2nd edn., Raven Press, New York, pp. 527–601.
34. Sontag, J.M., Page, N.P. and Saffiotti, U. (1976) Guidelines for carcinogen bioassay in small rodents. *DHHS Publication (NIH)*, Washington DC, pp. 76–801.
35. Environmental Protection Agency (1986) Draft policy chemical Regulation, *Reporter*, May 9, 158.
36. Muhle, H., Bellmann, B., Creutzenberg, O., Fuhst, R., Koch, W., Mohr, U., Takenaka, S., Morrow, P., Kilpper, R., MacKenzie, J. and Mermelstein, R. (1990) Subchronic inhalation study of toner in rats, *Inhal. Toxicol.*, **2**, 341–360.
37. McClellan, R.O., Miller, J.F., Hesterberg, T.W., Warheit, D.B., Bunn, W.B., Kane, A.G., Lippmann, M., Mast, R.W., McConnell, E.E. and Reinhardt, C.F. (1992) Approaches to evaluating the toxicity and carcinogenicity of man-made fibres: Summary of a workshop held 11–13 November 1991, Durham, NC, *Regul. Toxicol. Pharmacol.*, **16**, 321–364.
38. Davis, J.M.G. (1986) A review of experimental evidence for the carcinogenicity of man-made vitreous fibres, *Scand. J. Work Envrion. Health*, **12** (Suppl. 1), 12–17.
39. Pott, F., Roller, M., Ziem, U., Reiffer, F.J., Bellmann, B., Rosenbruch, M. and Huth, F. (1989) Carcinogenicity studies on natural and man-made fibres with the intraperitoneal test in rats. In, 'Non-occupational Exposure to Mineral Fibres', Bignon, J., Peto, J., Saracci, R., eds. IARC Scientific Publication No. 90, International Agency for Research on Cancer, Lyon, pp. 173–179.
40. Pott, F. (1987) Problems in defining carcinogenic fibres, *Ann. Occup. Hyg.*, **31**, 799–802.
41. McClellan, R.O. and Hesterberg, T.W. (1994) Role of biopersistence in the pathogenicity of man-made fibres and methods for evaluating biopersistence – a summary of two round-table discussions, *Environ. Health Perspect.*, **102** (Suppl. 5), 277–283.
42. Muhle, H., Bellmann, B. and Pott, F. (1994) Comparative investigations of the biodurability of mineral fibres in the rat lung, *Environ. Health Perspect.*, **102** (Suppl. 5), 163–168.
43. Morgan, A. (1980) Effect of length on the clearance of fibres from the lung and on body formation. In, 'Biological Effects of Mineral Fibres', Wagner, J.C., ed., IARC, Scientific Publication No. 30, International Agency for Research on Cancer, Lyon, pp. 329–335.
44. Bolton, R.E., Vincent, J.H., Jones, A.D., Addison, J. and Beckett, S.T. (1983) An overload hypothesis for pulmonary clearance of UICC amosite fibres inhaled by rats, *Br. J. Ind. Med.*, **40**, 264–272.
45. Morgan, A., Davis, J.A., Mattson, S.M. and Morris, K.J. (1995) Effect of chemical composition on the solubility of glass fibres *in vivo* and *in vitro*, *Ann. Occup. Hyg.* (in press).
46. Musselman, R.P., Miller, W.C., Easters, W., Hadley, J.G., Kamstrup, O., Thevenaz, P. and Hesterberg, T.W. (1994) Biopersistence of man-made vitreous fibres (MMVF) and crocidolite fibres in rat lungs following short-term exposures, *Environ. Health Perspect.*, 139–143.
47. Bellmann, B. and Muhle, H. (1994) Bioberständigkeit verschiedener Mineralfasertypen in der Rattenlunge nach intratrachealer Applikation. Wirtschaftsverlag NW, Bremerhaven, Germany, pp. 1–108.
48. Bellmann, B., Muhle, H., Kamstrup, O. and Draeger, U.F. (1994) Investigation on the durability of man-made vitreous fibres in rat lungs, *Environ. Health Perspect.*, **102** (Suppl. 5), 185–189.

49. Davis, J.M.G., Addison, J., Bolton, R.E., Donaldson, K., Jones, A.D. and Smith, T. (1986) The pathogenicity of long versus short fibre sample of amosite asbestos administered to rats by inhalation and intraperitoneal injection, *Br. J. Exp. Pathol.*, **67**, 415–430.
50. Yu, C.P., Zhang, L., Oberdörster, G., Mast, R.W., Glass, L.R. and Utell, M.J. (1994) Clearance of refractory ceramic fibres (RCF) from the rat lung: Development of a model, *Environ. Res.*, **65**, 243–253.
51. Muhle, M., Takenaka, S., Mohr, U., Dasenbrock, C. and Mermelstein, R. (1989) Lung tumor induction upon long-term low-level inhalation of crystalline silica, *Am. J. Ind. Med.*, **15**, 343–346.
52. Ferin, J., Oberdörster, G., Penney, D.P. (1992) Pulmonary retention of ultrafine and fine particles in rats, *Am. J. Respir. Cell Mol. Biol.*, **6**, 535.
53. Oberdörster, G., Ferin, J., Gelein, R., Soderholm, S.C. and Finkelstein, J. (1992) Role of the alveolar macrophage in lung injury: Studies with ultrafine particles, *Environ. Health Perspect.*, **97**, 193–199.
54. Oberdörster, G., Ferin, J. and Lehnert, B.E. (1994) Correlation between particle size, *in vivo* particle persistence and lung injury, *Environ. Health Perspect.*, **102** (Suppl. 5), 173–179.
55. Oberdörster, G., Gelein, R., Ferin, J. and Wiss, B. (1995) Association of particulate air pollution and acute mortality: invaluement of ultrafine particles? *Inhal. Toxicol.*, **7**, 111–124.
56. Heinrich, U., Fuhst, R., Rittinghausen, S., Creutzenberg, O., Bellmann, B., Koch, W. and Levsen, K. (1995) Chronic inhalation exposure of Wistar rats and two different strains of mice to diesel engine exhaust, carbon black and titanium dioxide, *Inhal. Toxicol.*, **7**, 539–556.
57. Muhle, H., Bellmann, B. and Creutzenberg, O. (1994) Toxicokinetics of solid particles in chronic rat studies using diesel soot, carbon black, toner, titanium dioxide and quartz. In, 'Toxic and carcinogenic effects of solid particles in the respiratory tract,' Mohr, U., ed., ILSI Press, Washington, 29–41.

Mechanisms and Significance of Particle Overload

P.E. MORROW

DEPARTMENT OF ENVIRONMENTAL MEDICINE, UNIVERSITY OF ROCHESTER, USA

1 Background

Over thirty years have passed since Professors Theodore Hatch and Paul Gross published their monograph on Pulmonary Deposition and Retention of Inhaled Aerosols.[1] Ted Hatch was a specialist in industrial hygiene; Paul Gross was a pulmonary pathologist. Both carried out extensive experimental studies of dusts. These astute investigators provided us with a remarkable compendium of experimental dust studies undertaken before the 1960s, as well as discussions of the basic requirements and concepts of inhalation toxicology. There are several passages from their monograph that are very appropriate to our topic today.

First, Hatch and Gross state: 'In order to establish a quantitative dose–response relationship at the critical site, one must first estimate how much of the inhaled aerosol (toxicant) is initially deposited and at what sites within the respiratory system, how rapidly and to what degree the deposited particles are cleared from the respiratory tract and lungs and, finally, what fraction of the retained material reaches the critical site within the lungs or other parts of the body to produce damage ... '.[2]

This statement may seem to be both dated and self evident, but one only needs to recall the common usage of exposure concentration as a surrogate for 'dose' in our experimental and epidemiological reporting. Moreover, exposure concentrations serve as the basis for all of our air quality standards for worker and public health protection. That we cannot achieve, except in rare instances, what Hatch and Gross termed 'a quantitative dose–response relationship' is a testimonial to the persisting difficulties of performing quantitative inhalation toxicology.

In a second statement, Hatch and Gross, point out: 'With light dust burdens, a reasonable balance is maintained in healthy lungs between the rate of initial deposition and pulmonary clearance rate. A slight excess of particles may be translocated from alveoli to the lymph nodes which drain the lungs. If the load of particulates initially deposited upon the alveolar surface is large, however, or if the lung clearance mechanism is impaired, a sequestering tissue reaction (pneumoconiosis) occurs which may be followed by other pathological events'.[3] Their statement is amplified by a further comment: 'Overloading of the pulmonary clearance mechanism will reduce its effectiveness according to findings from certain animal studies. It is not clear, however, how far these findings can be applied to human lungs under conditions of real industrial exposures'.[4]

These perceptive statements from Hatch and Gross about dust overloading remain correct in 1995 in every respect. It is only that we have recently rediscovered the phenomenon, we have postulated some mechanisms for it, and have found some additional reasons to be concerned about it. That is why I have been invited to review our current understanding of dust overload.

2 A Definition of Dust Overloading

Dust overloading is the outcome of excessive dust exposures whereby the normally linear kinetics of alveolar dust clearance become nonlinear. The prolongation of alveolar retention is the earliest sign of dust overloading and the impairment of dust clearance becomes greater as the extent of overload increases. This clearance impairment is believed to be due to the loss of alveolar macrophage (AM) mobility that is initially associated with a persistent inflammatory state, which can progress to induce chronic pathological changes, including tumours. The clearest demonstrations of dust overload involve the excessive exposure of Fischer 344 rats to insoluble dusts, which are both relatively benign and persistently retained.

At a symposium devoted to dust overload, Witschi[5] emphasized that overloading a biological system was not unique to the lungs, for it is well-established that the ability of many biological systems to function can be impaired by being subjected to excessive levels of substrate administration. Nonlinear kinetics are commonplace in toxicokinetic analyses. In the case of the lungs, the inhalation of particles that have an intrinsic toxicity may result in a dose-related reduction in AM particle removal. Toxic particle exposures can usually be distinguished from overload by the elicitation of this clearance reduction at much lower lung burdens, the concomitant association of increased AM lysis and the presence of other cytotoxic manifestations.

3 Overload Studies

Since the review of early experimental studies by Hatch and Gross[1] there have been a number of studies that provided further evidence of lung overload and its sequelae; these include Le Bouffant;[6] Klosterkötter and Gono;[7] Ferin;[8,9] Ferin and Feldstein;[10] Davis et al;[11] Middleton et al.;[12] Adamson and Bowden;[13,14] White and Bhagwan;[15] Lee and co-workers;[16] Green et al.;[17] Wehner et al.;[18] Bowden and Adamson;[19] Chan et al.;[20] Matsuno et al.;[21] Wolff et al.;[22,23] and Muhle et al.[24,25]

Just over ten years ago, Bolton et al.[26] published a paper on the pulmonary clearance of amosite fibres inhaled by rats, in which these investigators noted a prolongation of pulmonary clearance with increasing lung burdens that they hypothesized was due to 'overload'. This paper was followed by a second authored by Vincent et al.[27] in which they further discussed overload as the basis for the nonlinear clearance kinetics observed in the amosite study in rats. These two papers were widely cited in the subsequent literature on the subject of dust overload and they are the major reason dust overload was rediscovered.

4 Particle Clearance Kinetics

Some, perhaps many of you, may not be familiar with the usual kinetic descriptions used to describe alveolar dust clearance in experimental animals and man. It is important to review briefly our current views of dust clearance kinetics because they are fundamental to an understanding of dust overload.[28] Most 'insoluble' particles, characterised by a relatively long alveolar retention time, are usually found to follow first order kinetics, involving one, sometimes two, clearance rate constant, provided the dust particles are relatively benign and the amount of dust in the lungs is, as Hatch and Gross described, 'light'.

If we examine a simple case in Figure 1, we see that in the lungs, shortly after such a dust exposure, there follows a fairly rapid clearance phase. In most species, including man, this early, rapid clearance is attributable to mucociliary transport of particles within and ultimately out of the tracheobronchial (TB) airways. Looked at closely, it is found to be a nonlinear clearance process that requires about two or three days to run its course. It is nonlinear because the apparent clearance rate changes over time. This time dependency is based on the fact that mucociliary function varies in different regions of the respiratory tract. The velocity of mucociliary transport of dust in the trachea is in centimetres per minute, whereas in the smallest bronchioles, the average transport velocity is up to 100 times slower so as the overall dust clearance of the bronchial tree progresses, the apparent clearance rate becomes slower. To be sure, if you examine how investigators describe TB dust clearance, you will find a considerable variation. Often it is described by one or two or three exponential terms, but this is merely a curve fitting manoeuvre that approximates the reality. In contrast, the alveolar clearance of the insoluble particle usually appears to follow a single rate constant over time: within reasonable variations in absolute lung burdens, the same clearance rate prevails. This is indicative of a linear, simple first order clearance process (Figure 2).

In Figure 2, we have an example of simple first order kinetics involving a single compartment lung model during continuous dust exposure, where d is the rate of dust deposition (mg day^{-1}), k is fractional clearance rate of the lung compartment expressed as the ln $(2/t_{1/2})$, where $T_{1/2}$ is the retention half life of the alveolar dust in days. Thus k is expressed as the fraction per day (day^{-1}). The dust content of the lung, L, is expressed in milligrams. In this scheme, the build up rate of dust in the lungs, $L = d - kL$ (mg day^{-1}). With a persistently retained dust, the value of L keeps increasing until kL is equal to d, the daily deposition rate. At that time, L approaches zero and the value of L becomes constant, hence a steady-state lung burden, Lss, has been reached. This achievement of a steady-state lung

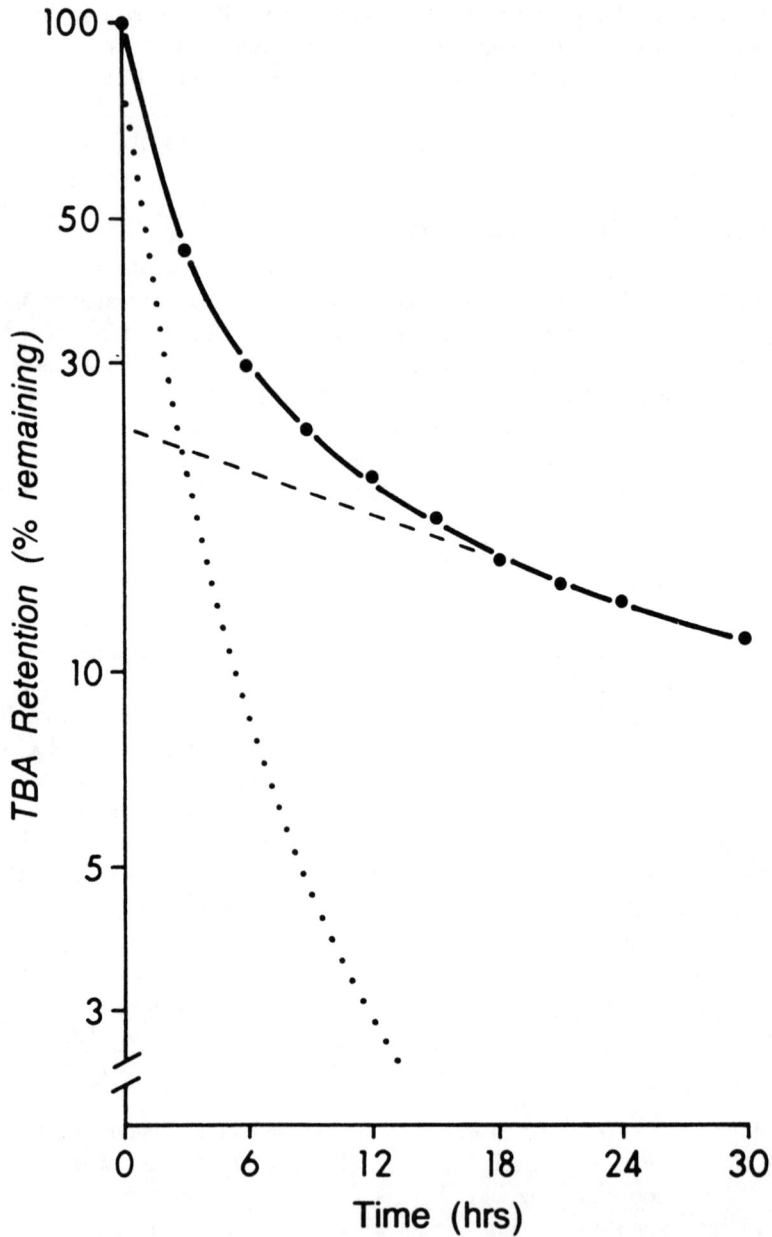

Figure 1 *The removal of an acutely administered insoluble dust from the tracheobronchial tree is depicted over 30 hours. Stripping the slowest clearance rate component, accounting for approximately 25% of the dust deposited (---), leaves a curvilinear residual (....), which could be stripped further thereby allowing the nonlinear TB retention curve to be approximated by three first order rate constants. Adapted from Morrow and Yu.[28]*

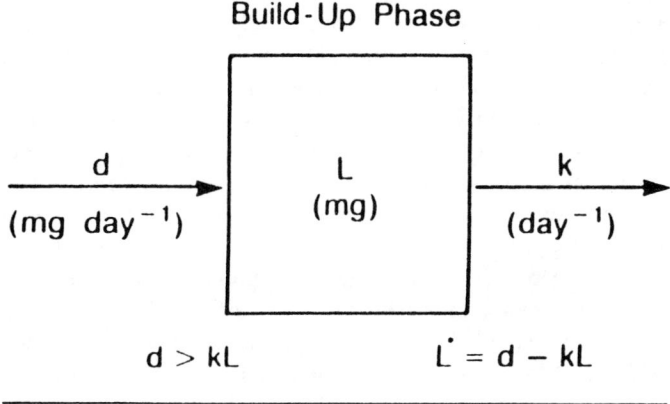

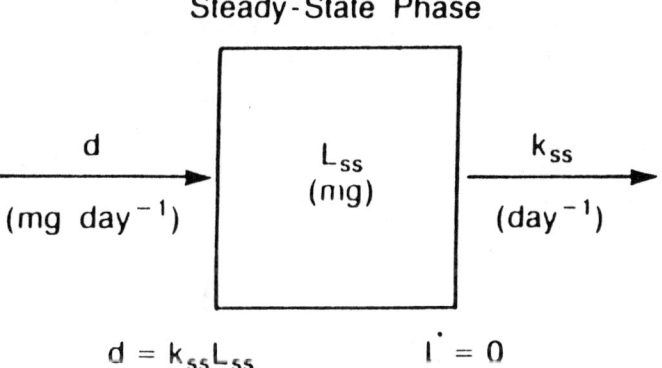

Figure 2 *Buildup and steady-state kinetics are depicted in a single lung compartment model utilizing first order clearance kinetics. The relationships between the daily dust deposition rate (d) and the daily dust elimination rate (k) are shown during the build-up phase (upper) and after achieving the steady-state phase. Adapted from Morrow.[29]*

burden in approximately five half lives ($t_{1/2}$), is the normal expectation of a first order dust clearance rate and a relatively constant dust deposition rate.

The time to achieve a steady-state lung burden obviously depends on the pulmonary dust retention time for the dust, other things remaining the same. For example, if the retention half life is 60 days for dust A and 200 days for dust B, both having the same value for the lung deposition rate, d, we expect that the steady state will occur around 300 and 1000 exposure days, respectively. On the other hand, if we compare two exposures to dust A with different particle size distributions such that the respective deposition fractions changed by a factor of ten, say from 0.01 to 0.10, then the value of the deposition rate, d, will likewise increase by a factor of ten. Since the value of k is expected to be unchanged (0.693/60 day), the magnitude of the respective steady-state lung burdens (L_{ss}) must also have the ratio 1 to 10 at the end of 300 days.

Let us consider one additional scenario. If, in the latter case of a deposition fraction of 0.10 for dust A, we found that a steady state did not occur in 300 days and the build up rate seemed to increase after the lung dust concentration reached about 1 mg dust per g lung tissue, we would be entitled to believe our loss of linear kinetics was due to dust overloading. With a constant particle size distribution and a constant dust exposure concentration throughout the exposure, the change in the compartmental kinetics must be due to the inconstant nature of k, the dust clearance rate. I hasten to add that all of the qualifying conditions used in this example are important, because, as I shall emphasize further, the development of nonlinear dust clearance kinetics is not a condition uniquely due to overload.

In Figure 3, we see examples of dust exposures performed by eight investigative teams where the value of k was measured for several different dusts.[30] Notice how k varied as a function of the lung dust concentration, and that below 1 or 2 mg dust per g of lung (log M_a = −2.5 to 0.3), the value of k was relatively constant, but as it exceeded the 1–2 mg dust per g lung tissue lung concentration, the value of k progressively decreased. At 10 mg dust pre g lung tissue, the value of k became so small as to suggest dust clearance had essentially ceased or was no greater than it would be if dust dissolution was the only dust removal mechanism operating.

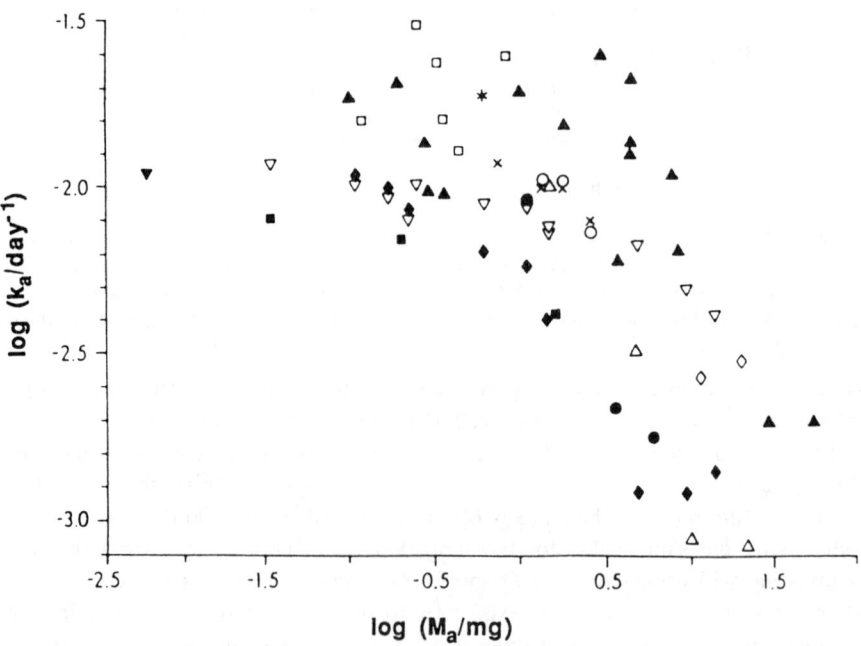

Figure 3 *Alveolar dust retention data from ten different investigations are summarized wherein each dust clearance rate (k_a) is plotted as a function of the total alveolar particulate burden (M_a/mg). Adapted from, Yu, Chen and Morrow.*[30]

5 Mechanistic Concepts

While these kinds of exposure and clearance data were being accumulated, it also became evident that in the lungs of rats with dust concentrations that induced decreased clearance, aggregations of particle laden macrophages were found more or less in relation to the lung dust concentration and, concurrently, the lymph nodal uptake of particles also increased in the same relative way. Collectively, these observations strongly incriminated the failure of the alveolar macrophage to transport the insoluble particles out of the lungs. For the rat, particulate transport by the AM from the alveolar region along the TB airways into the gastrointestinal tract, is known to be the dominant dust removal mechanism for insoluble particles. In general, when these static AM aggregations were seen, there did not seem to be a major problem with phagocytic uptake of particles, as the AMs were often engorged with particles, but there was an apparent loss of macrophage mobility to leave the alveoli. With decreased AM mobility, not only would dust clearance from the lungs be reduced, but the opportunity for interstitial penetration by particles to occur would be enhanced. This would be expected to result in increased particle uptake by the drainage lymph nodes. Additionally, in the view of Adamson and Bowden,[14,31] particles penetrating into the *interstitium* increase the toxicity of the particles probably through increased interactions with interstitial cells, *e.g.* fibroblasts and endothelial cells.

In 1988, in a paper with Mermelstein,[32] we pointed out that in studies using insoluble dusts with higher physical densities, more than 1 or 2 mg dust per g lung tissue was required to elicit a reduction in AM particle removal. However, when these dusts, *e.g.* titanium dioxide, $p = 4.2$ g cm^{-3}, were treated as unit density equivalents, their conformity to the clearance trend seen in Figure 3 improved. We proposed, consequently, that the volume of dust present in the lungs was more relevant to the impairment of clearance than was its mass[33] (Figure 4).

In a subsequent paper,[29] it was pointed out that the AM population in rat lungs activated by dust exposures was approximately 2.5×10^7 cells; therefore, it was possible to express the beginning of an overload condition in the Fischer 344 rat, the principal animal used in these studies, as the volumetric equivalent of 1 mg of spherical unit density particles per gram of lung allocated to 2.5×10^{-7} AM. In an adult F-344 rat with a lung weight of about 1.5 g, 1.5 mg of unit density dust constitutes a 1500 nl dust volume, thereby producing a volumetric concentration of 1000 nl g^{-1} lung associated with the beginning of overload. At the cellular level, 1500 nl, re-expressed as a 1.5×10^9 μm^3 dust volume, would result in an average 60 μm^3 dust per AM. If we accept 1000 μm^3 as the average volume of the AM, the onset of overload constituted about a 6% volumetric loading of dust, whereas the progression of overload until AM particle transport virtually ceases, implied an average of 60% increase in AM volume (600 μm^3) by phagocytosed dust.

These particulate loadings of the AM became the basis of a volumetric hypothesis.[29]

Oberdörster *et al.*[34] designed a study specifically to examine the volumetric hypothesis by administering 100 μg of radiolabelled 10.3 μm diameter plastic

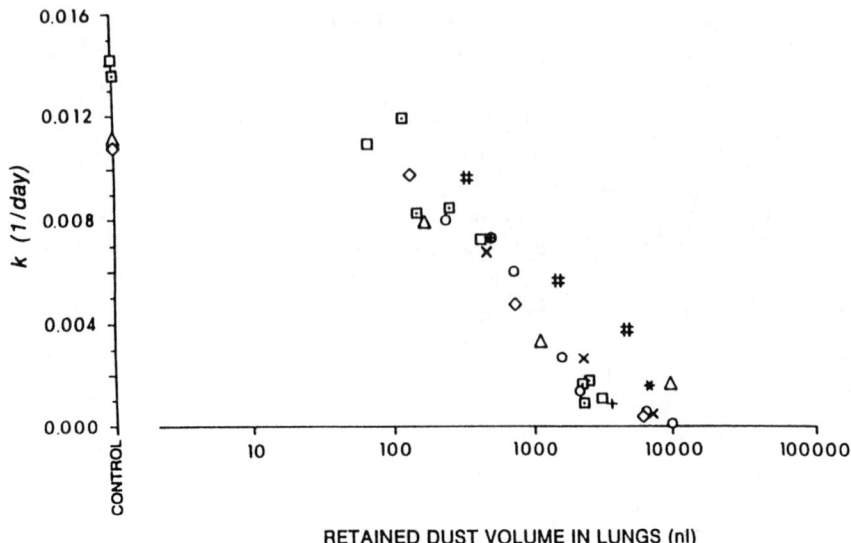

Figure 4 *Retention studies with a co-polymer toner (▪ □ ○ ◊ ∆), diesel particles (×), carbon black(*), poly(vinyl chloride) (#), and titanium dioxide (+, o) performed at Fraunhofer[24,25] in F-344 rats are shown where the volumetric particulate burden (nl) is correlated with the measured alveolar clearance rate (k). Adapted from Oberdörster.[33]*

microspheres to F-344 rats, each microsphere having a volume of ~600 μm³. Although the microspheres were promptly and completely phagocytized by AM, they underwent negligible clearance, thus providing support of the hypothesis.

Were these volumetric loadings interpretable in terms of macrophage migratory function? It seemed possible. For example, a volume-related mechanical interference with the cytoskeletal structures of the cell could be responsible.[35,36] Unfortunately, cell mobility has not been studied in these terms. AM mobility is clearly better understood in relation to chemotactic and chemokinetic factors whereby AM–particle interactions are facilitated. This facilitation can be seen in enhanced phagocytosis, but it does not necessarily translate into an action on AM migration from the alveolar region to the TB epithelial surface and out of the lungs.

Particle–cell interactions in the alveoli, especially those involving the AM, result in the elaboration of many kinds of factors, including those that are responsible for a major component of the onset and progression of dust overload *viz.* the appearance and persistence of inflammatory cells.[37] So many cytokines, enzymes, chemotactic factors, proteases and other mediators have been described in the condition of dust overload that it is tempting to consider that the loss of AM mobility may be both initiated and maintained by the excessive and persistent release of these endogenous materials. What arguments favour such a hypothesis?

In the Fraunhofer studies depicted in Figure 4, and in several subsequent studies, Muhle *et al.*[24,25] used insoluble test dusts and serial sacrifice procedures to establish lung burdens. Partly to reduce the number of sacrifices, small groups of rats were serially given a brief exposure to an insoluble radioactive particle, such as radiolabelled polystyrene (*PS) microspheres, and the alveolar retention of the *PS followed for several months. By measuring the retention times for the *PS microspheres during the course of both 90 day and 2 year studies, it was found that when an overload condition developed with the test dust, the test particle and *PS microsphere clearance were similarly affected. In other words, if at 10 months in the study the test particle burden in the lung began to exceed 1000 nl dust per g lung, rats given a brief, low level *PS exposure, showed evidence of reduced alveolar clearance when compared with concurrent control rats given the same *PS exposure. Independent measurements of the test dust clearance showed a similar reduction.

Lehnert et al.[38] showed by intratracheally-administered plastic microspheres that they could acutely produce overload in their rats when the engulfed particulate volume exceeded 500 nl, but perhaps more importantly, they showed that the distribution of particles per AM was extremely non-uniform. More recently, Morgan and co-workers[39] showed the same type of non-uniformity of particles per AM following the inhalation of fluorescent microspheres in rats. Their study and subsequent studies of overloaded lungs in rats have shown that many of the AM recovered in the bronchoalveolar lavagate were without particles. In the face of this realisation, how was it possible that the technique of using a radioactive surrogate particle clearly demonstrated that a newly inhaled dust was cleared at the same rate as dust which had been retained in the lungs for many months? The implication that the entire AM population was uniformly impaired in its particulate transport function despite highly irregular AM particle loading can best be explained by the elaboration of factors in some type of nonspecific dust dose–response system.

Obviously, this compelling view of the dust overload mechanism relegates the 60 μm^3 dust per AM to 600 μm^3 dust per AM range calculated from early overload studies to mere indices of the overload condition. The 1000 nl per g lung to 10,000 nl per g lung range of lung volumetric loading does remain a credible 'dose' for the dose–response system just alluded to. Thus, at the present time, there is a highly presumptive basis for considering the persistent, and possibly excessive elaboration of chemotactic and chemokinetic growth factors responsible for the reduction of AM-mediated particle transport from the alveolar region of the rat lung associated with dust overload.

6 The Reversibility of Overload

Another critical consideration of overload that we have not discussed pertains to the reversibility of overload. Here again, all of the evidence is derived from rat studies. As long as some AM-mediated particle transport from the alveolar region continues and parallel dust clearance mechanisms continue to function, *e.g.* dissolution and lymphatic uptake, it seems logical that the ensuing reduction of alveolar dust burden would eventually allow the AM to recover its normal clearance capacity. Creutzenberg et al.[40] and Bellmann et al.[41] undertook a reversibility study in F-344 rats up to 15 months after a subchronic exposure to toner particles, which basically confirmed the foregoing expectation. Their study showed that the condition of overload was completely reversible if the degree of overload was low, but with overload sufficient effectively to eliminate AM particle transport, there was no clear evidence of recovery.

7 Overload in Other Species

The references previously cited on inhalation studies that have described dust overload (*op. cit.*) obviously have an international character; nevertheless, with few exceptions, all of these studies were performed in the rat. In the United States, the National Toxicology Program set high animal quality standards for pre-chronic and chronic toxicity testing and these were subsequently met by several breeders of the Fischer-344 rat. As a consequence, virtually every toxicity study in the US since 1976 has depended on the F-344 rat. Less frequently, the Wistar or some other rat strain has been used. When a second species has been employed in toxicity studies, the Syrian Golden hamster or one of several strains of mice was selected. Muhle et al.[24] have reviewed many of these studies indicating a condition of dust overload during chronic dust exposures of the latter two species. The findings in hamster and mice are limited and the results have a greater variability, but there is a comparable picture of clearance retardation and sustained inflammatory changes, supporting the viewpoint that the rat is not unique.

A more extensive testing based upon non-rodent species is needed, but since experimentation with larger laboratory animals is decreasing, the prospects for this are poor. The ultimate question of species distinctions regarding dust overload, clearly focuses on humans.

8 Does Dust Overload Pertain to Humans?

With some conservative assumptions, it is easy to calculate the buildup characteristics of many kinds of dusts in human lungs. For instance, using the dust concentrations and particle sizes reported in US and European coal mines, the steady-state lung burdens that should occur (assuming a retention half life of about 1000 days) are well above the 1–2 mg per g lung tissue concentration, which portends the beginning of overload in the rat lung. Also, autopsy material from coal miners confirms that such lung dust concentrations are often exceeded in miners with many years of underground experience. Histopathological and pulmonary function assessments both indicate that lung injury does occur in coal miners in relation to their years of occupational exposure. This general conclusion is complicated by the fact that the rank of coal and the presence of other minerals in the mine dust play important pathogenic roles as do the smoking habits of miners. Despite these confounders, there is reason to believe that coal mining, like most dusty trades, produces lung injury. When we try to compare what is known about coal miners' lungs and lung coal burdens, we find there are no criteria that exactly match what we know about lung function and lung dust burdens in the rat. There are no data on lung retention as a function of lung burden in coal miners, we have no comparable pulmonary function tests for rats that we have for miners. Histopathological data do point out the comparative absence of lung tumours in coal miners, but granulomatous and fibrotic changes are known in both species.

Recent studies of particulate retention in human subjects by Bailey et al.[42] and Foster et al.[43] collectively suggest that alveolar macrophage function may not be a very important clearance mechanism in human lungs. Using a radio-labelled cobalt oxide, Foster et al. deduced, for example, that three-quarters of the alveolar dust clearance in humans was by simple dissolution. They also deduced that AM-mediated alveolar dust clearance had a half life of about 640 days! Bailey et al. arrived at similar values using labelled, fused aluminium silicate particles, but concluded that AM-mediated transport in the human lung was normally a nonlinear clearance process that could be empirically described by halflives ranging from 174 to 700 days. No explanation was offered for a time-dependent dust removal by AM.

Thus, at the present time, we can conclude that the circumstances for dust overload to occur certainly prevail in many coal mines, and possibly in other dusty trades, but at present there are no functional or pathological comparisons that permit us to conclude that the consequences of high lung dust concentrations are comparable in rodent and human lungs. In fact, two studies minimize the potential impact of dust overload in human lungs even if it occurred, owing to the apparently limited role of the human AM in dust removal.

9 The Implications of Overload

A conservative appraisal of the implications of dust overload would focus almost exclusively on two issues, viz. the confounding of pre-chronic and chronic inhalation exposure findings in rodents and the nongenotoxic tumour induction seen in F-344 rats, which appears to be both sex-specific and species-specific.

It is generally agreed that the condition of dust overload gives rise to non-specific dysfunctional and pathological states. Otherwise, how can one explain that toner, an organic copolymer; talc, a magnesium silicate; titanium dioxide, a prototypical benign, inert dust; carbon black, a chemically inert material; poly(vinyl chloride), a stable chlorethylene polymer; and volcanic ash, a complex carbon- aceous and vitreous material, all give rise to a similar progression of respiratory responses and effects as overload increases? Without any dosimetric distinctions, these different dusts presumably induce their respiratory system actions on some physical basis as we previously inferred.

Recent studies with ultrafine (< 100 nm diameter) inert particles, e.g. TiO_2 and carbon, have shown 'overload type' effects in the lungs of rats, but at lower lung dust concentrations. For example, consider the comparison of ultrafine and fine TiO_2 particles given in Figure 5, where alveolar retention and lymphatic uptake is seen to be greater for ultrafine particles at the same mass burden.[44] Microscopically, the penetration of ultrafine particle into the pulmonary interstitium is strikingly greater. In addition, the inflammatory response in rat lungs was found by Oberdörster and Yu[45] to correlate better with the particulate surface area than to mass or volume in the case of ultrafine particles. These findings and conclusions also suggest a physical basis for the biological activity of ultrafine particles and some important distinctions relative to the condition of dust overload vis a vis fine particles.

The rather consistent finding of tumours in rats with substantial degrees of dust overload had led to a consensus that diverse, virtually inert dusts are inducing tumours through a nongenotoxic mechanism.[46] At this time, there are no mechanistic explanations for this type of tumourigenesis, but the concept is widely accepted. This international seminar is an ideal forum for establishing the basis for the non-specific, particle induction of lung tumours.

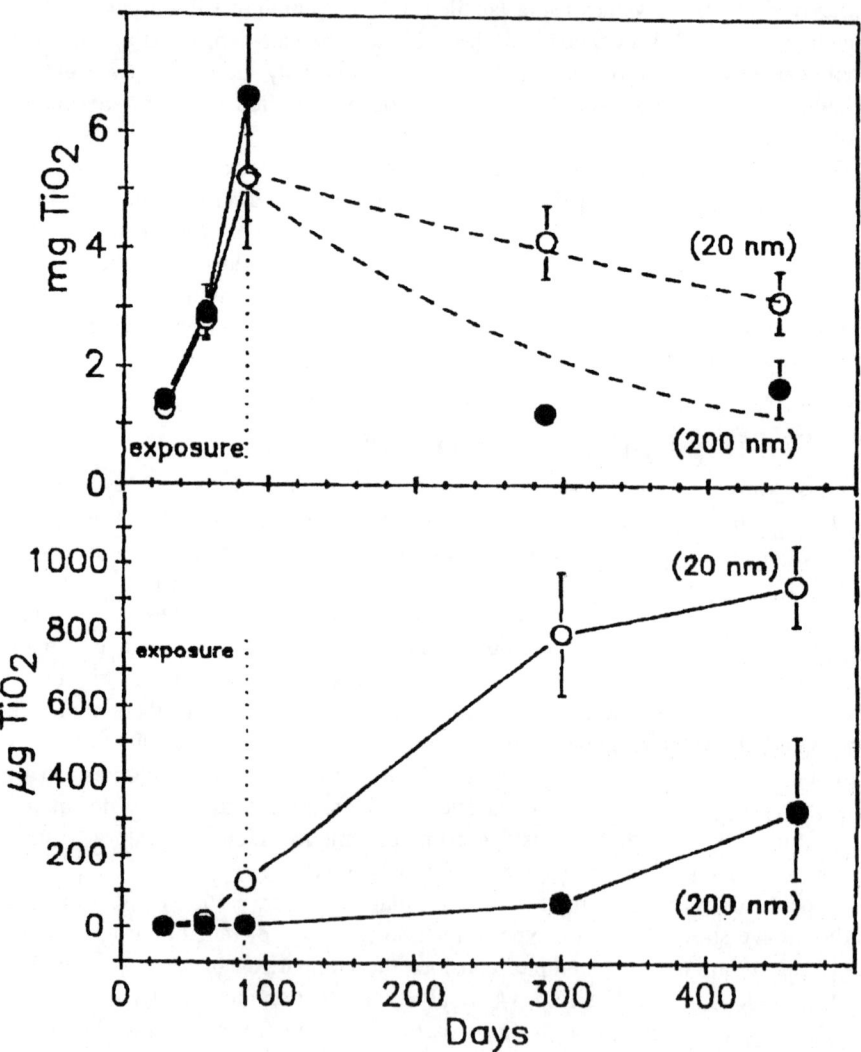

Figure 5 *Comparisons of similar fine (•) and ultrafine (o) titanium dioxide exposures on the subsequent lung retention (upper curves) and lymph nodal build-up (lower curves) measured in male F-344 rats. Adapted from Oberdörster et al.*[44]

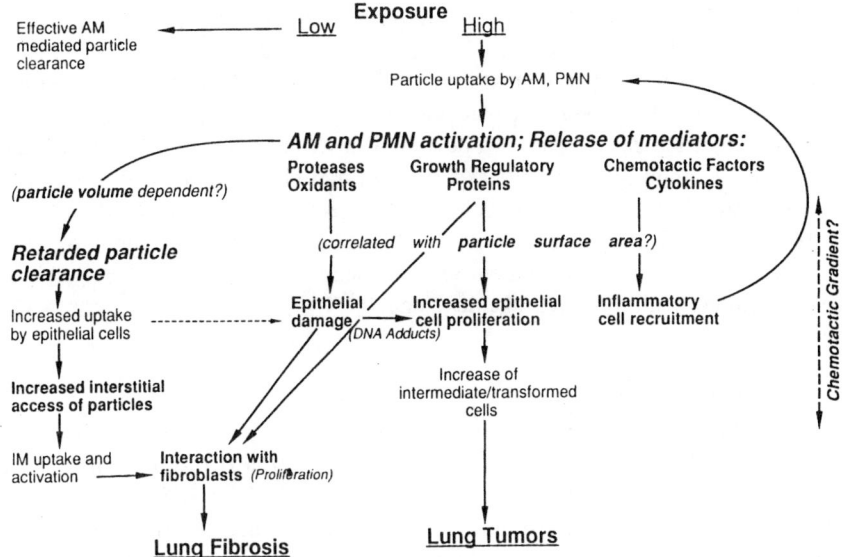

Figure 6 *A schematic depiction of the presumed relationships and effects induced by high and low particulate exposures. Adapted from Oberdörster and Yu.*[45]

With a less conservative appraisal of overload implications, one can speculate that excessive dust exposure levels, which lead to high dust deposition rates in the alveoli, always have the potential to overwhelm extant dust clearance processes. Default penetration of the respiratory epithelium by particles due to inefficient macrophage activity presents particles to interstitial cells that may be much more vulnerable to particle interactions.[14] The likelihood of overload effects in other species must continue to be considered, even if they are manifest in some actions other than those occurring in the rat prolongation of alveolar clearance. A hypothetical pathogenic sequence for particle effects in rats is given in Figure 6.

10 Relatable Issues

There are other issues that are tempting to include in this discussion of dust overload. The recent retrospective epidemiological studies by Pope, Dockery and Schwartz and their co-workers[47] in the US challenge the so-called PM_{10} standard for air quality. They have reported associations between air pollution levels, especially between the particulate concentrations in the PM_{10} range, and mortality among the aged in several US communities. Increments of 10 to 20 μg m^{-3} in urban dust levels have been correlated with significant, but small, increments in mortality! Moreover, the associations have involved communities with quite different particulate pollution patterns, so they postulate the effects seen that are both non-specific and particle-induced. There is not a shred of suspicion that dust overload is involved, but the apparent non-specificity of the particle effect gains one's attention.

In addition, there are the diesel particle studies[45] in which tumours were found, but were complicated in their interpretation because of the use of very high exposure levels. There was the related matter of adsorbed organic materials on the diesel particles, as it was known that these could be extracted, which gave a strong mutagenic signal and provided a plausible basis for tumourigenesis. Subsequent studies with purified carbon particles devoid of any mutagenic materials were also found to be as tumourigenic in rats as the diesel particles were at similar levels of dust overload. These carbon particle studies recall the historic difficulty of experimentally inducing cancers in the lungs of laboratory animals with carcinogenic agents, but the relative ease of accomplishing this when particles were given along with the carcinogenic agents.[48]

11 Conclusions

From the perspective of this seminar, lung overload can presently be regarded as an experimentally-encountered condition that both perplexes and excites those concerned with toxicity testing of insoluble particles in laboratory animals. The most reasonable explanation available for the effects of dust overload appears to depend upon nonspecific particulate properties that can induce a (1) decreased alveolar particle clearance by AM, (2) sustained pulmonary inflammatory response, (3) increased interstitial and lymph nodal particulate uptake and (4) long-term pathological sequelae, *e.g.* pulmonary fibrosis and/or nongenotoxic tumourigenesis, depending on the degree of dust overload. Efforts to identify initiating events, separate cause-and-effect relationships and understand the interrelated particle-induced responses are continuing. Our current understanding of these matters is schematically depicted in Figure 6.

It is especially important to determine whether the tumours produced by dust overload of rodent lungs will be found in other non-rodent species. From the perspective of understanding tumourigenesis, those involved in experimental investigations should seek a mechanistic basis for this overload effect and concurrently provide rational bases for selecting (or rejecting) specific animal models for determining the carcinogenic potential of inhaled materials in humans.

12 Acknowledgement

The author would like to express his appreciation to Joyce Morgan, Environmental Health Sciences Center, for her important help in the preparation of the manuscript, and to Dr Mark Frampton, pulmonologist in the Department of Medicine, and colleague for his valuable suggestions.

References

1. Hatch, T.F. and Gross, P. (1964) 'Pulmonary Deposition and Retention of Inhaled Aerosols', Academic Press, New York.
2. Ibid. p. 3.
3. Ibid. p. 70.

4. Ibid. p. 120.
5. Witschi, H. (1990) Lung overload: A challenge for toxicology, *J Aerosol Med.*, **3** (Suppl. 1), 189–196.
6. Le Bouffant, L. (1971) Influence de la nature des poussieres et de le charge pulmonaire sur l'epuration. In, 'Inhaled Particles III', Walton W.H., ed., Unwin, Old Woking, England, pp. 227–237.
7. Klosterkötter, W.S. and Gono, F. (1971) Long-term storage, migration and elimination of dust in the lungs of animals with special respect to the influence of polyvinyl-pyridine-N-oxide. In, 'Inhaled Particles III', Walton W.H., ed., Unwin, Old Woking, England, pp. 273-280.
8. Ferin, J. (1972) Observations concerning alveolar dust clearance, *Ann. New York Acad. Sci.*, **200**, 66–72.
9. Ferin, J. (1977) Effect of particle content of lung on clearance pathway. In, 'Pulmonary Macrophage and Epithelial Cells. Proceedings of the 16th Hanford Biology Syposium', ERDA Report CONF–760927, Technical Information Center, Energy Research and Development Administration, Washington, DC, pp. 414–423.
10. Ferin, J. and Feldstein, M.L. (1978) Pulmonary clearance and hilar lymph node content in rats after particle exposure, *Environ. Res.*, **16**, 342–352.
11. Davis, J.M.G., Beckett, S.T., Bolton, R.E., Collings, P. and Middleton, A.P. (1978) Mass and number of fibres in the pathogenesis of asbestos-related lung disease in rats, *Br. J. Cancer*, **37**, 673–688.
12. Middleton, A.P., Beckett, S.T. and Davis, J.M.G. (1979) Further observations on the short-term retention and clearance of asbestos by rats using UICC reference samples, *Ann. Occup. Hyg.*, **22**, 141–152.
13. Adamson, I.Y.R. and Bowden, D.A. (1980) Role of monocytes and interstitial cells in the generation of alveolar macrophages. II. Kinetic studies after carbon loading, *Lab. Invest.*, **42**, 518–524.
14. Adamson, I.Y.R. and Bowden, D.A. (1981) Dose response of the pulmonary macrophagic system to various particulates and its relationship to trans-epithelial passage of free particles, *Exp. Lung Res.*, **2**, 165–175.
15. White, H.J. and Bhagwan, D.G. (1981) Early pulmonary response of the rat lung to inhalation of high concentration of diesel particles, *J. Appl. Toxicol.*, **1**, 104–110.
16. Lee, K.P., Trochimowicz, H.J. and Reinhardt, C.F. (1985) Pulmonary response of rats exposed to titanium dioxide (TiO_2) by inhalation for two years, *Toxicol. Appl. Pharmacol.*, **79**, 179–182.
17. Green, F.H.Y., Boyd, R.L., Danner-Rabovsky, J., Fisher, M.J., Moorman, W.J., Ong, T. et al. (1983) Inhalation studies of diesel exhaust and coal dust in rats, *Scand. J. Work Environ. Health*, **9**, 181–188.
18. Wehner, A.P., Dagle, G.E. and Clark, M.L. (1983) Lung changes in rats inhaling volcanic ash for one year, *Am. Rev. Respir. Dis.*, **128**, 926–932.
19. Bowden, D.A. and Adamson, I.Y.R. (1984) Pathways of cellular efflux and particulate clearance after instillation to the lung, *J. Pathol.*, **143**, 117–125.
20. Chan, T.L., Lee, P.S. and Hering, W.E. (1984) Pulmonary retention of inhaled diesel particles after prolonged exposure to diesel exhaust, *Fundam. Appl. Toxicol.*, **4**, 614–631.
21. Matsuno, K., Tanaka, I. and Kodama, Y. (1986) Pulmonary deposition and clearance of a coal fly ash aerosol by inhalation, *Environ. Res.*, **41**, 195–200.
22. Wolff, R.K., Henderson, R.F., Snipes, M.B., Mauderly, J.L. and McClellan, R.O. (1985) Particle Clearance and Accumulated Lung Burdens in Rats. In, '*Abstracts of Annual Meeting, Albuquerque, NM*, American Association for Aerosol Research', p. 9.P11.

23. Wolff, R.K., Henderson, R.F., Snipes, M.B., Mauderly, J.L. and McClellan, R.O. (1987) Alterations in particle accumulation and clearance in lungs of rats chronically exposed to diesel exhaust, *Fundam. Appl. Toxicol.*, **9**, 154–156.
24. Muhle, H., Creutzenberg, O., Bellmann, B., Heinrich, U. and Mermelstein, R. (1990) Dust overloading of lungs: Investigations of various materials, species differences and irreversibility of effects, *J. Aerosol Med.*, **3** (Suppl. 1), 111–128.
25. Muhle, H., Bellmann, B., Creutzenberg, O., Dasenbrock, C., Ernst, H., Kilpper, R. *et al.* (1991) Pulmonary response to toner upon chronic inhalation exposure in rats, *Fundam. Appl. Toxicol.*, **17**, 280–299.
26. Bolton, R.E., Vincent, J.H., Jones, A.D., Addison, J. and Beckett, S.T. (1983) An overload hypothesis or pulmonary clearance of UICC amosite fibres inhaled by rats, *Br. J. Ind. Med.*, **40**, 264–272.
27. Vincent, J.H., Johnston, A.M., Jones, A.D., Bolton, R.E. and Addison, J. (1985) Kinetics of deposition and clearance of inhaled mineral dusts during chronic exposure, *Br. J. of Ind. Med.*, **42**, 707–715.
28. Morrow, P.E. and Yu, C.P. (1993) Models of aerosol behavior in airways and alveoli. In, 'Aerosols in Medicine', Moren, F., Dolovitch, M.B., Newhouse, M.T., and Neuman, S.P., eds., Amsterdam, Elsevier, pp. 157–193.
29. Morrow, P.E. (1988) Possible mechanisms to explain dust overloading of the lungs, *Fundam. Appl. Toxicol.*, **10**, 369–384.
30. Yu, C.P., Chen, Y.K. and Morrow, P.E. (1989) An analysis of alveolar macrophage mobility kinetics at dust overloading of the lungs, *Fundam. Appl. Toxicol.*, **13**, 452–459.
31. Adamson, I.Y.R. and Bowden, D.A. (1982) Effects of irradiation on macrophage resopnse and transport of particles across the alveolar epithelium, *Am. J. Pathol.*, **106**, 40–46.
32. Morrow, P.E. and Mermelstein, R. (1988) Chronic inhalation toxicity studies. In, 'Inhalation Toxicology: The Design and Interpretation of Inhalation Studies and Their Use in Risk Assessment', Mohr U., Dungworth, D., Kimmerle, G., Lewkowski, J., McClellan, R., Störer, W., eds., Springer-Verlag, New York/Berlin, pp. 103–117.
33. Oberdörster, G. (1995) Lung particle overload: Implications for occupational exposure to particles, *Regul. Toxicol. Pharmacol.*, **27**, 123–135. (Fig. 4).
34. Oberdörster, G., Gerin, J. and Morrow, P.E. (1991) Volumetric loading of alveolar macrophages (AM): A possible basis for diminished AM-mediated particle clearance, *Exp. Lung Res.*, **18**, 87–104.
35. Trotter, J.A. (1981) The organization of actin in spreading macrophages, *Exp. Cell Res.*, **132**, 235–248.
36. Cain, H. and Kraus, B. (1981) Cytoskeleton in cells of the mononuclear phagocyte system, *Virchows Archive B. Cell Pathol.*, **36**, 159–176.
37. Oberdörster, G. (1988) Lung clearance of inhaled insoluble and soluble particles. *J. Aerosol Med.*, **4**, 289–330.
38. Lehnert, B.E., Valdez, Y.E. and Bomalaski, S.H. (1985d) Lung and pleural "free-cell responses" to the intrapulmonary deposition of particles in the rat, *J. Toxicol. Environ. Health*, **16**, 823–839.
39. Morgan, A., Kellington, J.B., Morris, K.J., Collier, C.G. and Hodgson, A. (1995) Labeling rat alveolar macrophages with fluorescent microspheres, *Inhal. Toxicol.*, **7**, 255–268.
40. Creutzenberg, O., Muhle, H., Bellmann, B.L., Kilpper, R., Morrow, P.E. and Mermelstein, R. (1989) Reversibility of biochemical alterations in bronchoalveolar lavagate upon cessation of dust exposure, *Exp. Pathol.*, **37**, 243–247.

41. Bellmann, B., Muhle, H., Creutzenberg, O. and Mermelstein, R. (1990) Recovery behavior after dust overloading of lungs in rats, *J. Aerosol Sci.*, **21**, 377–380.
42. Bailey, M.R., Fry, R.A. and James, A.C. (1985) Long-term retention of particles in the human respiratory tract, *J. Aerosol Sci.*, **16**, 295–305.
43. Foster, P.P., Pearman, I. and Ramsden, D. (1989) An interspecies comparison of the lung clearance of inhaled monodisperse cobalt oxide particles. Part II. Lung clearance of inhaled cobalt oxide in man, *J. Aerosol Sci.*, **20**, 189–204.
44. Oberdörster, G., Ferin, J. and Lehnert, B.E. (1994) Correlation between particle size, *in vivo* particle persistence and lung injury, *Environ. Health Perspect.*, **102**, 173–179.
45. Oberdörster, G. and Yu, C.P. (1990) The carcinogenic potential of inhaled diesel exhaust: a particle effect? *J. Aerosol Sci.*, **21** (Suppl. 1), 5397–5401.
46. US Environmental Protection Agency (1993) *Report EPA/600/R-93/104. OHEA Workshop: Research Needs in Risk Assessment of Inhaled Particulate Matter*.
47. Pope, C.A., Dockery, D.W. and Schwartz, J. (1995) Review of epidemiological evidence of health effects of particulate air pollution, *Inhal. Toxicol.*, **7**, 1–18.
48. Stinson, S.F. and Saffiotti, U. (1983) Experimental respiratory carcinogenesis with polycyclic aromatic hydrocarbons. In, 'Comparative Respiratory Tract Carcinogenesis', Vol. II, Reznik-Schüller, T., ed., CRC Press, Boca Raton, FL, pp. 76–93.

Evaluating the Carcinogenicity of Crystalline Silica and Other Mineral Particles: Human, Animal, Cellular and Molecular Studies

U. SAFFIOTTI

LABORATORY OF EXPERIMENTAL PATHOLOGY, NATIONAL CANCER INSTITUTE, BETHESDA, MD, USA

1 An Historical Introduction

Criteria for the evaluation of the carcinogenicity of environmental chemicals have developed historically on the basis of preceding studies of experimental carcinogenesis in animal models and, when available, on epidemiological studies of exposed population groups, especially for occupational exposures. The study of the carcinogenicity of crystalline silica and other mineral particles has developed only recently. The specific physico-chemical properties of particulate minerals – e.g. their lack of solubility, their particle size and shape, their crystalline structure and the presence of mineral impurities – require that criteria for the evaluation of their carcinogenicity be considered with special attention.

In the introductory paper of this seminar, John Duffus[1] outlined the development of criteria for the evaluation of environmental carcinogens, in the 1960s and 1970s. I add here some recollections extending further back in time, as I was involved in the early development of methods, concepts and criteria for carcinogen evaluation since the 1950s. In the 1930s and 1940s, experimental animal studies provided increasing evidence for the carcinogenic activity of organic compounds from several different chemical classes, including categories of compounds for which there was also human evidence from occupational exposure, such as polycyclic aromatic hydrocarbons from tar products and aromatic amines used in dyestuff manufacture.[2] In 1942, Hueper[3] published a pioneering treatise on occupational cancer. Gradually, more concern developed about the environmental effects of carcinogens on the population at large, and ways to prevent exposures. A particular area of early concern was for the use of intentional food additives that may be carcinogenic, since their use could be stopped. A major turning point in the criteria for the evaluation of carcinogens resulted from a symposium on potential cancer hazards from chemical additives and contaminants to foodstuffs, convened by the International Union Against Cancer (UICC),[4] held in Rome in August 1956, with the participation of many leading scientists in carcinogenesis. Prior to this conference, most countries had legislation or regulations that singled out a limited list of toxic substances banned from intentional use in foodstuffs, while everything else, whether tested or untested, was permitted. This conference recommended that this approach be reversed, so that substances intended for human use be tested for chronic toxicity, and that only those substances that had been tested and found not to be carcinogenic should be allowed as intentional food additives. This policy implied the development of lists of permitted compounds (those tested and found negative) to the exclusion of all others (whether tested with positive results or not tested). Extension of these criteria to environmental contaminants, particularly pesticides, was also recommended. In subsequent years, legislation along these lines was enacted in many countries and supported by international organizations. These criteria were further defined in a series of meetings, held since the late 1950s by the European Committee for the Protection of Populations from Risks of Long-term Toxicity (Eurotox), the World Health Organization, the Food and Agriculture Organization, the UICC and later the International Agency for Research on Cancer (IARC).

In the United States, new concern over the toxicity of pesticides[5,6] led to a Presidential Commission's recommendation for systematic carcinogenicity testing. The National Cancer Institute (NCI) developed an initial testing project on 150 pesticides and industrial chemicals[7] and established in 1968 an expanded program on chemical carcinogenesis (which I was then called to direct). This program was developed along three interacting lines: basic research on carcinogenesis mechanisms, development of biological models at the animal and cellular level and bio-assays for carcinogenic activity.[8,9]

Nearly thirty years later, I still think that the interaction of these three areas of research is essential. Recent great progress in the study of molecular mechanisms of carcinogenesis needs to be related to organ models for various types of cancer and to relevant cellular models using animal and human cells. The development of appropriate cellular models is important for the understanding of specific pathways of carcinogenicity and their molecular controls, and therefore for the interpretation of animal carcinogenesis bio-assays. The NCI carcinogenesis bio-assay program was separated from research programs in the late 1970s and then moved to the National Institute of Environmental Health Sciences to form the National Toxicology Program (NTP).

The guidelines for carcinogen bio-assays in small rodents,[10] which we prepared in the 1970s on the basis of previously developed laboratory methods, are still currently used. They were based on the need to provide a common baseline of test data on which to make at least a first assessment of the likelihood of carcinogenic effects of given test agents, in other words a *minimum* of testing compatible with available resources. Tests in two rodent species were selected, as being a step better than tests in only one species, although additional species and strains might have provided a wider range of susceptibilities, but at prohibitive costs. For the test doses, we selected two doses, namely the maximum tolerated dose (MTD) and a fraction of it (usually one-half of MTD), mostly because we wanted to avoid the uncertainty of testing *only* at lower doses, which would have left out the observation of possible carcinogenic effects at the highest *tolerated* dose. Obviously this and other areas of toxicology need the benefit of mechanism research and the study of host factors, which is why we envisaged a parallel development of studies on basic research and biological models (experimental pathology and cell biology). The results of research and bioassays in the last two decades have provided a much greater understanding of carcinogenic effects and their mechanisms.

2 Studies on the Carcinogenicity of Crystalline Silica

The carcinogenic activity of crystalline silica and other mineral particles needs to be studied at the human, animal, cellular and molecular levels in order to elucidate the pathogenetic mechanisms and their relationship to human disease. Early suspicions on the possible carcinogenicity of crystalline silica and observations of lung cancer in silicotic patients were given little consideration because of the prevailing belief that there was no correlation between silicosis and cancer. I remember, as a young investigator in the 1950s, being discouraged from pursuing research in this direction. Even after evidence was established that *fibrous* mineral particles (asbestos) were carcinogenic, and that crystalline silica particles were carcinogenic by intrapleural injection in rats, inducing malignant histiocytic lymphomas,[11] this scepticism prevailed. By a fortunate coincidence, in the early 1980s, three major long-term animal studies were conducted, which included exposures of rats to quartz dust by inhalation or by intratracheal instillation, resulting in lung cancer induction in all experiments.[12–14] At about the same time, the first adequately designed epidemiological studies also indicated that human silicosis was associated with increased risk for lung cancer.[15] In addition, assays for the neoplastic transformation of Syrian hamster embryo cells in culture also showed transforming activity of quartz dust.[16] These and related studies were presented together in 1984 at the first International Symposium on Silica, Silicosis and Cancer.[15] To that symposium I contributed new evidence, obtained from re-examination of slides from my previous experiments on the histogenesis of silicosis in rats, that showed early and marked hyperplastic reactions of the alveolar epithelium adjacent to silicotic granulomas, which I interpreted as precursor lesions of lung carcinomas.[17] Interestingly, quartz dust treatments induced no lung fibrosis and no lung tumours in hamsters.[12,18] Hamsters, however, were the species of choice for the induction of respiratory carcinomas (from the larynx, trachea and bronchi) following intratracheal administration of mineral dusts (*e.g.*, hematite) to which organic carcinogens, such as benzo-[a]pyrene, were adsorbed.[19,20] With this type of combined administration, it was found that among four test samples of carrier dust, quartz (Min-U-Sil 5) resulted in the highest incidence of lung tumours, predominantly adenocarcinomas.[18] The studies of Miller and Hook[21] clearly demonstrated that the early response of the rat alveolar epithelium to quartz dust consisted of hypertrophy and hyperplasia of alvolar type

II cells. A critical review of the data on silica carcinogenesis was conducted in 1986 by the IARC.[22] Since then, several additional animal studies provided new evidence for lung carcinogenesis by quartz, including inhalation exposure at relatively low levels.[23–26]

We conducted a series of experiments in rats treated with a single intratracheal instillation of quartz (Min-U-Sil 5, abbreviated MQZ) or hydrofluoric acid etched MQZ (HFMQZ), examined at progressive time intervals for histogenesis and for long-term carcinogenesis. We reported the progressive development of alveolar type II cell hyperplasia, adenomatoid lesions, adenomas and carcinomas of the peripheral lung, which were predominantly adenocarcinomas.[26,27] Histogenesis experiments by single intratracheal instillation were also conducted in Syrian golden hamsters and mice of three strains. In hamsters, crystalline silica dust was stored in pulmonary macrophages, but failed to induce fibrosis or alveolar hyperplasia; in mice, it induced only moderate fibrosis, but again no alveolar hyperplasia and (in limited tests) no significant induction of lung tumours. These marked species differences indicated the critical role of host factors in lung carcinogenesis by quartz.[26–28]

In developing our research, we followed a working hypothesis on carcinogenesis by crystalline silica,[17] which implies combined pathogenetic mechanisms, involving direct DNA damage by crystalline silica to alveolar epithelial cells, as well as the role of cellular mediators linking fibrogenesis and carcinogenesis. The cellular mediators include those identified in silicotic tissues, $i.e.$ transforming growth factor-beta1 (TGF-β1), tumour necrosis factor-alpha (TNF-α), interleukin-1 (IL-1) and IL-6, as well as mast cell products and oxygen free radicals.[26] We investigated properties of defined crystalline silica samples and their pathogenetic mechanisms in selected organ and cellular models, including: (1) physicochemical characteristics of the samples (purity, particle size, surface area, charge); (2) generation of free radicals (oxygen radicals and silicon radicals) and the role of metal ions; (3) cytotoxicity and neoplastic transformation in cells in culture, and the role of inhibitors; (4) DNA binding and DNA damage by silica $in\ vitro$; (5) species and organ-dependent host factor mechanisms; (6) role of TGF-β1 in organ and cellular models; (7) correlations of experimental and human pathology of silica-induced lung lesions.

3 Neoplastic Transformation by Quartz

Transformation of cells in culture in appropriate cell systems, within its obvious limitations, represents an important alternative method to long-term animal bio-assays for the evaluation of toxicity and carcinogenicity of particulate minerals. Each bio-assay method has to be evaluated in its own context. Tests on cells in culture are of short duration (weeks) and incomparably less cumbersome and less costly than long-term animal bio-assays (years). On the one hand, cell culture assays exclude factors involved in the penetration and distribution of particles in the airways and other tissues, they also exclude the interplay of different cell types ($e.g.$ between cells of inflammatory reactions and target epithelial cells), and they do not allow histogenetic studies of organ-specific tumour types. On the other hand, tests on cells in culture offer specific advantages for determining the direct effects of test materials on single target cell types, for identifying the role of individual modifying factors (enhancers and inhibitors of cytotoxicity and transformation) and for studies on chromosomal alterations, gene damage, pathways of gene expression and molecular controls in the specific target cells. In addition, they offer unique opportunities for ultrastructural studies of intracellular particle localization and interaction with cellular organelles. The neoplastic nature of the transformed cells is demonstrable by subcutaneous inoculation in immunosuppressed athymic nude mice, where neoplastic cells grow as solid tumours (tumourigenicity assays).

Further development of cellular models for neoplastic transformation, especially for specific epithelial target cells representing major forms of cancer, is needed, but there are already several established systems for culture and transformation of epithelial cells from rodents and even some from human tissues. For respiratory tract epithelia, initial emphasis was given in the 1970s and 1980s to developing culture and transformation models for the tracheobronchial epithelium, because it represented the major site of origin of bronchogenic carcinomas (especially epidermoid), which were then the prevalent type of human lung cancers.[29] In the last decade, the changed prevalence of human lung cancer types (showing a relative decline in epidermoid carcinomas and a relative prevalence of adenocarcinomas), and the increasing interest in peripheral lung tumours, including fibrosis-associated adenocarcinomas, raised a new challenge to establish cellular models for the growth and transformation of pulmonary alveolar type II epithelial cells, a current goal of our laboratory.

Neoplastic transformation by crystalline silica has been studied so far in the following three different cell systems.

(a) *Syrian hamster embryo (SHE) cells*. Two quartz samples of different toxicity induced dose-dependent morphological transformation; of these, one sample was notable for inducing transformation at doses that were not cytotoxic, and the other, a Min-U-Sil sample, was more active in transformation and also induced increased frequencies of binuclei and micronuclei.[16,30] Preliminary data, presented in a poster at this seminar, showed the induction of dose-dependent morphological transformation of SHE cells by four different preparations of quartz and two of cristobalite.[31]

(b) *Mouse embryo cell line, BALB/3T3/A31-1-1*. Five different samples of quartz were tested in this system: MQZ, HFMQZ, Chinese standard quartz (CSQZ), DQ-12 and F600. All these samples induced dose-dependent morphological transformation, confirmed as neoplastic by tumourigenicity assays in nude mice. The transformation frequency reached a plateau level at the higher tested doses. The transformed cells showed chromosomal alterations (translocations, amplifications) suggestive of genomic instability, and several quartz-transformed cell lines showed altered expression of several oncogenes and of the tumour suppressor gene, *p53*.[32,33] In this cell system, other minerals, when mixed with quartz (Min-U-Sil 5) at equal weight, showed remarkable modifying effects: two non-toxic dusts, hematite (Fe_2O_3) and anatase (TiO_2), inhibited both cytotoxicity and transformation by quartz, whereas rutile (another TiO_2 polymorph), itself more cytotoxic than quartz, markedly enhanced quartz toxicity but did not alter its transforming activity (except at the highest dose).[33] These studies suggest a possibly important role for associated minerals in the evaluation of silica toxicity and carcinogenicity. In another study with this cell system, we found that TNF-α, a growth factor that markedly enhanced transformation initiated by the carcinogen 3-methylcholanthrene, resulted in the inhibition of transformation of BALB/3T3/A31-1-1 cells treated with quartz, indicating different underlying mechanisms.[34]

(c) *Fetal rat lung epithelial cell line, FRLE.* As a cellular model representing the alveolar type II cell (the stem cell of the alveolar epithelium and the target cell for quartz-induced rat lung carcinogenesis *in vivo*) we selected the FRLE cell line, established from lung epithelial cells of a Sprague-Dawley fetal rat at day 20 of gestation[35], as recently discussed.[26] We first obtained neoplastic transformation in the FRLE cell line by transfection with a plasmid containing a mutated c-K-*ras* insert, which allowed characterisation of morphological transformants and their tumourigenicity in nude mice and nude rats.[36] FRLE cells showed marked dose-dependent toxicity for quartz and very low toxicity for anatase; the antioxidants 2-PVPNO [poly(2-vinyl-pyridine-*N*-oxide)] and 4-PVPNO were shown to inhibit quartz toxicity in this cell line.[37] Three separate experiments on FRLE cells, treated with quartz (MQZ) at single doses varying from 12.5 to 100 µg cm^{-2} in serum-free MEM medium for 24 h, resulted in the appearance of colonies or foci with altered morphology, whereas none developed in untreated or DMSO treated controls. The morphologically altered cells, when subcultured and injected subcutaneously in nude mice, gave rise to undifferentiated or poorly differentiated carcinomas.[38,39] These findings show that quartz particles can induce neoplastic transformation directly in lung alveolar epithelial cells, without the mediation of mesenchymal cells that was hypothesized to occur *in vivo*. Quartz-treated transformed cells from the first experiment, the tumours derived from these cells in nude mice, and the cells subcultured from those tumours were studied for mechanisms involving TGF-βs.[38] Current studies are aimed at establishing cell lines from fully differentiated alveolar type II cells from young adult rats, as models for transformation by silica and other mineral particles.

4 Studies on the Role TGF-β

TGF-β is a multifunctional growth factor, involved in cell proliferation, differentiation, inflammation, fibrogenesis and carcinogenesis.[40] It is one of the cytokines identified in silicotic tissues.[41,42] By immunohistochemistry, mature TGF-β1 was localized in fibroblasts and mononuclear cells of the rat silicotic granulomas and in the extracellular matrix.[42,43] In contrast, TGF-β1/ latency-associated peptide (LAP), indicative of the site of intracellular formation of TGF-β1, was localized in silicotic rat lungs, in hyperplastic alveolar type II cells adjacent to silicotic granulomas, in adenomatoid proliferations and in adenomas, but not in carcinomas.[42,43] In human lungs, TGF-β1/LAP was similarly localized in silicosis-associated alveolar hyperplastic cells.[43] In the FRLE cell system, TGF-β1/LAP localization was weak in control cells, but strong in cells transformed either by quartz or by a mutated *ras* plasmid.[38] In contrast, carcinomas in nude mice, derived from transformed FRLE cells, did not show TGF-β1/LAP, in analogy with carcinomas induced by quartz *in vivo*. Increased expression of TGF-β1, TGF-β2 and TGF-β receptor type II mRNAs was found in the transformed cells as compared with untreated FRLE cells.[38,43] Parallel studies in our laboratory using another organ model, the liver, showed that quartz induced marked fibrosis and cirrhosis in hamster liver, in striking contrast with the lack of fibrogenesis in hamster lungs, discussed above.[44] These results indicate the importance of organ-specific factors in the cellular responses to quartz.

5 DNA Binding and DNA Damage by Crystalline Silica

The carcinogenic activity of quartz particles *in vivo* and in cell cultures needs to be interpreted on the basis of plausible molecular mechanisms leading to DNA damage. A number of studies has focussed on the generation of free radicals by quartz and on the important role of metals, especially iron. The reactive oxygen species include superoxide radical ($O_2^{\cdot-}$), singlet oxygen ($^1O_2$), hydroxyl radical ($\cdot OH$) and hydrogen peroxide (H_2O_2), as shown by analysis of induced DNA damage mechanisms[45] as well as by chemical methods and by electron spin resonance spin trapping.[46] The metalloenzyme superoxide dismutase (SOD) specifically catalyses the conversion of $O_2^{\cdot-}$ into H_2O_2 and O_2. Catalase dismutes H_2O_2 into H_2O and O_2. The Haber Weiss reaction produces the highly reactive $\cdot OH$ radical:

$$H_2O_2 + O_2^{\cdot-} \rightarrow \cdot OH + OH^- + O_2$$

Trivalent Fe accelerates this reaction by a mechanism involving reduction of Fe^{3+} to Fe^{2+} by action of the superoxide radical:

$$Fe^{3+} + O_2^{\cdot-} \rightarrow Fe^{2+} + O_2$$

and the Fenton reaction:

$$Fe^{2+} + H_2O_2 \rightarrow Fe^{3+} + \cdot OH + OH^-$$

which regenerates Fe^{3+} and yields the highly reactive hydroxyl radical.

Using a simple DNA electrophoretic assay *in vitro*, we showed the induction of DNA double strand breakage by quartz particles and its dependence on the presence of oxygen and the formation of the oxygen radicals outlined above.[45] The formation of double strand breaks accrued from overlapping single strand breaks over long incubation times was markedly accelerated by the presence of iron.[47] The formation of oxidized DNA bases by incubation of DNA *in vitro* with crystalline silica was demonstrated by the production of thymine glycol.[47] The formation of silicon-based free radicals has also been reported[48] and may contribute to the generation of ˙OH and H_2O_2.[46]

The hydroxyl radical, however, has a very short half-life, with a reaction distance of approximately 15 Å, less than the width of the DNA helix. We examined a possible mechanism that provides for the release of ˙OH in close proximity to DNA target sites. The interaction of DNA with the quartz surface in aqueous solutions at physiological pH was investigated by Fourier-transform infrared spectroscopy (FTIR). The results showed specific dose-dependent changes in the DNA spectrum induced by quartz, and changes of the silica spectrum induced by DNA. Analysis of the spectral changes was indicative of the binding of the DNA phosphate backbone to the silanol groups through hydrogen bonding. This binding provides an anchoring mechanism that allows generation of free radicals on the silica surface to take place within a few angstroms from the target DNA bases.[49,50]

6 Concluding Comments

We studied the reactions of crystalline silica with target macromolecules *in vitro* in order to understand their direct interaction. In whole cells and *in vivo*, other factors need to be considered, particularly DNA repair, which has not yet been studied for silica-induced damage. Binding of DNA to the silica surface may occur at preferential sites of DNA unwinding and it may be interesting to determine whether and how it impairs DNA repair. Electron microscopic observation of quartz-treated cultured cells (BALB/3T3/A31-1-1 and FRLE) revealed the localization of quartz particles, not only in the cytoplasm, but also inside the nucleus and in mitotic spindles.[38,51] Much remains to be learnt about the mechanisms leading to the very high carcinogenic responses observed even after a single intratracheal dose of quartz in rats.[26] The evidence of neoplastic transformation of cells in culture by quartz clearly shows that quartz can induce this effect directly in target cells without mediation by the cells of adjacent inflammatory tissues. The more complex reactions occurring *in vivo* may be representative of important modulating factors, including host factors responsible for species and organ susceptibility.

Quartz is ubiquitously deposited and retained in human lungs of the general population,[52] implying, in a sense, that there are no unexposed control populations. Of course, occupational exposure levels leading to silicosis and lung cancer are of much greater magnitude. The experimental results, discussed above, suggest plausible pathways for carcinogenesis by quartz. The mechanisms involved in the biological effects of quartz and other minerals on the cells of the lung, especially on the alveolar type II cells, need to be further investigated. In order to identify and evaluate the complex biological effects of mineral particles, their carcinogenic properties and their molecular mechanisms, this new and wide-open research field needs to be further explored.

References

1. Duffus, J.H. (1995) Setting the scene: the scientific context. In, 'International Seminar on Assessment of Carcinogenic Risk from Occupational Exposure to Inorganic Substances', Luxembourg, 17–20 October 1995, The Royal Society of Chemistry, Cambridge, pp. 1–5.
2. National Cancer Institute. Survey of compounds which have been tested for carcinogenic activity, *U.S. Public Health Service Publication No. 149*, Washington, DC, US Gov. Printing Office, 17 Vol., from 1951 to 1992.
3. Hueper, W.C. (1942) 'Occupational Tumors and Allied Diseases', Charles C. Thomas, Springfield, IL.
4. International Union Against Cancer (1957) Symposium on potential cancer hazards from chemical additives and contaminants to foodstuffs. Resolutions. *Acta Unio Internationalis Contra Cancrum*, **13**, 187–193.
5. Hueper, W.C. (1957) The potential role of non-nutritive food additives and contaminants as environmental carcinogens, *Acta Unio Internationalis Contra Cancrum*, **13**, 220–252.
6. Carson, R. (1962) 'Silent Spring', Houghton Mifflin, Boston.
7. Inness, J.R.M, Ulland, B.M., Valerio, M.G., *et al*. (1969) Bioassay of pesticides and industrial chemicals for tumourigenicity in mice: preliminary note, *J. Natl. Cancer Inst.*, **42**, 1101–1114.
8. Saffiotti, U. (1977) Identifying and defining chemical carcinogens. In, 'Origins of Human Cancer – Cold Spring Harbor Conferences on Cell Proliferation', Vol. 4, Hiatt, H., Watson, J.D., and Winsten, J., eds., Cold Spring Harbor, NY, Cold Spring Harbor Laboratory, pp. 1311–1326.
9. Saffiotti, U. (1977) Carcinogenesis, 1957–77: Notes for a historical review, *J. Natl. Cancer Inst.*, **59** (Suppl. 2), 617–622.
10. Sontag, J.M., Page, N.P. and Saffiotti, U. (1976) Guidelines for carcinogen bioassays in small rodents. *National Cancer Institute, DHEW Publication No. (NIH)*76–801, Washington, DC.
11. Wagner, M.M.F. and Wagner, J.C. (1972) Lymphomas in the Wistar rat after intrapleural inoculation of silica, *J. Natl. Cancer Inst.*, **49**, 81–91.
12. Holland, L.M., Wilson, J.S., Tillery, M.I. and Smith, D.M. (1986) Lung cancer in rats exposed to fibrogenic dusts. In, 'Silica, Silicosis, and Cancer', Goldsmith, D.F., Winn, D.M., Shy, C.M., eds., Praeger, New York, pp. 267–279.
13. Dagle, G.E., Wehner, A.P., Clark, M.L. and Buschbom, R.L. (1986) Chronic inhalation exposure of rats to quartz. In, 'Silica, Silicosis, and Cancer', Goldsmith, D.F., Winn, D.M., Shy, C.M., eds., Praeger, New York, pp. 255–266.
14. Groth, D.H., Stettler, L.E., Platek, S.F., Lal, J.B. and Burg, J.R. (1986) Lung tumours in rats treated with quartz by intratracheal instillation. In, 'Silica, Silicosis, and Cancer', Goldsmith, D.F., Winn, D.M., Shy, C.M., eds., Praeger, New York, pp. 243–253.
15. Goldsmith, D.F., Winn, D.M. Shy, C.M., eds. (1986) 'Silica, Silicosis, and Cancer', Praeger, New York.
16. Hesterberg, T.W. and Barrett, J.C. (1984) Dependence of asbestos- and mineral dust-induced transformation of mammalian cells in culture on fibre dimension, *Cancer Res.*, **44**, 2170–2180.
17. Saffiotti, U. (1986) The pathology induced by silica in relation to fibrogenesis and carcinogenesis. In, 'Silica, Silicosis, and Cancer', Goldsmith, D.F., Winn, D.M., Shy, C.M., eds., Praeger, New York, pp. 287–307.
18. Niemeier, R.W., Mulligan, L.T. and Rowland, J. (1986) Cocarcinogenicity of foundry silica sand in hamsters. In, 'Silica, Silicosis, and Cancer', Goldsmith, D.F., Winn, D.M., Shy, C.M., eds., Praeger, New York, pp. 215–227.

19. Saffiotti, U., Cefis, F. and Kolb, L. (1968) A method for the experimental induction of bronchogenic carcinoma, *Cancer Res.*, **28**, 104–124.
20. Keenan, K.P., Saffiotti, U., Stinson, S.F., Riggs, C.W. and McDowell, E.M. (1989) Multifactorial hamster respiratory carcinogenesis with interdependent effects of cannula-induced mucosal wounding, saline, ferric oxide, benzo[a]pyrene and N-methyl-N-nitrosourea, *Cancer Res.*, **49**, 1528–1540.
21. Miller, B.E. and Hook, G.E.R. (1988) Isolation and characterization of hypertrophic type II cells from the lungs of silica-treated rats, *Lab. Invest.*, **58**, 565–575.
22. International Agency for Research on Cancer (1987) Silica. In, 'IARC Monographs on the evaluation of the carcinogenic risk of chemicals to humans', International Agency for Research on Cancer, Lyon, Vol. 42, pp. 39–143.
23. Muhle, H., Takenaka, S., Mohr, U., Dasenbrock, C. and Mermelstein, R. (1989) Lung tumour induction upon long-term low-level inhalation of crystalline silica, *Am. J. Ind. Med.*, **15**, 343–346.
24. Muhle, H., Bellmann, B., Creutzenberg, O., Dasenbrock, C., Ernst, H., Kilpper, R., MacKenzie, J.C., Morrow, P., Mohr, U., Takenaka, S. and Mermelstein, R. (1991) Pulmonary response to toner upon chronic inhalation exposure in rats, *Fundam. Appl. Toxicol.*, **17**, 280–299.
25. Spiethoff, A., Wesch, H., Wegener, K. and Klimisch, H.-J. (1992) The effects of thorotrast and quartz on the induction of lung tumours in rats, *Health Phys.*, **63**, 101–110.
26. Saffiotti, U., Williams, A.O., Daniel, L.N., Kaighn, M.E., Mao, Y. and Shi, X. (1996) Carcinogenesis by crystalline silica: animal, cellular, and molecular studies. In, 'Silica and Silica-induced Diseases', Castranova, V., Vallyathan, V. Wallace, W.E., eds., CRC Press, Boca Raton, FL, pp. 345–381.
27. Saffiotti, U., Daniel, L.N., Mao, Y., Williams, A.O., Kaighn, M.E., Ahmed, N. and Knapton, A.D. (1993) Biological studies on the carcinogenic mechanisms of quartz, *Rev. Mineral.*, **28**, 523–544.
28. Saffiotti, U. (1990) Lung cancer induction by silica in rats, but not in mice and hamsters: Species differences in epithelial and granulomatous reactions. In, 'Environmental Hygiene II', Seemayer, N.H., Hadnagy, W., eds., Springer-Verlag, New York, pp. 235–238.
29. Saffiotti, U. (1987) Human lung cancer and experimental pathogenetic models: An increasingly close connection. In, 'Lung Carcinomas', McDowell, E.M., ed., Churchill Livingstone, Edinburgh, pp. 370–393.
30. Hesterberg, T.W., Oshimura, M., Brody, A.R. and Barrett, J.C. (1986) Asbestos and silica induce morphological transformation of mammalian cells in culture: a possible mechanism. In, 'Silica, Silicosis, and Cancer', Goldsmith, D.F., Winn, D.M., Shy, C.M., eds., Praeger, New York, pp. 177–190.
31. Elias, Z., Poirot, O., Daniere, M.C., Terzetti, F., Marande, A.M., Dzwigaj, S., Pezerat, H. and Fubini, B. (1995) Study of genotoxicity in Chinese hamster V79 cells and morphological transformation of Syrian hamster embryo (SHE) cells treated with physico-chemically different types of silica (abstr.). In, 'International Seminar on Assessment of Carcinogenic Risk from Occupational Exposure to Inorganic Substances', Luxembourg, 17–20 October 1995.
32. Ahmed, N. and Saffiotti, U. (1992) Crystalline silica-induced cytotoxicity, transformation, tumourigenicity, chromosomal translocations and oncogene expression (abstr.) *Proc. Am. Assoc. Cancer Res.*, **33**, 119.
33. Saffiotti, U. and Ahmed, N. Neoplastic transformation by quartz in the BALB/3T3/A31-1-1 cell line and the effects of associated minerals, *Teratog., Carcinog. Mutagen.* (in press).

34. Mao, Y., Saffiotti, U., Daniel, L.N., Shi, X. and Ahmed, N. (1995) Tumor necrosis factor-a (TNF-α) inhibits neoplastic transformation (Tf) induced by quartz in BALB/3T3/A31-1-1 cells, *Proc. Am. Assoc. Cancer Res.*, **36**, 177.
35. Leheup, B.P., Federspiel, S.J., Guerry-Force, M.L., Wetherall, N.T., Commers, P.A., DiMari, S.J. and Haralson, M.A. (1989) Extracellular matrix biosynthesis by cultured fetal rat epithelial cells. I. Characterization of the clone and the major genetic types of collagen produced, *Lab. Invest.*, **60**, 791–807.
36. Saffiotti, U., Kaighn, M.E., Knapton, A.D. and Williams, A.O. (1993) Transformation of the fetal rat alveolar type II cell line, FRLE, by lipofection with a mutated K-*ras* gene (abstr.). *Proc. Am. Assoc. Cancer Res.*, **34**, 102.
37. Mao, Y., Daniel, L.N., Knapton, A.D., Shi, X. and Saffiotti, U. (1995) Protective effects of silanol group binding agents on quartz toxicity to rat lung alveolar cells, *Appl. Occup. Environ. Hyg.*, **10**, 1132–1137.
38. Williams, A.O., Knapton, A.D., Ifon, E.T. and Saffiotti, U. (1996) Transforming growth factor beta expression and transformation of rat lung epithelial cells by crystalline silica (quartz), *Int. J. Cancer*, **65**, 639–649.
39. Mao, Y. and Saffiotti, U., unpublished observations.
40. Roberts, A.B. and Sporn, M.B. (1990) The Transforming growth factor-b's. In, 'Peptide Growth Factors and their Receptors', Sporn, M.B., Roberts, A.B., eds., Springer-Verlag, New York, pp. 419–472.
41. Piguet, P.F., Collart, M.A., Grau, G.E., Sappino, A.-P. and Vassalli, P. (1990) Requirement of tumour necrosis factor for development of silica-induced pulmonary fibrosis, *Nature*, **344**, 245–247.
42. Williams, A.O., Flanders, K.C. and Saffiotti, U. (1993) Immunohistochemical localization of transforming growth factor-b1 in rats with experimental silicosis, alveolar type II hyperplasia, and lung cancer, *Am. J. Pathol.*, **142**, 1831–1840.
43. Williams, A.O., Knapton, A.D. and Saffiotti, U. (1995) Growth factors and gene expression in silica-induced fibrogenesis and carcinogenesis, *Appl. Occup. Environ. Hyg.*, **10**, 1089–1098.
44. Williams, A.O. and Knapton, A.D. Silicosis, cirrhosis and liver tumours in mice and hamsters: 1. Studies on transforming growth factor-b expression, *Hepatology*, (in press).
45. Daniel, L.N., Mao, Y. and Saffiotti, U. (1993) Oxidative DNA damage by crystalline silica, *Free Radicals Biol. Med.*, **14**, 463–472.
46. Shi, X., Mao, Y., Daniel, L.N., Saffiotti, U., Dalal, N.S. and Vallyathan, V. (1994) Silica radical-induced DNA damage and lipid peroxidation, *Environ. Health Perspect.*, **102** (Suppl. 10), 149–154.
47. Daniel, L.N., Mao, Y., Wang, T.-C.L, Markey, C.L., Markey, S.P., Shi, X. and Saffiotti, U. DNA strand breakage, thymine glycol production and hydroxyl radical generation induced by different samples of crystalline silica *in vitro* (submitted).
48. Fubini, B., Giamello, E., Volante, M. and Bolis, V. (1990) Chemical functionalities at the silica surface determining its reactivity when inhaled. Formation and reactivity of surface radicals, *Toxicol. Ind. Health*, **6**, 571–598.
49. Mao, Y., Daniel, L.N., Whittaker, N.F. and Saffiotti, U. (1994) DNA binding to crystalline silica characterized by Fourier-transform infrared spectroscopy, *Environ. Health Perspect.*, **102** (Suppl. 10), 165–171.
50. Saffiotti, U., Daniel, L.N., Mao, Y., Shi, X., Williams, A.O. and Kaighn, M.E. (1994) Mechanisms of carcinogenesis by crystalline silica in relation to oxygen radicals, *Environ. Health Perspect.*, **102** (Suppl. 10), 159–163.

51. Daniel, L.N., Mao, Y., Williams, A.O. and Saffiotti, U. Direct interaction of crystalline silica and DNA: A proposed model for silica carcinogenesis, *Scand. J. Work Environ. Health* (in press).
52. Churg, A., Wright, J.T., Stevens, B. and Wiggs, B. (1992) Mineral particles in the human bronchial mucosa and lung parenchyma. II. Cigarette smokers without emphysema, *Exp. Lung Res.*, **18**, 687–714.

Overview of Epidemiological Studies on the Carcinogenicity of Metals

A. BERNARD

UNIT OF INDUSTRIAL TOXICOLOGY AND OCCUPATIONAL MEDICINE, CATHOLIC UNIVERSITY OF LOUVAIN, BRUSSELS, BELGIUM

1 Introduction

Study of cancers caused by industrial chemicals has contributed to the discovery and knowledge of chemical carcinogenesis probably more than any other inquiry line. The first chemical agents identified as carcinogens were polycyclic hydrocarbons present in soots, tars and mineral oils which were found to be associated with skin cancer already from the 18th century.[1] With the exception of arsenic, inorganic chemicals and in particular metals were recognised as carcinogens much later. The first case of cancer linked to a metal exposure was reported in 1890 by Newman who described an adenocarcinoma of the nares in a man who had worked for more than 20 years in the chromate industry in Scotland.[2] Although many other cases of cancer were reported in the chromate industry afterwards, the possibility that chromium compounds can cause respiratory tract cancers was denied for many years.[3] It was accepted only in the early 1950s when several studies demonstrated that the incidence of lung cancers in the chromate industry was greatly increased at up to 31 times the normal rate.[4] At about the same time, an excess of respiratory tract cancer was also recorded in the nickel refinery at Clydach[5] but the responsibility of nickel in these cancers was not immediately recognised because arsenic (present in sulfuric acid till 1923) was more under suspicion.[6]

A systematic analysis of cancer mortality in the metal industry began really after the 1950s. Initially, these studies were focused mainly on nickel and chromium compounds,[7] but they were rapidly extended to other metals in different industries since at that time – and this is still the case – there was no mechanistic reason to suspect only a certain category of metals of carcinogenicity.

Reviewing the epidemiological studies on metal carcinogenicity is a challenging task, not only because of the number of different metal compounds to which humans can be excessively exposed but mainly because very few studies have succeeded in linking the cancer risk to a specific metal without having to make some reserves about a possible confounding by tobacco smoking or other chemicals (*e.g.* arsenic in smelters, radon and silica in mines). To reduce the task, the present review will be focused on metals, especially transition metals, and, apart from antimony, non-metals and semi-metals will not be considered.

2 Antimony

The main applications of antimony are as flame retardants in plastics and textiles, components of specialized alloys and in the manufacture of glass and pigments. Antimony has been evaluated by an IARC working group in 1988, which concluded that there was sufficient evidence of the carcinogenicity of antimony trioxide in animals but inadequate evidence for classification in humans.[8] The only human data available to the IARC working group were those provided by the study of Davies[9] in an UK antimony processing plant. Nine lung cancer cases were found among workers employed solely on antimony smelting against 5.7 expected.[9] An update of the cohort in 1983 confirmed the excess of lung cancer.[10] The results of a mortality study on 1,014 men employed between 1937 and 1971 in a Texas antimony smelter have been recently published.[11] Mortality from lung cancer was significantly elevated in exposed workers and a significant positive trend with duration of employment was apparent. Data from both the UK and the US plant suggest thus that occupational exposure to antimony trioxide may increase the risk of lung cancer. This conclusion is, however, limited by the fact that these studies provided no data allowing to exclude a possible contribution from the arsenic present in the antimony ore.

3 Cadmium

The case of cadmium is very illustrative of the difficulties and uncertainties in the epidemiological evaluation of metal carcinogenicity. Occupational exposure to cadmium may be found in a variety of industrial settings or operations, such as zinc ore smelting and refining, production of nickel–cadmium batteries, welding and electroplating. The general population living in some polluted areas may be also excessively exposed to cadmium contaminating the food. In 1993, a working group of experts convened by the IARC classified cadmium and cadmium compounds as Group I human carcinogens, having concluded that there was sufficient evidence of cadmium being carcinogenic to humans and animals.[12] Epidemiological data reviewed by the IARC was derived from a total of six industrial cohorts: two from nickel-battery plants in UK and Sweden, two from copper–cadmium alloy plants also in UK and Sweden, one from a cadmium recovery plant in USA, one from a cadmium processing plant in UK, and one from a smelter in China. It is important to stress that in most of these studies exposure levels to cadmium were considerably higher than those in present day factories, and that alterations of renal or lung function were common findings. The weight of evidence of cadmium carcinogenicity varies substantially according to the type of cohort.

In nickel–cadmium battery plants, a significant excess of prostatic cancers was reported in one cohort examined in the 1960s, but this finding was not corroborated by follow up of the same cohort or by studies on other cohorts. A significant increase in lung cancer was detected in the UK nickel-battery plant cohort with a nonsignificant trend as a function of years of exposure. In the Swedish nickel–cadmium battery plant, a nonsignificant increase of lung cancer risk was detected among workers with more than five years of exposure and a latency period of more than 20 years. The evidence provided by these data, however, is limited by the lack of information on the possible exposure to nickel compounds.

In copper–cadmium alloy plants, the results from the UK cohort were inconsistent, with an increase in one plant and a decrease in another, and inconclusive because of the concurrent exposure to arsenic. A nested case-control study showed that the association was stronger with arsenic exposure than with cadmium exposure. In the Swedish copper–cadmium alloy plant, no significant increase was found for both lung and prostatic cancer (although the latter was slightly in excess).

The UK cadmium processing cohort studied by Kazantzis et al.[13] included workers from 17 plants where cadmium was produced or used in the manufacture of pigments, stabilisers and alloys. A significantly increased mortality from lung cancer was observed with some evidence of a dose–response relationship. The authors concluded, however, that this increase cannot be attributed to cadmium alone, owing to the presence of multiple confounding factors, especially arsenic but also nickel, beryllium, chromium and heated oil. A tentative breakdown of cause-specific mortality in those workers for which arsenic exposure was not known to have occurred and those with known arsenic exposure showed that the lung cancer risk was increased only in the latter.

The cadmium recovery plant in the USA has been the object of several mortality studies that have either expanded the initial cohort or attempted to control for contributions from cigarette smoking or arsenic. The feedstock indeed contained arsenic at levels that were particularly elevated before 1940. The interference of tobacco smoking appears in that half of the cohort was comprised of Hispanics who were light cigarette smokers and who did not show any excess of lung cancer, in contrast to the other half of the cohort composed of non-Hispanics. The latter showed a significant increase of lung cancer risk, which correlated with cumulative exposure to cadmium and persisted after controlling for tobacco smoking and period of hire (a surrogate estimate of arsenic exposure). The dose–response was even stronger after 1940 when arsenic exposure was low.[14] One important argument retained by the IARC working group in favour of the carcinogenicity of cadmium in humans was the existence of a dose–response relation between cumulative cadmium exposure and lung cancer risk, which in the opinion of the experts would not have been observed if data had been confounded by arsenic. This argument is, however, debatable, since for a given feedstock one can expect that airborne concentrations of cadmium correlate with that of other components of the ore, including arsenic. In addition, a cautionary note published by Sorahan and Lancashire[15] drew attention to the possibility of a misclassification of workers in the US cohort study, which might have distorted the analysis of confounding by arsenic.

In conclusion, the early suspicion that inhalation of cadmium may increase the risk of prostate cancer has been completely dispelled by subsequent studies. With respect to the lung, the interpretation is complicated by the fact that when statistically significant excesses were found it was nearly always in situations in which cadmium exposure could be confounded with exposure to nickel (in nickel-battery plant) or arsenic (smelters).[16] The only study in which the possibility of confounding was considered as unlikely is that of the cohort hired after 1940 in the US cadmium recovery plant. But even if after that date the feedstock contained only 1–2% of arsenic; the possibility that exposure to this potent carcinogen persisted to a level sufficient to account for the hazard cannot be entirely excluded owing to the lack of accurate information on exposure levels. Also intriguing is the fact that no excess of lung cancer was found in Hispanics who were known to be light smokers. This suggests the possibility of an interaction between tobacco smoking and an occupational factor.

Although environmental exposure to cadmium has been sufficiently high in some polluted areas to cause renal effects, there is to date no sound epidemiological evidence that cadmium may represent a carcinogenic risk (at the level of the prostate or at another site) for the general population exposed mainly via the oral route.[17] For instance, an association between cadmium levels in drinking water and prostatic cancer has been observed,[18] but since cadmium intake via drinking water is negligible compared with the intake from food or tobacco smoking, such a relation might well be not causal as were relations reported in the past between cadmium in drinking water and cardiovascular diseases.[19]

4 Chromium

Chromium and its compounds are widely used in industry as plating or alloying agents (steel industry), pigments (paint industry), preservatives (wood industry) or tanning agents (leather industry). As for nickel, chemical speciation is an important clue to the understanding of the toxic effects of chromium. The main distinction to be made for chromium is between hexavalent compounds, which readily cross cell membranes, and trivalent compounds which cannot and whose cell bioavailability is consequently much lower. These two classes of chromium compounds contain many individual species, each may present different persistence and biological reactivity in the respiratory tract, depending on the solubility, degree of hydration, crystalline structure, exact stereochemistry, surface properties, *etc*.

Occupational exposure to chromium compounds can be encountered in a number of industries or occupations. The most important exposures to hexavalent compounds are found during chromate production (sodium, potassium, calcium and ammonium salts), chrome plating (chromium trioxide), pigment production and spray painting (zinc and lead chromate). There may also be a significant exposure to chromium compounds during operations of welding, cutting and grinding of chromium alloys. Exposure to trivalent compounds occurs during chromate production (chromite ore) and in the ferrochromium industry.

There is no doubt that several hexavalent chromium compounds can produce lung or nasal cavity cancers in humans. Many cases of nasal cancers were reported in the chromate-producing industry in Germany, France, UK and USA. Epidemiological studies carried out in several countries have consistently shown an excess of lung cancer mortality among workers in both the chromate-producing industry and during chromate-pigment manufacturing. An excess risk of sinonasal cancer was also reported in these studies. But not all chromate salts present the same carcinogenic potency. In its last evaluation, IARC[20] classified all hexavalent compounds as carcinogenic in humans (Group I), but by stressing that this evaluation applies to the group of chemicals and not necessarily to all individual chemicals within the group. An evaluation of selected individual chromium compounds has been attempted by a CEC working group.[10] The only specific chromate salt to which the excess of lung cancer could be specifically linked was zinc chromate. An excess of cancers could also be linked to lead chromate but in association with zinc chromate. No epidemiological data were available to evaluate *per se* calcium chromate, strontium chromate and other chromium compounds.

An increased risk of lung cancer was also reported in platers who were exposed mainly to soluble chromium compounds (chromium trioxide). Although a concomitant exposure to other substances (*e.g.* nickel and polycyclic hydrocarbons) cannot be completely ruled out, hexavalent chromium is the most probable cause of the increased risk of respiratory tract cancers reported experienced by platers. Welders of stainless steel also showed an excess of respiratory cancers but no conclusion about the specific contribution of hexavalent chromium compounds can be drawn because of the simultaneous exposure to other carcinogenic compounds.

The results of epidemiological studies on cohorts of workers exposed predominantly to trivalent chromium (ferrochromium) have shown an excess of cancers in the respiratory tract and at other sites, but again the responsibility of trivalent chromium compounds in these cancers cannot be proven owing to the probable exposure to other carcinogens (benzo[*a*]pyrene, asbestos and hexavalent chromium).

In conclusion, there is ample epidemiological evidence that exposure by inhalation to hexavalent chromium in the production of chromate and of chromate pigments and in the chromium plating industry increases the risk of respiratory cancers. Zinc chromate and chromium trioxide are the only individual chromium compounds that have been specifically associated with these cancers by epidemiological studies. The carcinogenic potential of other individual compounds of chromium cannot be evaluated on the basis of available human data.

5 Cobalt

The main uses of cobalt metal are in the production of high temperature and other specialized alloys and as a cement for tungsten carbide in hard metal alloys for cutting tools. Cobalt compounds are also used in the production of plastics, ceramics, rubber, paints, *etc.*[20] Occupational exposure to cobalt-containing dust can cause pulmonary fibrosis (the so-called hard metal pneumoconiosis) and an occupational asthma. Since the first report of lung cancers in hard-metal workers in 1962,[21] two epidemiological studies have shown an elevated risk of lung cancer, one among hard-metal workers and the other for workers in the production of cobalt.[21] However, the follow up of the latter cohort did not confirm the excess risk of cancers,[22] which anyway could not be attributed specifically to cobalt in view of the possible exposure to other carcinogens (arsenic, nickel or chromium) during cobalt production. The Swedish study remains, from which no conclusion can be drawn because of the concurrent exposure to other components of the hard metal (tungsten carbide).

It is should be noted that cases of local sarcoma have been reported in patients with cobalt-containing implants, but when metal analysis was done on the tissues it revealed also the presence of other metals such as chromium or nickel.[21] Cobalt and cobalt compounds were considered by an IARC working group in 1990 which concluded that there was sufficient evidence for the carcinogenicity of cobalt metal powder in experimental animals, but inadequate evidence for their carcinogenicty in humans (Group IIB).[23]

6 Copper

A number of studies has been reported on the mortality of workers engaged in the production or the use of copper compounds, but in very few of them exposure was exclusively to copper. In copper smelters, workers are exposed to a complex mixture of gaseous or particulate chemicals likely to have a carcinogenic or co-carcinogenic activity (*e.g.* arsenic, sulfur dioxide) so that the contribution of copper and a fortiori of a specific copper compound cannot be evaluated. The same problem is encountered in copper miners who are exposed to silica, radon and other components of the ore.[24]

In copper refineries, the risk of confounding by other exposures, notably arsenic and sulfur dioxide, is lower than in smelters. Workers are probably exposed mainly to copper sulfate. No increase of cancer was found in those refineries. A rather specific exposure to copper sulfate has been described in Portugal, in vineyard sprayers using the Bordeaux mixture. Many cases of liver and lung diseases associated with copper accumulation have been recorded in these sprayers, among which were one liver angiosarcoma and three cases of lung cancer. However, the exposure history of these workers was insufficiently documented to exclude other possible causes.[21] Copper and its compounds have never been examined by a working group convened by the IARC.

7 Iron

Cases of lung cancer were reported in iron-ore miners in the 1950 and 1960s.[25,26] Several epidemiological studies carried out in iron-ore mines in UK, Sweden and USSR have confirmed the increased incidence of lung cancer, which in some mines was as high as 12 times the predicted value.[27] The risk was attributed to exposure to radon decay products, a conclusion supported by the demonstration of dose–response relations with the level of radioactivity.[28] Iron-ore mines in Lorraine, France, are characterised by low levels of radioactivity. Yet, a significant excess of lung cancer was demonstrated in both a proportional mortality study[29] and in a cohort study.[30] In the latter study, all cases were smokers, whereas in the former, information on smoking habits was fragmentary. Despite these uncertainties about smoking habits, it is unlikely that the higher rate of lung cancer in French iron-ore miners can be explained by differences in smoking habits only. The contribution of a workplace factor, perhaps acting in synergy with tobacco smoke, seems more plausible. Whether this factor is hematite, another component of the ore, the excessive inhalation of dust or any combination of these pollutants cannot be determined.

8 Lead

Lead is one of the metals the longest used by man and a major occupational pollutant. The main industrial activities with potential exposure to lead are lead mining, smelting and refining, manufacture of lead batteries, welding and cutting of metal coated with lead-based paints. The highest levels of exposure are found in smelting and refining operations, which may result in air concentrations largely exceeding 1000 µg m^{-3}. Lead and inorganic lead compounds have been classified by the IARC in the Group IIB as possibly carcinogenic to humans.[31]

More than 15 cohort studies have been published since 1936 on the cancer mortality of lead-exposed workers in different industries.[32] In several of these studies, lead exposure, assessed by air or biological measurements, was very high and largely exceeded the current permissible levels. No consistent pattern of cancer mortality emerged from these studies. Excesses from all malignant neoplasms, largely contributed by respiratory and digestive cancers, were seen in several cohorts of smelters and battery plant workers but no relation could be established between cancer incidence and the intensity or duration of exposure. A cancer site of special interest is the kidney, since animal data have demonstrated that lead can produce renal tumours. The possibility of a carcinogenic effect of lead on human kidney is suggested also by case-reports of kidney cancer in two workers who had been highly exposed to lead.[33,34] Positive findings on kidney cancer in epidemiological studies include a significant excess of deaths in a subcohort with high exposure to lead,[35] an excess approaching statistical significance in the study by Selevan et al.[36] and a significant excess in the PMR in plumbers.[37] Other studies found no change or even a slight nonsignificant deficit.[32]

Case-control studies based on industrial workers or on the general population have not provided any consistent evidence of a cancer risk related to previous exposure to lead. A meta-analysis of all case-control and cohort studies available till 1990 was recently carried out by Fu and Bofetta.[32] The combined results showed a significantly increased risk of overall cancer, stomach cancer, lung cancer and bladder cancer. The relative risk for kidney cancer was increased but did not reach statistical significance. When the analysis was restricted to most heavily exposed workers, significantly increased relative risks were found for stomach and lung cancer. The authors recognise that their analysis may present several limitations linked to potential confounders that could no be controlled for, to which one has to add the well-known publication bias inherent to any meta-analysis study.

In conclusion, cohorts of lead-exposed workers examined were sufficiently large and exposure levels sufficiently high to detect excesses of cancers but they all lacked accurate information on the past lead exposure and, which is more critical, on the importance of confounding by tobacco smoking and diet. In smelters and glass factories, workers were probably exposed to carcinogenic chemicals such as arsenic, hexavalent chromium or polycyclic hydrocarbons. Under these conditions, it is difficult to establish the exact contribution of lead in the excesses of lung or stomach cancer reported in several studies. Observations on kidney cancer are perhaps less influenced by these confounders. These observations, supported by animal data, suggest that heavy exposure to lead may increase a risk of kidney cancer. However, present standards in occupational hygiene have made such exposures quite unlikely.

9 Mercury

Mercury occurs in several chemical forms with clearly distinct patterns of toxicity: elemental mercury (mercury vapour), inorganic mercury salts and organomercury compounds. Only the first two forms will be considered here. Human exposure to mercury vapour or inorganic mercury is mainly occupational (*e.g.* chloralkali plant, dentistry). This form of mercury is readily absorbed by inhalation and distributes mainly in the CNS (central nervous system) (as Hg^0) and the kidneys (after oxidation into Hg^{2+}). A possible source of exposure to these two forms of mercury for the general population is via dental amalgams (Hg^0) or the use by dark-skinned people of skin lightening creams or soaps (Hg^{2+}).

Studies on industrial workers have largely documented the toxic effects of inorganic mercury on the CNS and kidney, which are the two main target organs. The epidemiological studies on the carcinogenicity of mercury vapour and inorganic mercury have been reviewed in 1993 by an IARC working group, which concluded that these two forms are unclassifiable as to their carcinogenicity in humans (Group III).[12] Epidemiological studies have investigated four main occupational groups: miners (one study), chloralkali workers (three studies), hat workers (one study), workers in the nuclear weapon industry (one study) and dentists (four studies). The study in miners[38] and in hat workers,[39] together with two studies in chloralkali workers,[40,41] suggest an association between mercury and an increased risk of lung cancer, but no trend was found of an increased risk with duration or intensity of exposure. The concomitant exposure to other carcinogens (radon or silica) in miners, the confounding by tobacco smoking and the lack of association in dentists do not allow to draw any conclusion from these data. Some studies in dentists are suggestive of a carcinogenic effect on the CNS or the pancreas,[42–44] but the studies on other occupational groups with potentially higher exposures do not support these associations. Two population-based case-control studies were also published but they do not provide any additional observation consistent with above studies (one reported an increased risk of prostate cancer) (see review by Bofetta et al.[45]). As with lead, animal data point to the kidney as a possible target organ for the carcinogenicity of mercury. An excess of kidney cancers was found in one study of dentists[46] but no evidence has been reported in other occupational groups, even in chloralkali workers. It should be noted that kidney cancers have been produced only in male mice treated with doses of mercury sufficient to cause renal toxicity.

Epidemiological studies carried out so far do not provide a consistent cancer pattern of mercury or its inorganic compounds in humans. Since these studies have examined cohorts of workers who have experienced exposures much higher than present levels in industry, one can reasonably think that, below current permissible exposure levels, it is very unlikely that inorganic mercury and its compounds present a carcinogenic risk.

10 Nickel

Nickel and its compounds are chemicals of major economic importance. They are used for the production of steels, alloys, coins, electrical equipment, domestic utensils, *etc*. Human exposure can occur during mining and processing of nickel ores, the production and fabrication of nickel alloys, a variety of electrolytic procedures, in the manufacture of batteries and in welding procedures, in the chemical industry (use of nickel and nickel oxide as catalysts), and in the glass and ceramics industries. The chemistry of nickel compounds is highly complex. Man can be exposed to many different chemical species of nickel, whose biological properties may further change according to the degree of hydration or the stereochemistry, and this without mentioning the intermediates that may appear in industrial processes.

Epidemiological studies have demonstrated a strong association between lung and nasal cancers and exposure by inhalation to nickel coumpounds in nickel refineries in Canada, Norway and the UK (for reviews see Refs. 19 and 47). In some studies, the risk of of developing nasal cavity tumours was increased by up to 240 times. Exposure in these plants was in the past considerable with airborne concentrations of nickel largely in excess of 1 mg m^{-3}. Because of the widespread use of nickel, much effort has been devoted to the identification of chemical species responsible for these carcinogenic effects. In general there has been limited characterisation of the inhaled particles and the identification has relied mainly on the chemical processes. In refineries, the excess risk of cancer has been related primarily to the early stage of nickel refining, *i.e.* the oxidation of the nickel and copper sulfide matte, which led to the conclusion that exposure to nickel subsulfide (nickel matte) and nickel monoxide, two insoluble forms, has largely contributed to the excess of cancers in refineries. Limited epidemiological data suggest that there might be also an association between nickel sulfate (soluble nickel in the hydrometallurgy refining process) and respiratory tract cancers.

Available epidemiological studies (*e.g.* in nickel alloys plants) do not indicate that elemental nickel causes cancers in human.[48-51] There are no adequate epidemiological data to assess the risk of other nickel compounds (nickel carbonate, nickel dioxide, nickel hydroxide, nickel oxide and nickel sulfide).[47]

11 Zinc

Zinc is an essential element for man, playing an important role in a number of enzymatic systems, in the immune system, in gene and hormone expression. Epidemiological studies in China, Iran, Russia and South Africa have shown an association between zinc deficiency and oesophageal cancer rates, some of which were 50 times the worldrate. There is no evidence from case reports or epidemiological studies of cancers being specifically related to zinc or zinc compounds. Studies in the mining or smelting of zinc-ores are inadequate to evaluate the carcinogenicity of zinc or its compounds *per se* because of the simultaneous exposure to other established or suspected carcinogens or cocarcinogens (arsenic, cadmium, radon, silica, acid mist, polycyclic hydrocarbons, *etc.*).[21] Zinc and its compounds have not been considered by the IARC.

12 Other Metals

Manganese. No case report or epidemiological study of cancer in humans specifically exposed to manganese metal or compounds has been published.

Precious metals. No data are available to assess specifically the carcinogenicity of platinum, gold, silver, tin and their inorganic compounds in humans. Excesses of lung cancers have been reported in the mining of ores containing these metals but they were mainly concentrated among subjects who had experienced exposure to other carcinogens such as radon, arsenic or silica.[52]

Titanium. One case of lung adenoma was reported in a 53-year-old man who had been exposed to titanium oxide for 14 years.[53] No excess of cancer mortality or incidence was found in one study among workers who had been exposed to titanium dioxide[54] or titanium tetrachloride[55] between 1956 and 1985 in two US titanium dioxide producing plants.

13 Conclusions

A number of epidemiological studies has been carried out world-wide to identify metals or metallic compounds that might cause cancers in humans. The vast majority of these studies has examined the cancer mortality or incidence in cohorts exposed to metals at the workplace, mainly by inhalation. These studies usually present an adequate sensitivity essentially due to the high exposure levels in the past. In most cases, however, they suffer from a lack of accurate information allowing the effects of the metal to be clearly dissociated from those of other chemicals at the workplace or of tobacco smoking. Under these conditions, proving a causal association with a specific metal and a fortiori identifying the exact chemical species responsible for the risk is a difficult task, which cannot be achieved without introducing a component of personal judgement. This review and its conclusions are no exception to that rule.

Epidemiological studies have provided convincing evidence that exposure by inhalation to some nickel and chromium compounds is associated with an increased risk of respiratory tract cancers. For nickel the excess risk has been linked mainly to nickel monoxide and nickel subsulfide, and for chromium to hexavalent compounds and in particular zinc chromate, lead chromate in association with zinc chromate, and chromium trioxide. No inference can be made about the carcinogenicity of other nickel and chromium compounds on the basis of epidemiological data. It is important to stress that cancer excesses have been reported in industrial facilities where exposure levels were considerably higher than those encountered in present day factories and at a time when hygiene conditions were more than rudimentary. Chronic irritating effects on the respiratory tract, sometimes very severe (*e.g.* nasal septum perforation), were common and undoubtedly have contributed to promote the development of cancers. This is consistent with the fact that tumours induced by nickel and chromium compounds were primarily located in the regions which sustained predominant exposures, *i.e.* the respiratory sites.[56]

Epidemiological studies have also shown a significant excess of lung cancer in cohorts of workers exposed to cadmium. The difficulty with epidemiological data on cadmium carcinogenicity is to ascertain the absence of a significant exposure to other potential carcinogens such as arsenic and nickel. The recent classification of cadmium as human carcinogen by the IARC was mainly based on the assumption that in a US cadmium recovery plant, exposure to arsenic after 1940 was too low to account for the risk. This assumption, however, relied on a surrogate estimate of arsenic exposure and not on accurate measurements of airborne concentrations of arsenic. The case of antimony presents similarities to that of cadmium. Antimony trioxide is indeed an animal carcinogen that has been associated with an excess of lung cancer in several epidemiological studies, but, again, under conditions that do not allow a possible contribution from arsenic to be entirely eliminated.

Lead is one the rare metals for which a significant excess of cancer has been reported at a site other than the respiratory tract. Case reports and epidemiological studies strongly suggest that heavy exposures to lead as observed in the past have led to the development of kidney cancers.

With respect to other metals, epidemiological data are either inadequate or non-existent to assess their carcinogenicity in humans. There is also so far no consistent epidemiological evidence that metals reviewed here present a carcinogenic risk for the general population exposed via food or drinking water, even at exposure levels sufficiently high to produce early effects on the kidney (cadmium) or the CNS (lead).

References

1. Bernard, A. and Lauwerys, R. (1991) Relationship between cancer and occupational exposure to chemicals: an overview of the evidence. In, 'Recent Results in Cancer Research', Vol. 122, Eylenbosch, W. J. *et al.* eds., Springer-Verlag, Heidelberg, pp. 52–59.

2. Newman, D.A. (1890) A case of adenoma of the left inferior turbinated body, and perforation of the nasal septum, in the person of a worker in chrome pigments, *Glasgow Med. J.*, **33**, 469–470.
3. Legge, T.M. (1922), *Br. Med. J.*, **1**, 1110.
4. Machle, W. and Gregorius, F. (1948) Cancer of the respiratory tract in the United States chromate-producing industry, *US Public Health Report*, **63**, 1114–1127.
5. Amor, A.J., (1932), *J. Ind. Hyg.*, **14**, 216.
6. Hunter, D. (1978) 'The diseases of occupations', 6th edn., Hoddon and Stoughton, London.
7. Doll, R. (1958) Cancer of the lung and nose in nickel workers, *Br. J. Ind. Med.*, **15**, 217–223.
8. IARC (1989) 'Monograph on the evaluation of the carcinogenic risks to humans. Organic solvents, some resins monomers, some pigments and occupational exposures in paint manufacture and painting trades', Vol. 47, IARC, Lyon.
9. Davies, T.A.L. (1973) 'The health of workers engaged in antimony trioxide manufacture: a statement', Department of Employment, Medical Advisory Service, London, pp. 1–2.
10. Berlin, A., Draper, M.H., Krug, E., Roi, R. and van der Venne, M.Th., eds.(1989) 'The Toxicology of Chemicals', Series. 1, Carcinogenicity, Vol I, Commission of the European Communities, Luxembourg.
11. Schnorr, T.M., Steenland, K., Thun, M.J. and Rinsky, R.A. (1995) Mortality in a cohort of antimony smelters workers, *Am. J. Ind. Med.*, **27**, 759–770.
12. IARC (1993) 'Monograph on the evaluation of carcinogenic risks to humans. Beryllium, cadmium, mercury and exposures to glass manufacturing industry', Vol. 58, IARC, Lyon.
13. Kazantzis, G., Blanks, R.G. and Sullivan, K.R. (1992) Is cadmium a human carcinogen? In, 'Cadmium in the Human Environment: Toxicity and Carcinogenicity', Nordberg, G.F., Herber, R.F.M., Alessio, L., eds., IARC, Lyon, pp. 435–446.
14. Stayner, L., Smith, R., Thun, M., Schnorr, T. and Lemen, R. (1992) A quantitative assessment of lung cancer risk and occupational cadmium exposure. In, 'Cadmium in the Human Environment: Toxicity and Carcinogenicity', Nordberg, G.F., Herber, R.F.M., Alessio, L., eds., IARC, Lyon, pp. 447–455.
15. Sorahan, T. and Lancashire, R. (1994) Lung cancer findings from the NIOSH study of United States cadmium recover workers: a cautionary note. *Occup. Environ. Med.*, **51**, 139–140.
16. Doll, R. (1992) Cadmium in the human environment: closing remarks. In, 'Cadmium in the Human Environment: Toxicity and Carcinogenicity', Nordberg, G.F., Herber, R.F.M., Alessio, L., eds., IARC, Lyon, pp. 459–464.
17. Nishijo, M., Nakagawa, H., Morikawa, Y., Tabata, M., Senma, M., Miura, K., Takahara, H., Kawano, S., Nishi, M., Mizukoski, K., Kido, T. and Nogawa, K. (1995) Mortality of inhabitants in an area polluted by cadmium: 15 year follow up, *Occup. Environ. Med.*, **52**, 181–184.
18. Bako. G. *et al.* (1982) The geographical distribution of high cadmium concentration in the environment and prostatic cancer in Alberta, *Can. J. Publ. Health*, **73**, 92–94.
19. Bernard, A. and Lauwerys, R. (1986) Effects of cadmium exposure in humans. In, 'Handbook of Experimental Pharmacology', Foulkes, E.C., ed., Springer-Verlag, Heidelberg, Chapter V, pp. 135–177.
20. IARC (1990) 'Monograph on the evaluation of the carcinogenic risks to humans. Chromium, nickel and welding', Vol. 49, IARC, Lyon.

21. Aresini, G., Draper, M.H., Duffus, J.H. and Van der Venne, M.Th., eds. (1993) 'The Toxicology of Chemicals', Series. 1, Carcinogenicity, Vol. IV, Commission of the European Communities, Luxembourg.
22. Moulin, J.J., Wild, P., Mur, J.M., Fournier-Betz, M. and Mercier-Gallay, M.A. (1993) Mortality study of cobalt production workers: an extension of the follow up, Am. J. Ind. Med., 23, 281–288.
23. IARC (1991) 'Monograph on the evaluation of the carcinogenic risks to humans. Chlorinated drinking water, chlorination by-products, some other halogenated compounds, cobalt and cobalt compounds', Vol. 52, IARC, Lyon.
24. Chen, R., Wei, L. and Huang, H. (1993) Mortality from lung cancer among copper miners, Br. J. Ind. Med., 50, 505–509.
25. Faulds, J.S. and Stewart, M.J. (1956) Carcinoma of the lung in haematite miners, J. Pathol. Bacteriol., 72, 353–365.
26. Monlibert, L. and Roubille, R. (1960) Bronchial cancer in iron ore miners, J. Fr. Méd. Chir. Thor., 14, 435–439.
27. Cavelier, C., Mur, J.M. and Cericola, C. (1980) Le cancer bronchique chez les mineurs de fer: sélection bibliographique d'enquêtes épidémiologiques et d'études expérimentales, Cahiers de Notes Documentaires, 100, 363–371.
28. IARC (1988) 'Monograph on the evaluation of the carcinogenic risks to humans. Man-made mineral fibres and radon', Vol. 43, IARC, Lyon.
29. Mur, J.M., Meyer-Bisch, C., Pham, Q.T., Massin, N., Moulin, J.J., Cavelier, C. and Sadoul, P. (1987) Risk of lung cancer among iron ore miners: a proportional mortality study of 1,075 deceased miners in Lorraine, France, J. Occup Med., 29, 762–768.
30. Pham, Q.T., Gaetner, M., Mur, J.M., Braun, P., Gabiano, M. and Sadoul, P. (1983) Incidence of lung cancer among miners, Eur. J. Respir. Dis., 64, 534–540.
31. IARC (1980) 'Monograph on the evaluation of the carcinogenic risks to humans. Some metals and metallic compounds', Vol. 23, IARC, Lyon.
32. Fu, H. and Bofetta, P. (1995) 'Cancer and occupational exposure to inorganic lead compounds: a meta-analysis of published data'. Occup. Environ. Med., 52, 73–81.
33. Baker, E.L., Goyer, R.A., Fowler, R.A., Khetty, U., Bernard, D.B. and Adlers, S. et al. (1980) Occupational lead exposure, nephropathy and renal cancer: a case report, Am. J. Ind. Med., 1, 139–148.
34. Lilis, R. (1981) Long term occupational exposure, chronic nephropathy and renal cancer, Am. J. Ind. Med., 2, 293–297.
35. Steenland, K., Selevan, S. and Landrigan, P. (1992) The mortality of lead smelters workers: an update, Am. J. Public Health, 82, 1641–1644.
36. Selevan, S., Landrigan, P.J., Stern, F.B. and Jones, J.H. (1985) Mortality of lead smelters, Am. J. Epidemiol., 122, 673–683.
37. Cantor, K.P., Sontag, J.M. and Heid, M.F. (1986) Patterns of mortality among plumbers and pipefitters, Am. J. Ind. Med., 10, 73–89.
38. Amandus, H. and Costello, J. (1991) Silicosis and lung cancer mortality in US miners, Arch. Environ. Health, 46, 82–89.
39. Buiatti, E., Kriebel, D., Geddes, M., Santucci, M. and Pucci, N. (1985) A case control study of lung cancer in Florence, Italy. I. Occupational risk factors, J. Epidemiol. Community Health, 39, 244–250.
40. Barregard, L., Sallsten, G. and Jarvholm, B. (1990) Mortality and cancer incidence in chloralkali workers exposed to iorganic mercury, Br. J. Ind. Med., 47, 99–104.
41. Ellingsen, D., Andersen, A., Nordhagen, H.P., Efskind, J. and Kjuns, H. (1991) Cancer incidence and mortality among workers exposed to mercury in the Norwegian

chloralkali industry. Eighth International Symposium in Occupational Health, Paris 10–12 September 1991.
42. Ahlbom, A., Norell, S., Rodvall, Y. and Nylander, M. (1990) Dentists, dental nurses and brain tumours, *Br. Med. J.*, **292**, 662.
43. Walrath, J., Rogot, E., Murray, J. and Blair, A. (1985) 'Mortality patterns among US veterans by occupational and smoking status', Vol. 1, US Department of Health and Human Services, Bethesda, MD.
44. Petersen, G.R. and Milham, S. (1980) 'Occupational mortality in the State of California 1959–1961', NCI, Bethesda, MD.
45. Bofetta, P., Merler, E. and Vainio, H. (1993) Carcinogenicity of mercury and mercury compounds, *Scand. J. Work Environ. Health*, **19**, 1–7.
46. Gallagher, R.P., Threlfall, W.J., Band, P.R. and Spirelli, J.J. (1989) 'Occupational mortality in British Columbia', Cancer Control Agency.
47. Berlin, A., Draper, M.H., Krug, E., Roi, R., Van der Venne, M.Th., eds. (1990) 'The Toxicology of Chemicals', Series 1, Carcinogenicity, Vol. II. Commission of the European Communities, Luxembourg.
48. Godbold, J.H. and Tompkins, E.A. (1979) A long term mortality study of workers occupationally exposed to metallic nickel at the Oak Ridge Gaseous Diffusion plant, *J. Occup. Med.*, **21**, 799–806.
49. Cragle, D.L., Hollis, D.R., Newport, T.H. and Shy, C.M. (1984) A retrospective cohort mortality study among workers occupationally exposed to metallic nickel powder at the Oak Ridge Gaseous Diffusion Plant. In, 'Nickel in the Human Environment' Sunderman, F.W., Jr, ed., Vol. 53, IARC, Lyon, pp. 56–63.
50. Cox, J.E., Doll, R., Scott, W.A. and Smith, S. (1981) Mortality of nickel workers: experience of men working with metallic nickel, *Brit. J. Ind. Med.*, **38**, 235–239.
51. Redmond, C.K. (1984) Site specific cancer mortality among workers involved in the production of high nickel cadmium alloys plant. In, 'Nickel in the Human Environment', Sunderman, F.W., Jr, ed., Vol. 53, IARC, Lyon, pp. 73–86.
52. Simonato, L., Moulin, J.J., Javelaud, B., Ferro, G., Wild, P., Winkelman, R. and Saracci, R. (1994) A retroprospective mortality study of workers exposed to arsenic in a gold mine and refinery in France, *Am. J. Ind. Med.*, **25**, 625–634.
53. Yamadori, I., Oshumi, S., Taguchi, K. (1986) Titanium deposition and adenocarcinoma of the lung, *Acta Pathol. Jpn.*, **38**, 783–790.
54. Chen, J.L. and Fayerweather, W.E. (1988) Epidemiologic study of workers exposed to titanium oxide, *J. Occup. Med.*, **30**, 937–942.
55. Fayerweather, W.E., Karns, E., Gilby, P.G. and Chen, J.L. (1992) Epidemiologic study of lung cancer mortality in workers exposed to titanium tetrachloride, *J. Occup. Med.*, **34**, 164–168.
56. Klein, C.B., Frenkel, K. and Costa, M. (1991) The role of oxidative processes in metal carcinogenesis, *Chem. Res. Toxicol.*, **4**, 592–604.

Problems Encountered in Determining Metal Carcinogenesis through Epidemiological Studies

J.M. HARRINGTON

INSTITUTE OF OCCUPATIONAL HEALTH, UNIVERSITY OF BIRMINGHAM, UK

1 Introduction

Establishing whether occupational exposure to a substance or process leads to an increased risk of human cancer involves a number of approaches. These approaches include knowledge of the physical and chemical forms of exposure, biokinetics, bio-assays, *in vivo* and *in vitro* studies of the relevant compounds and, finally, epidemiological studies of exposed human populations. The purpose of this paper is to review the role of epidemiology in the elucidation of carcinogenic potential from workplace exposure to metals and their compounds.

The paper begins with a review of the epidemiological method with its advantages and disadvantages and is followed by some general considerations of the specific issues involved in studying metals for their carcinogenic potential. The problems are illustrated with particular reference to five metals – beryllium, cadmium, chromium, lead and nickel. The implications for industry and legislators regarding current knowledge and the way these metals are presently classified is discussed and some ways of improving the relationship between good science and appropriate action are suggested.

The literature cited is selective and, by and large, recent. The starting point has tended to be from the date of the latest relevant International Agency for Research on Cancer (IARC) Monograph and in general this means peer reviewed papers published between 1990 and early 1995.

2 The Epidemiological Method
2.1 Development

Epidemiological thinking probably dates back to John Graunt (1620–1674) and his analysis of the Bills of Mortality. Monson[1] summarizes Graunt's approach into four succinct pieces of advice, which could remain the model for modern epidemiologists:

(1) reduce voluminous data to a few pertinent tables,
(2) describe the data briefly,
(3) interpret the data conservatively,
(4) note that population-based data are an advantage.

Little development followed until the advent of studies to determine the causation of infectious disease in the 19th century. Chronic disease epidemiology was essentially a 20th century phenomenon, whilst occupational epidemiology did not really develop until after World War II. Nevertheless, the epidemiological method is now of paramount importance in the study of carcinogenesis, if only because it is the only type of study relies on real exposures in human populations. Its relevance to the issue of human cancer is not questioned, but its shortcomings as an investigative tool are considerable.

2.2 Uses

In the context of human cancer risk studies, its main uses are three-fold:[2]

(1) to discover risks that are overlooked or suggested only tentatively by laboratory based tests;
(2) to estimate the level of exposure that produces the highest additional risk of disease that is socially acceptable;
(3) to check the correctness of conclusions about the cause of a cancer hazard by monitoring the effect of its removal, to which could be added a fourth use:
(4) to estimate any interaction between the agent and other factors such as other workplace exposure or lifestyle factors.

The requirements of a good quality epidemiological study are that the results should not be explicable by bias, confounding factors or chance; that the data show a dose relationship and/or time dependency. In addition, it would be useful if the findings described have been observed in a number of previous studies.

2.3 Types of Study

For the purposes of occupational cancer epidemiology there are really only two studies of paramount importance – the cohort study and the case control study. Case series are of limited value and experimental studies rarely feasible. In occupational cancer studies, the accurate enumeration of a cohort of workers exposed to the suspect compound or working on a relevant process involving the substance is almost a *sine qua non*. Case control studies can be of value particularly as they are quicker and cheaper to undertake and, increasingly, the use of case control studies nested with a cohort is becoming the epidemiological study in vogue – combining as it shows the advantages of both studies whilst minimizing their disadvantages. The longitudinal aspect of both these analytical studies is crucial in the quest for exposure–health outcome relationship.[3] One additional technique that is gaining in popularity – partly because it might lead to earlier preventive measures – is the study of DNA adducts, point mutations, chromosomal derangement and the influence of oncogenes in the early phases of exposure. Such studies have been loosely termed 'molecular epidemiology'.[4]

2.4 Shortcomings of the Epidemiological Method

Epidemiology is not an easy science – it aims to be just as exact as chemistry or physics – but some of its major difficulties include:[5]

(1) the healthy worker effect – where the comparison group has a different *general* health status from the index population;
(2) low response rates in the study populations;
(3) high turnover of study population with selection bias operating on membership of exposed groups;
(4) latency between exposure and effect being longer than the study period;
(5) insufficient evidence of different effects by differing exposures;
(6) poor quality of the health effects data;
(7) poor quality of the exposure data;
(8) multiple exposures and;
(9) difficulty in interpreting low or no effect results – the 'non-positive' study.

2.5 Meta-analysis

The difficulties listed above occur in many published studies in various combinations. Well designed multinational, multicentric studies may overcome some of these but they bring formidable problems of logistics and finances. One recent development has been meta-analysis in which various studies are assembled, reviewed and their results pooled (various methods available) in an attempt to develop a summary result. The method is fashionable but is not without problems.

Recent debate on meta-analysis has highlighted some of the pitfalls in assuming that such an approach is the answer to the problems of reviewing conflicting studies to achieve a consensus view.[6-8] In essence, the technique is often employed to attempt to understand cause–effect relationships from weak associations. One danger is that spurious stability from meta-analytic techniques of low relative risk studies could discourage further research. Quality weighting of the studies does not eliminate bias or confounding and cannot overcome the possibility of publication bias in favour of 'positive' studies. It has been argued that meta-analysis is appropriate for pooled randomised clinical trials but less appropriate for non-experimental studies that have no generic set of quality criteria. Heterogeneity is the order of the day. There are currently no universally acceptable statistical techniques to compensate for input data limitations. Meta-analysis should not be abandoned, however, as it can be an aid in critical comparison of different studies and for finding patterns among studies. Care must be taken in interpretation, though, and even for randomised controlled clinical trials, serious mistakes have been made in interpreting small trends before a large-scale trial gave the answer.[9]

2.6 Exposure Data

The most serious defect in many otherwise well-designed epidemiological studies is the absence of good quality exposure data. The development of techniques for retrospective exposure assessment has received much attention of late and whole issues of journals[10] have been devoted to debating the best techniques for assessing exposure in the absence of good quality raw data collected at the time.

Improvements in retrospective exposure assessments[11] should enable epidemiologists to move beyond reporting an excess relative risk or odds ratio into the area of quantitative assessment of cancer risks to populations, which is much needed by policy makers who have to act on the 'science'. However, the complexity of exposures to a wide variety of physical and chemical forms of the element concerned remains a major problem and is well-illustrated by the problems of exposure to nickel and chromium compounds.

Finally, decisions on which agents should be studied should increasingly be driven by a practical 'need to know'. In this context there is much justification for concentrating future human studies on agents that have been shown to be animal carcinogens or *in vitro* mutagens but for which little human data exist.[12] A list of ten criteria to be used in establishing priorities for epidemiological research has been drawn up by the IARC.[13]

3 The Classification of Carcinogens

It is over a quarter of a century since IARC began its programme to evaluate the carcinogenic risk of chemicals to humans and to produce monographs on individual chemicals. The programme has expanded to include industrial processes which involve exposures to complex mixtures. Sixty monographs have been published and the process of evaluation adapted by IARC has become a model for others engaged in the evaluation process.

In essence, the IARC method is to assemble independent experts who evaluate published data on the chemical and physical properties of the agents under review, their human exposure circumstances, relevant animal studies, *in vitro* genotoxicity investigations and human epidemiological data. An overall evaluation is then agreed on a four point scale ranging from 1 (proven human carcinogenic risk) through probable (2A), possible (2B) and not classifiable (3) to, rarely, that an agent (mixture) is probably not carcinogenic to humans (4). [Exceptionally, an agent (mixture) may be placed in category 1 when evidence in humans is less than sufficient, but, where there is sufficient evidence of animal carcinogenicity and strong evidence in exposed humans that the agent (mixture) acts through a relevant mechanism of carcinogenicity.]

The criticism of the Agency, that it fails to quantify the risks it evaluates and fails to provide the practical lead for action to legislators, is somewhat unjust – it does not pretend to do that. Yet the IARC evaluations are often only read in summary and the designation of 1, 2A or 2B carries great weight in the world despite the fact that it is a qualitative rather than a quantitative assessment. For example, whilst a number of oestrogens are classified as Group 1 carcinogens, no consideration is possible in this classification to consider the therapeutic advantages of oestrogens or even their protective role against some human cancers. That is, the classification is based on the *strength* of the evidence – not the potency of the 'carcinogen'.

Other authorities such as the US Environmental Protection Agency (USEPA), and the European Union (in concert and as individual countries) also evaluate carcinogens and produce lists similar in nature and subdivision to IARC. A recent review of ten such organizations and their methods shows that all engage in carcinogen identification, but only some go on to extrapolate risk.[14] Risk assessments may be based on scientific judgement alone, involve an upper bound risk assessment, divide carcinogens into genotoxic and nongenotoxic and then use different extrapolation procedures for each whilst some consider nongenotoxic chemicals as threshold toxicants and genotoxic ones as having no threshold. The lists produced by the various agencies do not differ greatly for the known or probable human carcinogens, but there has been little attempt to harmonise strategies that would be of great value to the policy makers and of considerable help in communicating the results of any risk assessment for appropriate management of that risk. Even the purpose of these evaluative processes varies from country to country, with some concerned primarily with occupational or environmental exposures and others with more general labelling requirements.

It is, therefore, advisable to be cautious when scanning the different lists and to review the basis for the evaluation and the purpose behind the exercise. Nevertheless, no agency has developed the risk characterization procedure very far and few attempt to place such risks in the context of other (occupational or non-occupational) risks, which their national populations may face.

Finally, as will be discussed later in the context of individual metals and their compounds, a change in evaluation from say 2B to 2A or most significantly from 2A to 1 (using the IARC grading system) carries considerable legal and financial implications to the industries involved as well as their employees and their customers. Indeed, the scientific justification for such changes is often a matter of heated debate. This is even more problematical when the evaluation is concerned with a process – such as painting or aluminium refining – than it is with a specific metal such as nickel, but even here the carcinogenic risks from individual metallic compounds often remain unquantified.

4 Metal Carcinogenesis
4.1 Mechanistic Considerations
It is not the purpose of this paper to review the intricacies of metal carcinogenesis, but certain aspects of the literature on this subject are relevant to the process of epidemiological evaluation.

In general, occupational exposure to carcinogens tends to affect certain target organs (such as the lung, bladder, haemopoietic system, liver, nasal cavities and skin) more than others.[15] This biological fact needs to be borne in mind when studies show an excess risk for tumour sites that appear to be biologically implausible given the exposure characteristics and the agent(s) involved. Of the 20 chemicals classified by IARC as Group 1 carcinogens, five are metal/metal compounds (arsenic, beryllium, cadmium, chromium and nickel). Group 2A contains no metals, but Group 1 includes metal processing industries such as iron and steel founding. Most Group 1 carcinogens have been first reported in studies published before 1970.[16] The first good human data on arsenic dates back to 1947, but it has been a suspect human carcinogen for a century.[17]

Unfortunately, mechanistic aspects of carcinogenesis for metals are nowhere near as well understood as for some of the organic chemical carcinogens and the problems of exposure uncertainties are compounded by the different chemical and physical forms of the metals and their salts that are encountered in various occupational settings. Different metals act at different stages in the carcinogenic process with, for example, mutagenic and altered gene expression being a feature of exposure to chromium, nickel and cadmium. Some metals such as cadmium can interfere with calcium binding and regulation and thus affect intracellular communication and cell membrane permeability and this can influence the process of metastasis.[18] There is also the problem of metal/metal interaction to be considered.

Specifically, metals and their salts are, usually, ionic and thus may not require metabolic activation *in vivo*, they are often highly tissue-specific, and may follow metabolic pathways used by essential metals.[19] Indeed, some essential metals, such as chromium, are carcinogenic as well. These differences between metal compounds and the organic chemicals are obfuscated when one considers the organo-metal compounds.

The complex exposure characteristics of industrial processes can also lead to situations where interaction between different metals becomes important or where some concomitant metal exposures may act as anti-carcinogens – for example selenium or zinc. Furthermore, valency may be important in terms of bioavailability and the ability to cross the cell membrane (hexavalent *vs.* trivalent chromium) and the physical characteristics of the same compound may vary with the temperature of the reaction that produced it (nickel oxides).

In summary, the mechanistic aspects of metal carcinogenesis are complex and may not be analogous to organic chemicals. Little attention has been paid to this in the epidemiological literature and consideration of such aspects as valency, physical characteristics, route of administration (*vis a vis* the animal studies) and metal/metal interaction are rarely mentioned.[20] If specific metal compounds and their workplace concentrations can be characterized, that is usually as far as the epidemiologist goes though these other characteristics may be of crucial importance in assessing biological affect as opposed to chemical exposure.

4.2 Metal Industries

Apart from the consideration of occupational exposures to specific metals and their compounds, it is important, for completeness, to remember that a number of epidemiological studies have been concerned with metal alloys such as ferrochromium and stainless steel as well as processes involving mixed exposures in iron and steel founding and the highly complex exposure characteristics of welding. In the foundries, there is good evidence for a carcinogenic risk to the lung in some studies[21-23] but not in others.[24-26] For welders, respiratory tract cancer excesses have also been found, but not consistently so.[27-34] Specific metals are rarely invoked as the cause, but confounding exposures to smoking and polynuclear aromatics may well be important determinants of the excess cancer risks found.

The mining of metalliferous ores has long been known to be linked to excess cancer – usually of the respiratory or haemopoietic systems. Recent studies have confirmed these risks, but most authors suggest that the excesses are more likely to be due to radon and radon daughters, silica, exhaust gases or smoking or a combination of these with or without exposure to arsenic, which is commonly found in metal-bearing ore.[35-42] A further potential confounding factor in the metal processing industry is in the exposures to strong acid mists such as sulfuric acid, which has been strongly linked to laryngeal cancer.[43]

5 Specific Metals and their Compounds

Five metals have been chosen for review, the choice is not random. They illustrate the problems of determining carcinogenesis from epidemiological studies.

5.1 Beryllium

Beryllium is a rare and very light metal. Its main industrial uses relate to its ability to alloy to other metals to produce a hard, non-corrosive, non-magnetic metal of great tensile strength. The human health effects of beryllium are, in part, due to its ability to damage DNA *in vitro* and to inhibit enzymes such as alkaline phosphatase. It combines with phosphate-binding enzymes and has a strong cellular immune response *in vivo*. The soluble salts can cause acute berylliosis of the lungs and the effects are dose-dependent. A fifth of these cases are said to progress to chronic disease, which is characterized by granuloma formation. Chronic berylliosis is a hypersensitivity reaction that seems to be linked to exposure to high temperature calcined oxide. The latent period for the chronic disease can be but a few years but more commonly is 10–15 years with an upper limit of 30 years. Beryllium is a potent cause of lung cancer in rats[44] and its human toxicity appears to be related to particle deposition in the lungs.[45]

The epidemiological studies of cancer in beryllium-exposed workers are restricted to the US Beryllium Case Registry (of berylliosis) and seven US manufacturing plants. Up to 1991 there had been six studies of two US cohorts from some of these plants (4300 workers) and the Registry data. These studies showed a 50% excess of lung cancer 15 years after first exposure, an excess that reached statistical significance at 25 years. The excess was largely confined to those who had acute berylliosis and short exposure (less than five years often less than one year) with a two-fold excess in the Registry cases.

At this stage, the case for beryllium being a proven human carcinogen was in some doubt[46] as the excess seemed to be related to acute dose rather than length of exposure and more related to acute disease than chronic (fibrotic) disease.

Further studies – largely updates or reworkings of the original cohorts with additions or subtractions – have led to a heated debate. In particular, the update of the Registry data in 1991 – which included women for the first time but excluded cases who were diagnosed as having lung cancer *before* inclusion in the registry – confirmed the two-fold excess in acute berylliosis cases, but also found a 50% excess in the chronic cases.[47] A cohort study of 9225 workers from seven plants (including the Reading and Lorain plants of earlier reports), confirmed the excess at the Lorain plant but suggested that this excess also existed for workers hired after 1950 and at other plants. The Standardised Mortality Ratios (SMRs) ranged from 0.82 to 1.69.[48] A further report pointed out that all 17 lung cancer cases with acute berylliosis worked at the Lorain plant where exposures were known to be over 40 times higher than elsewhere,[49] but the cohort study had shown excesses at the other plants in workers who had not been in the Registry group.

In 1993, on the basis of these new studies, IARC reclassified beryllium from 2A to 1,[50] though the Monograph is less critical of the quality of the studies than MacMahon's review in 1994.[51] In MacMahon's view the early pre-1987 studies of two plants are 'unintrepretable' – owing partly to remarkable events, including the whittling down of the original cohort from 10 356 workers to 3055 workers and the resignation of the senior author on one paper following public criticism of the report. The later studies[47,48] are better studies, but smoking data of the workforce are still limited, the results still fail to show a dose–response relationship, exposures fell dramatically after 1950, the excess risks are around 50% and based on small numbers and the Lorain plant stands out as being greatly different in exposures and effects from the other plants.

The truth or otherwise of the carcinogenicity of beryllium hinges on whether one believes that high dose, short-term exposure can lead to an effect not seen in the lower dose, longer-term exposed workers. This would explain the absence of a dose–response effect but also implies that the workers selected out of the industry with acute disease are the ones at greatest risk of acquiring cancer later. It is also unfortunate that such limited exposed populations are available for study.

The dilemma here, and its effect on policy makers and the industry, is summed up by MacMahon:

> 'Even if such (high) exposures were carcinogenic, their relevance to occupational experience of the last three decades is nil. The situation exemplifies, perhaps, the futility, in the context of public health and public policy, of defining whether a chemical is or is not carcinogenic to humans without due consideration of dosages and routes and mechanisms by which humans are exposed'.

5.2 Cadmium

Cadmium is commonly mined as a cadmium/zinc-ore (less commonly with lead-ores) and is a by-product of zinc smelting. Its main uses are in pigments, batteries, electroplated products and as a plastic stabiliser. It is absorbed well from the lungs and has a half life in the human of ten years. Cadmium binds to sulfur-containing amino acids such as metallothionen, which is a zinc storage protein. It is toxic to the kidneys and on inhalation can cause emphysema.

Like beryllium, the epidemiological literature on cadmium and cancer is restricted to a relatively small number of cohorts of exposed workers, and the interpretation of the findings in these studies is still a source of much debate. The main study populations are a composite group of nearly 7000 UK workers at 17 cadmium processing plants, 3025 workers at a UK nickel–cadmium battery plant, 606 employees of a US smelter in Colorado and 522 Swedish battery workers.

The cancer story probably starts with a short report in 1965 of a cluster of four cases of prostate cancer (expected 0.58) in men working at the UK nickel–cadmium plant.[52] This report initiated the rest of the literature but no convincing confirmatory evidence of the prostate excess has been found since then,[2] despite that fact that some authors still cite the excess as real,[53] including the most recent (1994) edition of a major textbook of occupational medicine.[54]

The UK study of 17 processing plants[55] an excess of lung cancer reported which was highest in the ever high exposure group with ten years or more of exposure. Lung cancer was found in UK nickel–cadmium plant to be associated with duration of high or moderate-exposed employment with the excess confined to those first employed before 1947.[56] The small Swedish study[57] also found eight cases (4.1 expected) of lung cancer.

A crucially important cohort relates to the smelter at Globe, Colorado, which started in 1896 as a lead smelter, then as an arsenic smelter (1920–1926) and thereafter as a cadmium smelter. The study of this population has led to a series of reports by Lemen et al. (1976), Thun et al. (1985) and recently by Stayner et al. in 1992.[58] Stayner limited the study population to 606 workers employed after arsenic smelting ceased in 1926 and used the occupational exposure histories gleaned by Thun from the company records as well as some air monitoring data. The IARC Monograph of 1993[50] considers the Stayner report to have the best characterised population so far published. In this study a statistically significant excess of lung cancer was found (observed 21, expected 9.95) in non-Hispanics with high exposure and longest follow up. Various modelling techniques support a dose–response relationship. There was some attempt to control for smoking. These positive findings are disputed in a paper by Lamm et al. using a case control approach, although only 21 cases of lung cancer are common to both studies.[59]

Doll reviewed the literature up to and including the 1992 report on the Globe smelter[60] and noted that 500 cases of lung cancer had been found in five cohorts. The excess risk for lung cancer was mostly from 10–50% with some excess as high as 200%. However, smoking data were limited, and the workers' exposures often included nickel and arsenic and there are some differences between studies of the same populations regarding regional or national comparison rates for cancer and the use of arithmetic or geometric means for the exposure data.

Nevertheless, in 1992, the evidence seemed to be stronger for cadmium being a lung carcinogen for humans – indeed, in combination with the animal bio-assay results, it was sufficient for IARC to reclassify it to Group 1 in the 1993 Monograph.[50] One is led to the conclusion that the IARC Working Group were greatly influenced by the Stayner study with its detailed occupational exposure assessments.

However, some doubt must now be cast on the wisdom of that re-evaluation if it was based on the latest Globe smelter report. Sorahan and Lancashire have also had access to the Globe records and in a preliminary report[61] of their findings they note that the worker records are of two types: personnel and service records, which have limited job history data, and the detailed time sheets with different dollar rates for different jobs. They are probably correct in assuming that the latter records are the more accurate and they are certainly the more detailed. NIOSH did not use these records, but in a limited comparison of the job exposure data from the two sources for 149 workers for the period April to August 1949, Sorahan and Lancashire have found notable mismatches between them. The implications of this await full analysis but misclassification, if confirmed, could have a considerable influence on the exposure-effect relationship reported by Stayner.

5.3 Chromium

Chromium is a hard metal used for plating and special steels. A number of its salts are used in dyes, paints and as tanning agents. The commercially important valencies are Cr^{3+} and Cr^{6+}. Hexavalent chromium compounds are rapidly absorbed and excreted and are well-documented to be irritant, corrosive and potent skin sensitisers. Chromium is one of the 'essential' metals involved in carbohydrate metabolism and insulin production. Chromium compounds have been known to be carcinogenic to humans for 60 years. Indeed, the first case report of an adenocarcinoma of the nasal turbinate in a chrome pigment worker appeared in 1890.[62] Unlike the two preceding metals, the issue here is not one of carcinogenicity but relates to 'what chromium compounds in what amounts cause what diseases'.[63]

IARC[64] reviewed chromium and chromium compounds in 1990 and in the same year Langard published an authoritative review of the epidemiology of chromium carcinogenicity.[62] The literature to that date can be divided into the studies for different industries and processes. For chromate-producing plants in the UK and the US, a consistent excess risk for lung cancer had been noted with inconsistent findings for gastrointestinal cancers. For chrome pigment plants in the UK and Norway, the same findings apply to the same target organs. Some debate surrounds whether lead chromate is a human lung carcinogen, but only one UK plant produced lead chromate alone (the rest produced a mixture, including zinc and strontium chromates). No excess was found in the lead chromate plant except for those previously diagnosed as having lead poisoning. Studies of chrome plating populations in the UK, US, Italy, Scandinavia and the former USSR have consistently found lung and nasal cancer excesses but they were often small or statistically not significant. One Japanese study[65] of 265 workers in 40 small plating plants (not cited by Langard) found the excess of lung cancer was restricted to workers with a high degree of skin ulceration or nasal perforation. Excess lung cancer risks have also been noted in ferrochrome alloy workers.

Since then further studies of Japanese chrome platers[66] and chrome pigment workers[67] as well as UK and US chromate production workers[68,69] have been published. Some of the populations are too small or have been followed for too short a time to demonstrate any excess but the UK update reported an intriguing excess of nasal cancers (4 observed, 0.26 expected) in addition to confirmation of the lung cancer excess noted in earlier reports. The authors noted the major process changes that occurred between 1958 and 1960 and that the follow up of workers first employed since that time show no excess of lung cancer (14 observed, 13.7 expected). This one remaining UK plant has also been the source of an interesting study of ten chrome workers and ten controls looking for DNA strand breaks and eight hydroxydeoxy guanosine production in lymphocytes.[70] No excesses were found, although the workers were only exposed to airborne chromium concentrations at 20% of the current UK Occupational Exposure Limit.

There are four issues of interest concerning chrome carcinogenicity:

(1) the valency – hexavalent chromium alone appears to have the greatest carcinogenic potential;
(2) water solubility – soluble and sparingly soluble compounds are more likely to be carcinogenic than insoluble ones;
(3) industrial process – the data are stronger from pigments and dichromates than they are for chrome plating;
(4) potency – zinc chromate is more potent than lead chromate, whilst strontium chromate has not been studied separately in epidemiological studies.

Chromium compounds are convincingly genotoxic and are animal carcinogens.[17] Either the sparingly soluble compounds are carcinogenic or they are more carcinogenic than the soluble and finally, the difference between chromates and chromic compounds seems to reflect the penetrability of cell membranes for Cr^{6+} but not Cr^{3+} compounds whereas the kinetics of chromates and the *in vitro* reaction of Cr^{3+} with DNA suggest that the ultimate intracellular carcinogenic species is Cr^{3+}.

5.4 Lead

By contrast, lead has been used for 6000 years and its toxic properties to the haemopoietic, reproductive and nervous systems have been known for centuries. Lead and lead compounds are widely used in industry. Inhalation of lead dust or fume is the main route of entry and the metal ion has a strong affinity for sulfhydryl groups. It is therefore a potent enzyme inhibitor and competes with calcium in metabolic pathways.

Most mortality studies of lead-exposed workers have highlighted the effects on renal function leading to hypertension as a cause of cerebrovascular disease. Some studies have also shown modest (25–50%) excesses for some cancers, including lung, kidney and stomach, whereas others have not. The epidemiological literature has recently been well-reviewed by Fu and Boffetta.[71] They also applied meta-analysis to the assembled data and suggest some evidence to support the hypothesis of an association between stomach and lung cancer and exposure to lead. Notwithstanding the limitations of meta-analysis, a review of the literature does not provide convincing evidence for the carcinogenicity of lead compounds. The studies often show very modest excesses, others of equal quality are negative. Little attention has been paid to confounders such as smoking or diet, or indeed to competing occupational exposures. The studies linking lead with renal and urinary tract cancers may reflect a publication bias.

Even after such an analysis, given the long history of lead toxicity over the centuries and the considerable literature on large exposed populations, the weak or inconsistent data on carcinogenicity suggest that the lead is unlikely to be found subsequently to be a potent cause of human cancer.

5.5 Nickel

Nickel is a hard, silvery metal, which has the important properties of being malleable, ductile, non-corrosive and capable of taking a high polish. It is used as a catalyst, in alloys, coins, magnets and batteries. It is readily absorbed and excreted by humans, and is an element essential for normal haemopoiesis. It is a potent skin sensitiser. Nickel compounds may be soluble or insoluble and this property has an important bearing on biological effect.

The literature on nickel carcinogenesis in exposed working populations has been authoritatively reviewed and updated to 1990 in the report of the International Committee on Nickel Carcinogenesis in Man (ICNCM).[72] Nickel was first noted to cause lung and nasal sinus cancer over 50 years ago at the nickel refinery at Clydach in South Wales. This report was confirmed by an epidemiological study by Doll and Morgan twenty-five years later. Other reports followed from refinery populations in Canada, Norway, Germany and Russia. Whilst some excess cancer rates have been reported for larynx, stomach, kidney, prostate and buccal cavity, the consistent finding has been the high risk of lung cancer and an even higher risk for nasal cancer. The ICNCM report updates nine previously studied cohorts and one case control study. The respiratory cancer excess appears to be mainly associated with roasting, sintering and calcining of ore and is probably linked to oxidic and sulfidic nickel compounds. Mining and smelting operations do not appear to carry such risks in any consistent way, but the conflicting results on electrolysis workers leave an open question on the carcinogenic potential of soluble nickel salts. One study of metallic nickel exposure found no cancer excess and nickel alloys also seem to be relatively benign. Good dose–response data are sparse, but it appears that excess cancer risks require greater than 1 mg Ni m^{-3} (soluble) or greater than 10 mg Ni m^{-3} (insoluble).

The 1990 IARC Monograph[64] relies heavily on the ICNCM Report for its evaluation and concludes that there is sufficient evidence of carcinogenicity in humans for nickel sulfate and for combinations of nickel sulfides and oxides encountered in the refining industry.

Further studies of the Sudbury Basin workers in Canada suggest that although polynuclear aromatic hydrocarbon exposure may be significant,[73] the millers, miners and smelter workers do have a modest lung cancer excess (50%), although the absence of a dose–response effect, low nickel exposure and no data on smoking leave the question of the role of nickel unresolved.[74] Similarly, a small nickel refinery in Alberta of 716 workers followed for 6–30 years is small and may be it is too early to assess the effect of nickel exposure.[75]

In 1992, updates of the main nickel-exposed cohorts were published in book form in Nickel & Human Health. The main thrust of these reports was to look again at the practical issues surrounding nickel carcinogenicity. These are:

(1) Which nickel compounds/processes are the cause of cancer?
(2) What is the mechanism of nickel carcinogenesis?
(3) What are the biologically relevant routes *vis a vis* the animal studies?
(4) What 'doses' are required to produce a cancer excess?
(5) How can the occupational exposures be better characterised?

The reports cover the Clydach cohort,[76] as well as the Norwegian[77] and Canadian[78] cohorts. More nickel in air data were available and the cohort enumerations were more complete. Earlier excess cancer risks were confirmed with the suggestion that the risk has fallen in the more recently employed populations – indeed, the Clydach excess is confined to those employed before 1930.

Nevertheless, the impression gained from these careful studies is that one is no further on from the conclusions of the 1990 reports (apparently, the late publication of the book has confused the chronology. The reports in the book *predate* the ICNCM report). Further small studies from smelting and refinery populations in New Caledonia[79] and Finland[80] are, in essence, 'non positive' but further updates will become increasingly valuable. The role of nickel (and/or chromium) in lung cancer excesses in stainless steel welders has been well-reviewed by Langard,[81] but the issues remain unresolved. An interesting recent report of the Ontario sinter workers[82] suggests that the excess risks of lung cancer remain 30 years after cessation of exposure, but this study raises more questions to be answered about the persistence of nickel in body tissues.

The issue of nickel carcinogenicity is similar to the story of chromium. The excess cancer risks for certain tumours and certain process exposures is not in dispute. The practical problem for both industry, its employees and the legislators is which compounds, acting singly or together or in association with others, have been found to be so active and why. Perhaps clearance rates and relative toxicity could be important factors. Tissue response being more affected by time than dose. More fundamental is the problem of understanding the physico-chemical properties of nickel salts. It is even too simplistic to talk of nickel and nickel salts in a strict atom-for-atom stoichiometric way.[83] For example, the valency must be important and most nickel salts are crystalline, but there is an amorphous form of nickel sulfide, which is relatively inert biologically, whilst 'nickel oxide' has widely varying biological properties depending on the temperature of formation with black to grey to green crystalline structures. Similar complexities surround the nickel sulfides. The role of solubility in nickel carcinogenesis is unresolved. Some epidemiological evidence, backed by recent animal studies, suggests that soluble nickel compounds can affect cell proliferation but not genotoxicity. Insoluble nickel compounds may have contrasting properties. A *combination* of the two could thus be more potent than either alone, but it is clearly premature to classify soluble nickel salts such as nickel chloride as proven human carcinogens. In addition, nickel has a high affinity for cobalt and the influences of blood transport by type of nickel could be crucial in assessing biological effect.

Finally, the issue of the carcinogenic potential of nickel alloys has received little attention to date. It appears that high nickel alloy plants have a different mortality experience from the nickel producing plants. Studies of these plants are continuing, but the differing health effects between nickel producer and nickel user industries are of crucial importance in the future classifications of nickel carcinogenesis.

6 The Implications for Industry and the Legislators
6.1 The Epidemiological Debate

There is no doubt that the results of epidemiological studies of metal carcinogenesis have a profound influence on the evaluation process. Unfortunately, the unavoidable difficulties in the epidemiological method – being essentially non-experimental for occupational studies – means that the quality of the data is highly dependent on what is available for study. In many cases, one must conclude that the data bases are inadequate for firm conclusions to be drawn, whatever the results seem to imply. These shortcomings are compounded by variable interpretation of the data, which can be influenced by politics, industry interest, labour union views, personal career prospects and even frank mistakes or misjudgements.

6.2 The Classification of Carcinogens

The classification of carcinogens by international agencies such as IARC, supranational bodies such as the EC or national committees concentrates mainly on agent/process identification and is less consistent within and between agencies in risk extrapolation. Even the people involved in the process vary in expertise between agencies. Some are drawn strictly from the independent scientific community, whilst others rely on a tripartite grouping including representatives of industry and organized labour. There is a lamentable lack of consensus or consistency between the approaches used, even for carcinogen identification.

6.3 Risk Limitation and Control

The ultimate goal of all epidemiological studies should be to provide good quality data upon which to base strategies to control hazardous exposures and thereby prevent human ill health. In most cases, the poor quality of the data precludes confident predictions of dose–effect relationships, leaving the industry with little practical guidance on how to manufacture or process particular metal products for profit and yet protect the health of employees on whom that commercial success depends.

6.4 General Conclusions

The current state of knowledge on metal carcinogenesis is well-illustrated by the five metals reviewed. In essence the problems of interpretation are caused by several factors:

(1) limited data bases (beryllium);
(2) poorly defined exposure data (all five metals);
(3) inadequately recorded job history data (cadmium);
(4) inability to distinguish individual compounds or processes (chromium, nickel);
(5) ill defined risks from a ubiquitous agent (lead);
(6) problems of confounding exposures (cigarettes, arsenic, other metal interactions) and finally;
(7) the frustrating paradox of the undoubted relevance of epidemiological studies to the debate with the inherent complexities of the method. This compares with the relative clarity of the animal data and its inherent problems of extrapolation to the human.

The failure to resolve these issues leaves the industry not knowing how best to handle the problem, the employees confused and worried by the ill defined threat to their health and the legislators in a quandary on how to control badly characterised exposure hazards to limit an ill defined but life threatening health risk.

7. The Way Forward
7.1 Improve Epidemiological Quality

The development by the US chemical industry of Good Epidemiological Practice (GEP) is a step in the right direction for quality control of epidemiological studies. Peer reviewed journals are also beginning to adopt similar guidelines in assessing whether a paper should be published or not. However, such guidelines can only go so far. There is a danger that the assessment of whether a study is 'good' will be reduced to check lists and that would be a mistake. The guidelines should be just that, and the assessment of what constitutes quality science cannot be reduced to a slavish adherence to some listed points.

A fundamental defect in many studies is the exposure assessments. Techniques to improve the evaluation of even limited past exposures are being developed and such techniques should be encouraged and widely used. For the future it is essential that good quality exposure data are kept so that future cohort analyses can be better placed to answer the vital dose–response question.

7.2 Consider Mechanistic and Biological Plausibility

Too little attention has been paid by epidemiologists to the mechanistic issues of metal biotransformation and biological affect. It is rare for epidemiologists to join with biologists and chemists to consider, together, how they can solve the issues with which they all grapple. Frequently, specialists work in isolation from the others. The problem is well-illustrated by the complexity of nickel and chromium chemistry and the influence of different compounds by different routes of entry on *in vivo* human biology.

7.3 Promote 'Molecular Epidemiology'

Investigating early DNA adduct effects could enable researchers to discover more about the mechanisms of human carcinogenesis and at an early stage in the process. They could expand knowledge and simultaneously open the way for earlier protective measures for exposed human populations. Such studies are presently, however, at an early stage of development with no clear-cut evidence of beneficial effects. The ethics of undertaking such studies in the absence of such benefits is debatable but the possible gains could be substantial; the issue of ethics must be faced and debated openly.

7.4 Improve Lay Communication of Epidemiological Data

Whilst the non-scientific community seems to place increasing credence on the results of epidemiological studies, the scientific community has failed to communicate the shortcomings and caveats needed when interpreting the results of such studies. A major educational exercise is needed or the misinterpretation of the studies could have serious implications if the action of policy makers is misguided by their lack of understanding of the limitation of the science.

7.5 Extrapolation to Low Dose 'Real' Exposure

In most modern industrial societies, workplace exposures are relatively low. Extrapolating epidemiological data from past exposures to assess current worker risk is not easy but should be addressed. By the same token, there is a need to make risk assessments for the (usually) even lower environmental exposures. The USEPA and the WHO do this by using SMRs from the 'best' epidemiological studies, though how far this is justified needs further consideration.

8 Conclusions

The problems inherent in the execution and interpretation of epidemiological studies is no excuse for not improving matters where possible. Epidemiological studies are the most relevant type of study for investigating population-based human health effects. The objective of all carcinogenesis research is to learn more about the hazards, leading to accurate assessment of the risk, which in turn leads to the ultimate goal of all these actions: the protection of the health of human populations.

References

1. Monson, P.R. (1990) 'Occupational Epidemiology', 2nd edn., CRC Press, Boca Raton, FL.
2. Doll, R. (1984) Occupational Cancer: Problems in Interpreting Human Evidence, *Ann. Occup. Hyg.*, **28**, 291–305.
3. Harrington, J.M. and Saracci, R. (1994) Occupational Cancer: clinical and epidemiological aspects. Chapter in 'Hunters Disease of Occupations' 8th edn. Raffle P.A.B., Adams P.H., Baxter P.J., Lee W.R., eds., Edward Arnold, London, pp. 654–690.
4. Hamm, R.D. (1990) Occupational cancer in the oncogene era, *Br. J. Ind. Med.*, **47**, 217–220.
5. Harrington, J.M. (1995) 'Epidemiology chapter in Occupational Hygiene' 2nd edn., Harrington, J.M., Gardiner, K., eds., Blackwell Scientific Publications, Oxford (in press).
6. Shapiro, S. (1994) Meta-analysis/Schmeta-analysis, *Am. J. Epidemiol.*, **140**, 772–778 and 778–791.
7. Petitti, D. (1994) Of Babies and Bathwater, *Am. J. Epidemiol.*, **140**, 779–782.
8. Greenlands, S. (1994) Can Meta-analysis be salvaged? *Am. J. Epidemiol.*, **140**, 783–787.
9. Egger, M. and Davey-Smith, G. (1995) Misleading meta-analysis, *Br. Med. J.*, **310**, 752–754.
10. International Workshop on Retrospective Exposure Assessment for Occupational Epidemiologic Studies, (1991) *Appl. Occup. Environ. Hyg.*, **6**, 417–559.
11. Stayner, L., Smith, R., Bailer, J., Luebeck, E.G. and Moolgavkar, S.H. (1995) Modelling Epidemiologic Studies of Occupational Cohorts for the Quantitative Assessment of Carcinogenic Hazards, *Am. J. Ind. Med.*, **27**, 155–170.
12. Gardiner, M.J. (1991) Contribution of occupational exposure to cancer: recent developments, *Br. J. Ind. Med.*, **48**, 217–220.
13. IARC Internal Technical Report No 86/004. (1986) Priorities in Occupational Cancer Epidemiology, Joint IARC/CEC Working Group Report, IARC, Lyon.
14. Moolenaar, R.J. (1994) Carcinogen Risk Assessment: International Comparison, *Regul. Toxicol. Pharmacol.*, **20**, 302–336.
15. Kogevinas, M. (1995) An investigation of occupational cancer in the European Union, Medicina de Lavaro, (in press).
16. Vineis, P. and Blair, A. (1992) Problems and perspectives in the identification of new occupational carcinogens, *Scand. J Work Environ. Health*, **18**, 273–277.
17. Magos, L. (1991) Epidemiological and Experimental Aspects of Metal Carcinogenesis: Physico-chemical properties, kinetics, and the active species, *Environ. Health Perspect.*, **95**, 157–189.
18. Snow, E.T. (1992) Metal carcinogenesis: Mechanistic Implications, *Pharmacol. Ther.*, **53**, 31–65.
19. Waalkes, M.P., Coogan, T.P. and Barter, R.A. (1992) Toxicological Principles of Metal Carcinogenesis with special emphasis on Cadmium, *CRC Crit. Rev. Toxicol.*, **22**, 175–201.
20. Boffetta, P. (1993) Carcinogenicity of trace elements with reference to evaluations made by the International Agency for Research on Cancer, *Scand. J. Work Environ. Health*, **19** (Suppl. 1), 67–70.

21. Sorahan, T., Faux, A.M. and Cooke, M.A. (1994) Mortality among a cohort of United Kingdom steel foundry workers with special reference to cancers of the stomach and lung, 1946–90, *Occup. Environ. Med.*, **51**, 316–322.
22. Finklestein, M.M., Boulard, M. and Wilk, N. (1991) Increased risk of lung cancer in the melting department of a secondary Ontario Steel Manufacturer, *Am. J. Ind. Med.*, **19**, 183–194.
23. Finklestein, M.M. (1994) Lung cancer among steel workers in Ontario, *Am. J. Ind. Med.*, **26**, 549–557.
24. Andjelkovich, D.A., Mathew, R.M., Richardson, R.B. and Levine, R.J. (1990) Mortality of Iron Foundry Workers 1: overall findings, *J. Occup. Med.*, **32**, 529–540.
25. Andjelkovich, D.A., Mathew, R.M., Yu, R.C., Richardson, R.B. and Levine, R.J. (1992) Morality of Iron Foundry Workers II. Analysis by Work Area, *J. Occup. Med.*, **34**, 391–401.
26. Andjelkovich, D.A., Shy, S.M., Brown, M.H., Janszen, D.B., Levine, R.J. and Richardson, R.B. (1994) Mortality of Iron Foundry Workers III. Lung cancer case-control study, *J. Occup. Med.*, **36**, 1301–1309.
27. Melkild, A:, Langard, S., Anderson, A. and Tonnessen, J.N.S. (1989) Incidence of cancer among welders and other workers in a Norwegian shipyard, *Scand. J. Work Environ. Health*, **15**, 387–394.
28. Moulin, J.J., Portefaix, P., Wild, P., Mur, J.M., Smagghe, G. and Mantout, B. (1990) Mortality study among workers producing ferroalloys and stainless steel in France, *Br. J. Ind. Med.*, **47**, 537–543.
29. Moulin, J.J., Wild, P., Haguenoer, J.M., Falcon, D., De Gaudemarais, R., Mur, J.M. *et al.* (1993) A mortality study among mild steel and stainless steel welders, *Br. J. Ind. Med.*, **50**, 234–243.
30. Steenland, K., Beaumont, J. and Elliot, L. (1991) Lung cancer in mild steel workers, *Am. J. Epidemiol.*, **133**, 220–229.
31. Simonato, L., Fletcher, A.C., Andersen, A., Anderson, K., Becker, N., Chang-Claude, J. *et al.* (1991) A historical prospective study of European stainless steel, mild steel and shipyard welders, *Br. J. Ind. Med.*, **48**, 145–154.
32. Danielson, T.E., Langard, S., Andersen, A. and Knudsen, O. (1993) Incidence of cancer among welders of mild steel and other shipyard workers, *Br. J. Ind. Med.*, **50**, 1097–1103.
33. Jockel, K.-H., Ahrens, W. and Bulm-Audurff, U. (1994) Lung cancer risk and welding – Preliminary results from on-going case-control study, *Am. J. Ind. Med.*, **25**, 805–812.
34. Comba, P., Barbieri, P.G., Battista, G., Belli, S., Ponterio, F., Zanetti, D. *et al.* (1992) Cancer of the nose and para nasal sinuses in the metal industry: a case control study, *Br. J. Ind. Med.*, **49**, 193–196.
35. Ahlman, K., Koskela, R.S., Kuikka, P., Koponen, M. and Annanmaki, M. (1991) Mortality among sulfuric ore miners, *Am. J. Ind. Med.*, **19**, 603–617.
36. Ebihara, I., Shinokawa, E., Kawami, M. and Kurosawa, T. (1991) A retrospective cohort mortality study of copper miners, *J. Sci. Labour*, **67**, 7–13.
37. Hodson, J.T. and Jones, R.D. (1990) Mortality of a cohort of tin miners 1941–86, *Br. J. Ind. Med.*, **47**, 665–676.
38. Schuttmann, W. (1993) Schneeberg Lung Disease and uranium mining in the Saxon ore Mountains (Erzebirge), *Am. J. Ind. Med.*, **23**, 355–368.
39. Chen, R., Wei, L. and Huang, H. (1993) Mortality from lung cancer among copper miners, *Br. J. Ind. Med.*, **50**, 505–509.
40. Kabir, H. and Bilgi, C. (1993) Ontario Gold Miners with Lung Cancer, *J. Occup. Med.*, **35**, 1203–1207.

41. Tomasek, T., Swerdlow, A.J., Darby, S.C., Placek, V. and Kunz, E. (1994) Mortality in uranium miners in West Bohemia: a long term cohort study, *Occup. Environ. Med.*, **51**, 308–315.
42. Icso, J., Szollosova, H. and Sorahan, T. (1994) Lung cancer among iron ore miners in east Slovakia: a case control study, *Occup. Environ. Med.*, **51**, 642–643.
43. Soskolne, C.L., Jhangri, G.S., Siemiatycki, J., Lakhani, R., Dewar, R., Birch, J.D. et al. (1992) Occupational exposure to sulfuric acid in Southern Ontario, Canada, in association with laryngeal cancer, *Scand. J. Work Environ. Health*, **18**, 225–232.
44. Saracci, R. (1991) Beryllium and Lung Cancer: adding another piece to the puzzle of Epidemiologic Evidence, *J. Natl. Cancer Inst.*, **83**, 1362–1363.
45. Newman, L.S. (1992) Beryllium. Chapter in 'Hazardous Materials Toxicology, Clinical principles of Environmental Health', Sullivan, J.B., Kreiger, G.R., eds., Williams and Wilkins, Baltimore, MD, pp. 882–890.
46. International Agency For Research Against Cancer (1987) 'IARC Monograph on the Evaluation of Carcinogenic Risks to Humans. Overall Evaluation of Carcinogenicity: An updating of IARC Monographs' Volumes 1 to 42, IARC, Lyon.
47. Steenland, K. and Ward, E. (1991) Lung cancer incidence among patients with Beryllium disease: a cohort mortality study, *J. Natl. Cancer Inst.*, **83**, 1380–1385.
48. Ward, E., Okun, A., Ruder, A., Fingerhut, M. and Steenland, K. (1992) A mortality study of Workers at Seven Beryllium Processing Plants, *Am. J. Ind. Med.*, **22**, 885–904.
49. Eisenbud, M. (1993) Lung Cancer incidence among patients with Beryllium disease, *J. Natl. Cancer Inst.*, **85**, 1697–1698.
50. International Agency For Research On Cancer (1993) 'IARC Monographs on the evaluation of the Carcinogenic Risks to Humans. Beryllium, Cadmium, Mercury and exposures in the Glass Manufacturing Industry', Vol. 58, IARC, Lyon.
51. Macmahon, B. (1994) The epidemiological evidence on the carcinogenicity of Beryllium to Humans, *J. Occup. Med.*, **36**, 15–24.
52. Kipling, M.D. and Waterhouse, J.A.H. (1967) Cadmium and prostate carcinoma, *Lancet*, (i), 730–731.
53. Mateo, M.E.M., Rabadan, J. and Boustamante, J. (1990) Comparative Analysis of Certain Metals and Tumour Makers in Bronchopulmonary Cancer and Colorectal cancers, *Clin. Physiol. Biochem.*, **8**, 261–266.
54. Waldron, H.A. and Scott, A. (1994) Metals. Chapter in 'Hunters Diseases of Occupation', 8th edn., Raffle, P.A.B., Adams, P.H., Baxter, P.J., Lee, W.R., eds., Edward Arnold, London, Vol. 4, 90–138.
55. Armstrong, B.G. and Kazantzis, G. (1983) The Mortality of Cadmium Workers, *Lancet*, (i), 1424–1427.
56. Sorahan, T. and Waterhouse, J.A.H. (1983) Mortality study of nickel–cadmium battery workers by the method of regression analysis in Life tables, *Br. J. Occup. Med.*, **40**, 293–300.
57. Elinder, C.-S., Kjellstrom, T., Hogstedt, C., Andersson, K. and Spang G. (1985) Cancer mortality of cadmium workers, *Br. J. Int. Med.*, **42**, 651–655.
58. Stayner, L., Smith, R., Thun, M., Schnorr, T. and Lemen, R. (1992) A dose response analysis and quantitative assessment of lung cancer risk and occupational cadmium exposure, *Ann. Epidemiol.*, **2**, 177–194.
59. Lamm, S.H., Parkinson, M., Anderson, M. and Taylor, W. (1992) Determinants of lung cancer risk among cadmium exposed workers, *Ann. Epidemiol.*, **2**, 195–211.
60. Doll, R. (1992) Is cadmium a human carcinogen?, *Ann. Epidemiol.*, **2**, 335–337.
61. Sorahan, T. and Lancashire, R. (1994) Lung cancer findings from the NIOSH study of United States cadmium recovery workers: a cautionary note, *Occup. Environ. Med.*, **51**, 139–140.

62. Langard, S. (1990) One hundred years of chromium and cancer: A review of epidemiological evidence and selected case reports, *Am. J. Ind. Med.*, **17**, 189–215.
63. Lees, P.S.J. (1991) Chromium and Disease: Review of Epidemiological Studies with particular reference to Etiologic Information provided by Measures of Exposure, *Environ. Health Perspect.*, **92**, 93–104.
64. International Agency For Research On Cancer (1990) 'IARC Monographs on the evaluation of Carcinogenic Risks to Humans. Chromium, Nickel and Welding'. Vol. 49, IARC, Lyon.
65. Horiguchi, S., Morinaga, K., Endo, G. (1990) Epidemiological Study of Mortality from cancer among chromium platers, *Asia-Pacific J. Public Health*, **4**, 169–174.
66. Takahashi, K. and Okubo, T. (1990) A prospective Cohort Study of Chrome Plating Workers in Japan, *Arch. Environ. Health*, **45**, 107–111.
67. Kano, K., Horikawa, M., Utsunomiya, T., Tati, M., Satoh, K. and Yamaguchi, S. (1993) Lung cancer mortality among a cohort of male chromate pigment workers in Japan, *Int. J. Epidemiol.*, **22**, 16–22.
68. Davies, J.M., Easton, D.F. and Bidstrup, P.L. (1991) Mortality from respiratory cancer and other causes in United Kingdom chromate production workers, *Br. J. Int. Med.*, **48**, 299–313.
69. Pastides, H., Austin, R., Lemeshow, S., Klar, J. and Mundt, K.A. (1994) A retrospective – cohort study of occupational exposure to hexavalent chromium, *Am. J. Int. Med.*, **25**, 663–675.
70. Gao, M., Levy, L.S., Faux, S.P., Aw, T.C., Braithwaite, R.A. and Brown, S.A. (1994) Use of molecular epidemiological technique in a pilot study on workers exposed to chromium, *Occup. Environ. Med.*, **51**, 663–668.
71. Fu, H. and Boffetta, P. (1995) Cancer and occupational exposure to inorganic lead compounds: a meta-analysis of published data, *Occup. Environ. Med.*, **52**, 73–81.
72. International Committee Of Nickel Carcinogenesis In Man (1990) Report of the committee, *Scand. J. Work Environ. Health*, **16** (Suppl), 1–82.
73. Verma, D.K., Julian, J.A., Roberts, R.S., Muir, D.C.F., Jadon, N. and Shaw, D.S. (1992) Polycyclic Aromatic Hydrocarbons (PAHs): A possible cause of lung cancer mortality among nickel/copper smelter and refining workers, *Am. Ind. Hyg. Assoc. J.*, **53**, 317–324.
74. Shannon, H.S., Walsh, C., Jadon, N., Julian, J.A., Weglo, J.K. and Thornhill, P.G. et al. (1991) Mortality of 11,500 Nickel Workers – extended follow up and relationship to environmental conditions, *Toxicol. Ind. Health*, **7**, 277–293.
75. Egedahl, R.D., Coppock, E. and Homik, R. (1991) Mortality experience at a Hydrometallurgical Nickel refining in Fort Saskatchewen, Alberta between 1954 and 1984, *Occup. Med.*, **41**, 29–33.
76. Easton, D.F., Peto, J., Morgan, L.G., Metcalfe, L.P., Usher, V. and Doll, R. (1992) Respiratory Cancer in Welsh Nickel refiners: which nickel compounds are responsible? In, 'Nickel and Human Health: Current Perspectives' Nieboer, E., Nriagu, J.O., eds., Wiley, New York, pp. 603–619.
77. Andersen, A. (1992) Recent follow up of nickel refining workers in Norway and respiratory cancer. In, 'Nickel and Human Health: Current Perspectives,' Nieboer, E., Nriagu, J.O., eds., Wiley, New York, pp. 621–627.
78. Roberts, R.S., Julian, J.A., Jadon, N. and Muir, D.C.F. (1992) Cancer Mortality in Ontario Nickel Workers: 1950–1984. In, 'Nickel and Human Health: Current Perspectives' Nieboer, E., Nriagu, J.O., eds., Wiley, New York, pp. 629–648.
79. Goldberg, M., Goldberg, P., Leclerc, A., Chastang, J.F., Marne, M.J., Gueziec, J. et al. (1992) A seven year survey of respiratory cancers among nickel workers in New

Caledonia (1978–1984). In, 'Nickel and Human Health: Current Perspectives', Nieboer, E., Nriagu, J.O., eds., Wiley, New York, pp. 649–657.
80. Karjalainen, S., Kerttula, R. and Pukkala, E. (1992) Cancer risk among workers at a copper/nickel smelter and nickel refining in Finland, *Int. Arch. Occup. Environ. Health*, **63**, 547–551.
81. Langard, S. (1994) Nickel-related cancers in welders, *Sci. Total Environ.*, **148**, 303–309.
82. Muir, D.C.F., Jadon, N., Julian, J.A. and Roberts, R.S. (1994) Cancer of the respiratory tract in nickel sinter plant workers: effect of removal from sinter plant exposure, *Occup. Environ. Med.*, **51**, 19–22.
83. Berlin, A., Draper, M.H., Krug, E., Roi, R. and Van Der Venne, M.Th., eds (1990) 'The Toxicity of Chemicals' Series 1, Carcinogenicty, Vol. III, Summary Reviews of the Scientific Evidence, Commission Of The European Communities, Luxembourg, pp. 3–52.

A Re-assessment of Respiratory Cancers at the Clydach Nickel Refinery: New Evidence of Causation

M.H. DRAPER

EDINBURGH CENTRE FOR TOXICOLOGY, HERIOT-WATT UNIVERSITY, EDINBURGH, UK

1 Introduction

The investigations reported here were undertaken in 1990 in order to resolve some of the controversial issues that had arisen about the evaluations of nickel and simple nickel inorganic compounds as carcinogens. At that time there were similar concerns about other metals such as cadmium, cobalt, beryllium and chromium. However, it was believed that if the nickel problem was re-examined in-depth the new insights expected to arise could be applied to other metals where similar extrapolations from mixed exposure situations had been made. The focus of the re-examination was to address the situation that at one refinery in particular, Clydach in Wales, UK, there had been an astonishingly potent carcinogen in the environment, yet the extensive use of its products in industry appeared to offer little risk from cancer. There was an additional factor that aroused concern at this refinery, in that there was a well-documented presence of up to 10% of arsenic in some of the process materials that had never been satisfactorily accounted for. In the event, the study has continued for over five years because of the mass of data that has been revealed, including for example some 8000 years of detailed work histories of the 360 cases of lung and nasal cancers that have been attributed to exposure to the environment in the refinery complex.

In 1903 the Mond Nickel Refinery at Clydach in Wales, UK (now Clydach Refinery, INCO Europe Limited) was commissioned and continues to this day refining nickel by the Mond carbonyl process. Apart from the production of nickel metal and powder, throughout the years large quantities of a wide variety of simple nickel inorganic compounds have been produced such as oxides and salts, particularly nickel sulfate. The remarkable magnitude of the risks, especially for nasal cancers during the first three decades of operation of the refinery can be seen from the figures calculated by the ICNCM in 1989,[1] the salient features of which are set out in Table 1. Here it can be seen that in a population that never exceeded some 800 process workers, the standard mortality ratio (SMR) for nasal cancers in the early decades was estimated to exceed 30 000, but by the fourth decade it was effectively zero.

Table 1 *Standard mortality ratios (SMRs) for respiratory cancers at Clydach nickel refinery between 1903 and 1949*

Recruitment Years	Lung Cancers		Nasal Cancers	
	SMR	95% CI	SMR	95% CI
Before 1920	550	440 – 680	34,000	26,000 – 44,000
1920 – 1929	300	240 – 380	7,135	3,700 – 12,500
1930 – 1939	154	97 – 230	1,440	36 – 8,000
1940 – 1949	130	71 – 220	0	0 – 6,000

Adapted from Table 17 of the ICNCM report (1990).[1] The figures have mostly been rounded as the purpose of the table is to illustrate the magnitude of the SMRs and their decline through the decades.

The study of the first decades of operation of the Clydach refinery has revealed a surprisingly complex situation involving the feedstock supply, the work force and external factors, such as the influence of the First World War and the subsequent depression. A comparison of the times when subjects entered the evolving environment of the refinery, *i.e.* their recruitment times, and the fluctuations in feedstock supplies and their type, and the manpower required to operate the refinery at any one time has revealed some surprising discrepancies. Understanding the reasons for these discrepancies, together with the strong association of the occurrence of the cancers with exposures to process materials with a high arsenic content, has cast doubt on the accepted view that nickel and its simple inorganic derivatives were the causative agents of the respiratory cancers that occurred at Clydach.

In the course of this prolonged investigation many disciplines were called upon to help unravel this extremely complex web of factors that had some bearing on the understanding of why this cancer 'epidemic' arose, why it petered out after 1924, and what were the likely extremely toxic molecules in the environment that had such specific target areas as the ethmoid region in the nose and the bronchi in the lung, and why squamous carcinoma occurred and not adenocarcinoma. In the event, a molecule was identified, but what was more important for the future understanding of the possible toxic properties of other refinery environments was the specificity of the method of carriage to the target cells. This directs attention to the surface receptors of the target cells and how their function could be interfered with to produce eventual malignancy.

The systematic multidisciplinary approach to the problems at Clydach has been so successful that the thought has arisen that this approach could be generalized into a model for the investigation of other environments in the smelting and refining industries where the presence of carcinogens is suspected. To designate such a co-ordinated multidisciplinary approach, the term 'metademography' is suggested.

2 The Occurrence of Nasal and Lung Cancers at the Clydach Nickel Refinery Complex 1923 – 1990

In 1923 two workers died, one of an 'undifferentiated carcinoma' with 'nasal obstruction and epiphora' and the other of a 'squamous carcinoma' in the 'frontal sinus'. Their ages at death were, respectively, 50.3 and 36.7 years. By 1927 there were three more similar cases, aged 43.2, 48.7 and 58.1 years, and one 'carcinoma of the left lung' ('large mass extending from the left hilum') who died aged 44.0 years. Because nasal cancer is such a rare disease, suspicions were aroused, which were confirmed in a report from the Chief Inspector of Factories and Workshops (HMSO, 1933), and confirmation came in 1939 in a report prepared for the company by Bradford Hill that demonstrated that there was not only an excess of nasal cancers, but also an excess of lung cancers.[2] During the next 50 or so years there were to be, in all, 85 deaths from somewhat characteristic carcinomas of the upper nasal regions, predominantly ethmoidal, and 280 deaths from carcinomas of the lung, predominantly squamous carcinomas of the bronchi, associated with employment at the Clydach nickel refinery complex.

Figure 1 is a histogram showing the number of deaths from nasal cancer for each year from 1900 to 1990, and it can be seen that the first deaths occurred in 1923. Figure 2 shows the same for the lung cancer deaths. Here the first death occurred in 1927. A comparison of the two figures shows that from 1923 to 1944 the numbers of the nasal and lung cancer deaths were approximately in step, with a slight excess of lung cancers. Thereafter, the lung cancer deaths predominate. Figure 3 combines the nasal and lung cancer deaths, hereafter designated respiratory cancers for convenience, to show the envelope of the respiratory cancer 'epidemic' that occurred from 1923. It can be seen that there was an increase in deaths from 1923 to 1940, plateauing at about ten deaths per year for the next 25 years, and thereafter declining to an average of about three deaths per year, predominantly lung cancers. The first question that arises with any obvious

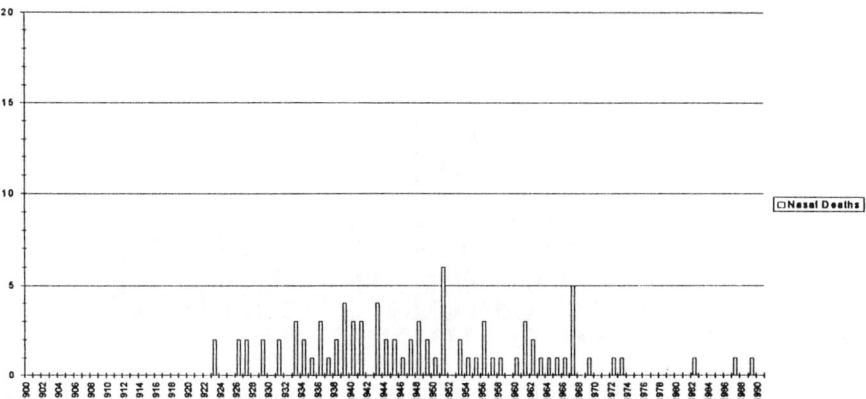

Figure 1 *Histogram of the number of deaths from nasal cancer each year among the process workers at the Clydach nickel refinery complex 1923 to 1989. x-axis: calendar year, 1900 to 1990; y-axis: number of deaths per year.*

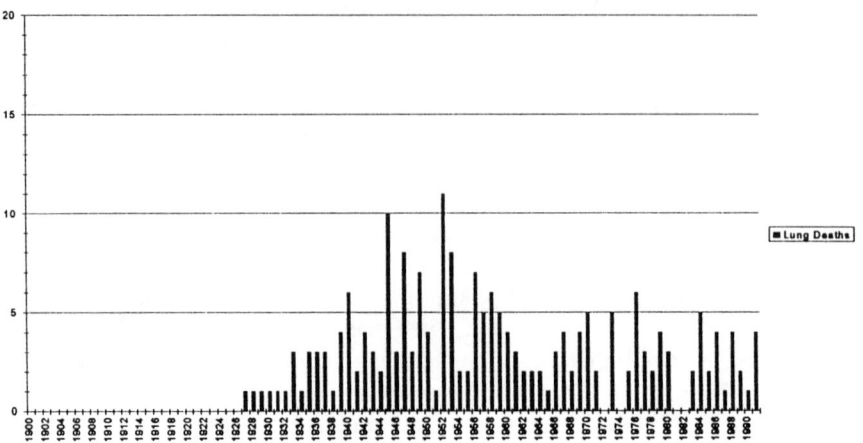

Figure 2 Histogram of the number of deaths from lung cancer each year among the process workers at the Clydach nickel refinery complex 1927 to 1991. x-axis: calendar year, 1900 to 1992; y-axis: number of deaths per year. The numbers plotted are from the reduced cohort of 207 'true' Clydach cases.

relationship with cancer and a working environment is that of duration, that is when did the workers enter the environment, how long did they remain and how long did they survive?

Figures 4, 5 and 6, are histograms of the annual recruitment numbers for the nasal cancer cases, the lung cancer cases and the two combined, respectively. These figures reveal a remarkable state of affairs in that only those workers who

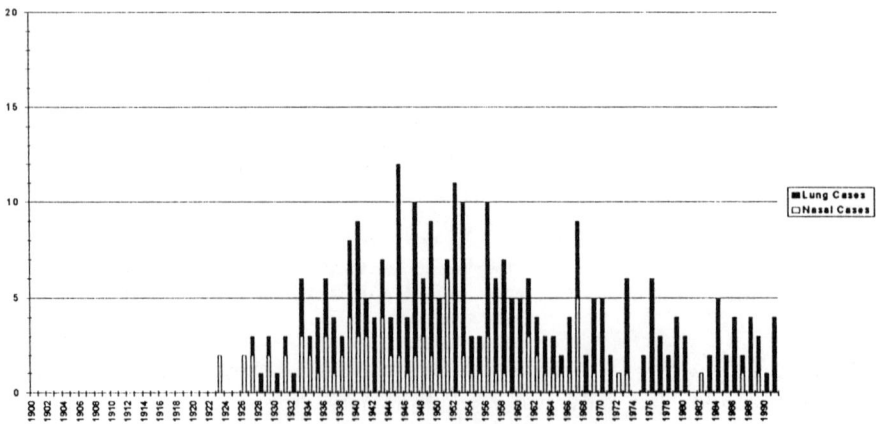

Figure 3 Histogram of the number of deaths from nasal and lung cancer each year among the process workers at the Clydach nickel refinery complex 1923 to 1991. x-axis: calendar year, 1900 to 1992; y-axis: number of deaths per year. Nasal cancers: unfilled lower part of columns; lung cancers: filled upper portion of column.

Figure 4 *Histogram of the recruitment per year of the nasal cancer cases among the process workers at the Clydach nickel refinery complex, 1903 to 1964. x-axis: calendar year, 1900 to 1992; y-axis: number recruited each calendar year. Note the decreased number of cases that were recruited between 1915 and 1921.*

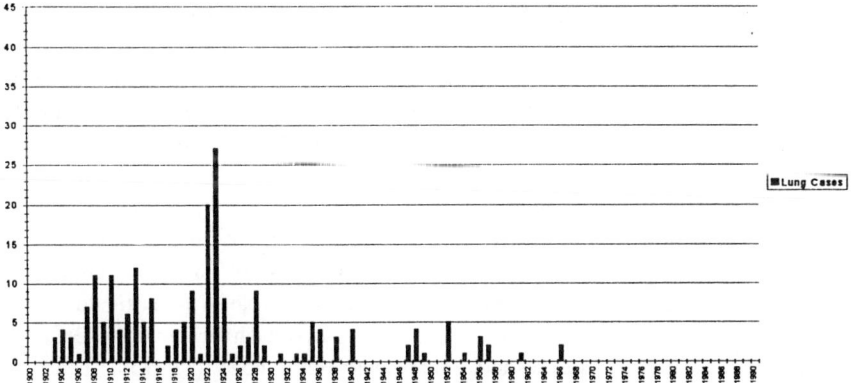

Figure 5 *Histogram of the recruitment per year of the lung cancer cases among the process workers at the Clydach nickel refinery complex, 1903 to 1966. x-axis: calendar year, 1900 to 1992; y-axis: number recruited each calendar year. The numbers plotted are from the reduced cohort of 207 'true' Clydach cases.*

started at Clydach before 1928 were at high risk; indeed, for nasal cancer the risk declined sharply after 1924. Reference back to Figure 1 shows that the first cancer deaths occurred after only two decades of nickel production. It can also be appreciated that up until 1955 it would have been difficult to establish the 1924/28 cut-off years. This possibility was first clearly pointed out in 1958.[2,3] One of these authors (Morgan) was the company's chief medical officer. Thus it becomes of considerable importance to try to establish what was so different during those early years of operation, because the basic processes involved did not change

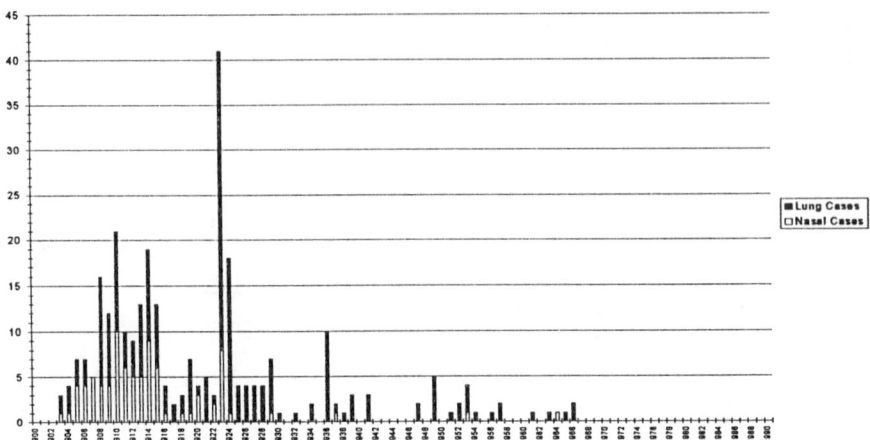

Figure 6 *Figures 4 and 5 combined to give a histogram of the recruitment per year of the nasal and lung cancer cases among the process workers at the Clydach nickel refinery complex, 1903 to 1964. x-axis: calendar year, 1900 to 1992; y-axis: number recruited each calendar year. Note that as in the nasal cancer cases there was a decrease in the number of cases that were recruited for a few years after 1915.*

substantially until the years 1932 to 1936. After this there were considerable changes to the initial processes, but the output of nickel mass metal continued and a new product, nickel metal powder was introduced, as well as a great increase in the output of nickel oxides and a variety of nickel salts for commercial use. Furthermore, all these changes were instituted without any notable attention, particularly during the World War II years, to matters of industrial hygiene.

Thus far the availability of two essential sets of information concerning the cases have been discussed, namely the deaths per year from respiratory cancers and their corresponding recruitment years. Unfortunately, the third set of essential information, the figures detailing the total numbers of workers and their dispositions for each year, are no longer available. Nevertheless, important relevant factors that bear on the size of the workforce each year can be deduced by a careful dissection of the refinery operations. Here, a key factor has been the availability of the records of the annual feedstock supplies.

The only nickel-containing feedstock from the commencement of refining operations in 1903 until 1930 was bessemer matte imported from Conniston in Canada. This sulfidic feed was then phased out to be replaced by low sulfur oxidic feeds. Figure 7 shows the annual intake of Canadian feedstocks from 1902 to 1960. It can be seen that the intake of bessemer matte increased from some 100 tonnes in 1903, peaking at some 27 500 tonnes in 1929 and falling to zero after 1934. From 1930 to 1936 there were considerable changes in the process, to be discussed later, and by 1940 the use of an oxidic sinter, known as SRS/PRS, peaked at 31 000 tonnes. The refining processes were again modified during the immediate post-World War II years to peak at some 35 000 tonnes in the 1950s with an improved feedstock known as R&S sinter. The figures for bessemer matte are replotted as a histogram in the lower part of the composite Figure 8, which for

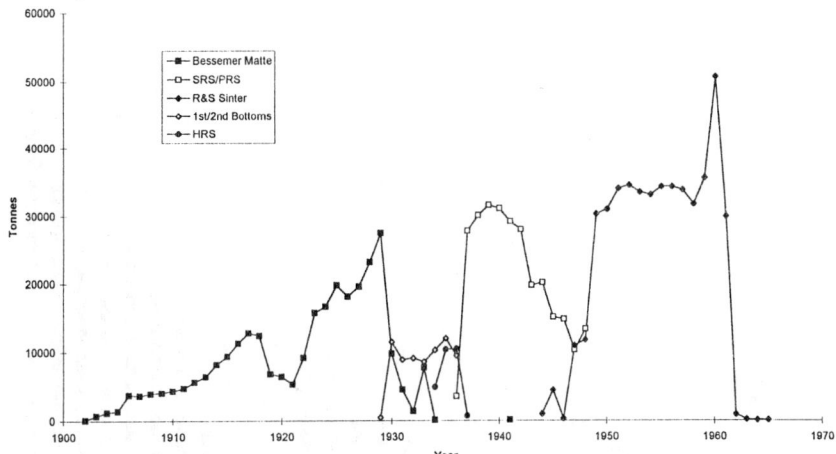

Figure 7 *A plot of the annual intake of various feedstocks from 1902 to 1965. x-axis: calendar year, 1900 to 1970; y-axis: tonnes of feed stock. The reduction in intakes associated with the ending and immediate postwar periods of WWs I and II should be noted. The period from 1930 to 1936 was when the leaching of copper from sulfidic bessemer matte by sulfuric acid was phased out, to be replaced by low copper oxidic feedstocks. The various feedstocks were known by short codes that refer to the technical aspects of their production. Thus SRS/PRS refers to a mixture of soft and partially roasted sinter, HRS refers to hard roasted sinter, bottoms refer to a feedstock prepared by Orford furnaces followed by very slow cooling and subsequent splitting off of the nickel rich lower part of the ingot.*

the critical period 1903 to 1930 sets out the relationship between feedstock levels and workforce establishment. In addition significant plant developments are indicated, to be discussed later, such as the times when each additional refining plant was commissioned, commencing in 1903, then 1907, 1911, 1916, 1918, 1928 and finally the seventh plant in 1930. In 1905 nickel sulfate production commenced, in 1907 the furnacing processes were commissioned and the linear calciners were considerably modified in 1911 and 1924 before being successively phased out between 1932 and 1936. As can be seen, each of these alterations was associated with changes in feedstock tonnages and manpower requirements.

Until 1936, the removal of copper from the bessemer matte feedstock was carried out at Clydach, whereas subsequently the copper was largely removed in Canada and the resulting sinters were shipped to Clydach. The copper was extracted from calcined matte by treatment with dilute sulfuric acid and the resultant solution of copper sulfate was processed to crystals and marketed. As bessemer matte contained about 35% of copper it can be appreciated that sulfuric acid was a major feedstock. Thus for every 1000 tonnes of bessemer matte, some 550 tonnes of commercial chamber sulfuric acid (75% acid) were imported. Unfortunately, this chamber acid was derived from iron pyrites and was known to be contaminated with arsenic, but it was cheap, and at the time no commercial disadvantage was apparent because of its presence, despite the fact that the presence of arsenic decreased the efficiency of the nickel extraction. A final important raw material with health implications was the tens of thousands of

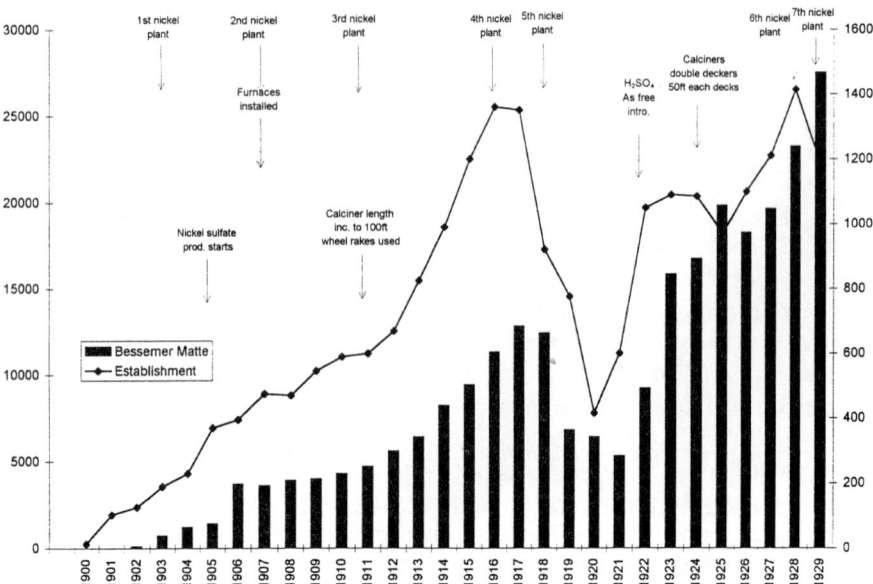

Figure 8 *A composite figure to show the relationship between annual feedstock input, set out as a histogram, and refinery annual establishment requirements plotted above. x-axis: calendar years, 1900 to 1929; y-axis, RHS: numbers of hourly paid workers; LHS: tonnes of bessemer matte imported each year. In addition, the years when a number of important refinery plant changes were made are indicated. The great influence of WW I and its economic aftermath is evident, as are the influences of plant change on the rates of increase in bessemer matte import.*

tonnes of high quality coal required to furnish the water and producer gases, needed for the supply of hydrogen for the reduction of nickel oxide and the carbon monoxide for the production of nickel carbonyl, as well as for general heating purposes throughout the plant.

3 Demographic Consideration

The above are the material facts of the complex of refining plants at Clydach. The next consideration is to describe the development and disposition of the workforce required to operate the processes leading to the production of nickel metal, nickel salts, copper sulfate and other products, such as precious metals. This is a surprisingly complex matter and to understand some of the unusual features associated with the workforce at the refinery, it is necessary to consider some of the demographic features of the industrial population in this part of Wales, before considering what process tasks were called for in the refinery operations.

At the turn of the century, South Wales was the greatest coal exporting region in the world. Prior to this the uplands north of the coal fields had been the greatest iron-producing region in the world. With the development of the South Wales ports to serve the enormous coal exporting industry, and in consequence the easy availability of high class coal, there developed around these ports, particularly Swansea, great metallurgical industries for the production of steel and the refining of copper, zinc and tin. The region also became the world's leading producer of tin plate and galvanized iron. Thus it was no surprise that when Ludwig Mond and Carl Langer developed a commercially feasible method of producing metallic nickel from a Canadian nickel–copper sulfide complex (bessemer matte), based on their discovery of the formation and easy decomposition of volatile nickel carbonyl, they decided to establish their refinery operations in South Wales. In 1899 a site of 33 acres was purchased at Clydach in the Swansea valley, about 9 km north-east of Swansea, and here under the direction of Carl Langer, a chemical engineering genius, the design and construction of the first refining plant was carried out. It was commissioned in late 1902.

The peak of South Wales coal, metal refining and coated steel production occurred in 1913/14 when the onset of World War I not only virtually eliminated export trade, but forced nations to develop new sources. Thus, after the War, South Wales not only had lost overseas markets, but, with the rest of the UK (and most of the world), entered a lengthy period of economic depression. Although the unique, high purity of the nickel produced by the carbonyl process ensured its value and desirability in world markets, as can be seen from Figure 7, these world events were faithfully reflected by decreased imports of feedstocks, and hence a considerable reduction of the workforce with little, if any, recruitment and fewer workers subsequently at risk.

Apart from these world trade influences during the 1920s and 1930s there is another possibly unique aspect about the workforce at Clydach that is of considerable importance in the interpretation of exposure data, and that is the simple fact that working at the Clydach refinery during the early years of the century could be exceedingly unpleasant, and a large number of recruits refused to 'stick it' for more than six or so months. Thus, as will be discussed, in some years some 500 might present for employment, but only a hundred or so might remain, even for one year.

With the above background in mind the manpower disposition at Clydach can be seen to be complex. Furthermore, a complete understanding is no longer possible as the Company's general attendance records were destroyed some years ago. Fortunately, the work records of the respiratory cancer cases were preserved. Figure 8 sets out, in the plot above the bessemer matte import histogram, what is known, namely the establishment figures for each year; that is, the number of workers required each year to operate the refinery for the critical period from 1900 to 1930.[3] Starting at 13 in 1900 it can be seen that there was a steady growth in the number of workers required as the number of separate nickel plants was increased, so that by 1916 when nickel plant 4 was commissioned the establishment required was 1350 workers. However, it can be seen that thereafter in the difficult years of the final stages of World War I and the first few years of peace, the years from 1917 to 1920 saw a reduction in the required workforce from 1 350 to some 400. The next decade saw an increase to nearly 1100 by 1924 followed by a fall off to 950 in 1926, the year of the General Strike, and it was not until 1929 that the establishment reached the levels of 1916/17.

A comparison with the annual bessemer matte imports from Canada, plotted in Figure 8, shows in general a parallel with the workforce requirements, reflecting the difficult economic times of the postwar years, particularly in Wales where there was a great retrenchment in the coal industry. What is surprising is the relationship up until 1916 of the greatly expanding production of nickel and other products with the recruitment of those workmen destined to succumb to respiratory cancers. Reference back to Figure 6 shows that from 1903 to 1915 there were on average about 15 future respiratory cancer cases recruited per year, despite the increase in the establishment from 200 to 1200. From 1916 to 1921 the number of future cases entering the refinery actually declined to about five per year, followed by a substantial increase in 1922/23. The majority of these latter recruits in fact succumbed to lung cancer, and in this connection it needs to be remembered that it was after World War I that cigarette smoking began to increase greatly.

The reason for this lack of association between future cases and the increases in establishment requirements lies in the fact referred to above, that working at the Clydach refinery was extremely unpleasant. In the discussion about the workforce so far, care has been taken to refer to the establishment, *i.e.* the managerial decision about how many workmen can be employed at any given time. However, the implementation of these decisions depends on workers actually turning up to be employed and, in the case of Clydach, having been recruited, remaining at the refinery. The scale of the turnover problem can be appreciated from the statistics that from 1900 to 1929, 5400 workers were recruited but 3310 left after less than one year's experience and a further 1240 left with less than 10 years of employment, leaving some 850 who 'stuck it out' for 10 years or more. As will be discussed, an understanding of this remarkably high turnover of workers has important implications for the interpretation of the recruitment pattern of those who subsequently became respiratory cancer cases. Figure 9 sets out a histogram of the year by year pattern of the rapid turnover of workers, the empty columns represent the actual numbers required to meet the establishment quota set each year by the management and the filled columns indicate the number of recruits that presented, but quit before the year was over.

The significance of these workforce turnover figures can be appreciated by considering the changes from 1911 to 1921. From Figure 8 it can be seen that the period from 1911 to 1917/18 was a period of rapid expansion, bessemer matte import increased from some 4000 tonnes to 12 000 tonnes with a corresponding increase in the established work force from 600 to 1350. However, reference to Figure 9 shows, for example, that in the period January to December 1913 in order to meet the labour force requirements of an additional 150 men, some 220 men needed to be recruited during that year. The situation in the next year was, in this respect much worse, in that nearly 500 were recruited in order to maintain an increment of some 200 to meet the total establishment need of some 1200 for the year 1915. This state of affairs continued through the 1914–1918 war years and into the immediate postwar years with a notable change in that from 1916/17 until 1920 there was apparently no net recruitment. On the contrary, a considerable contraction of the established workforce occurred, yet Figure 9 shows that some

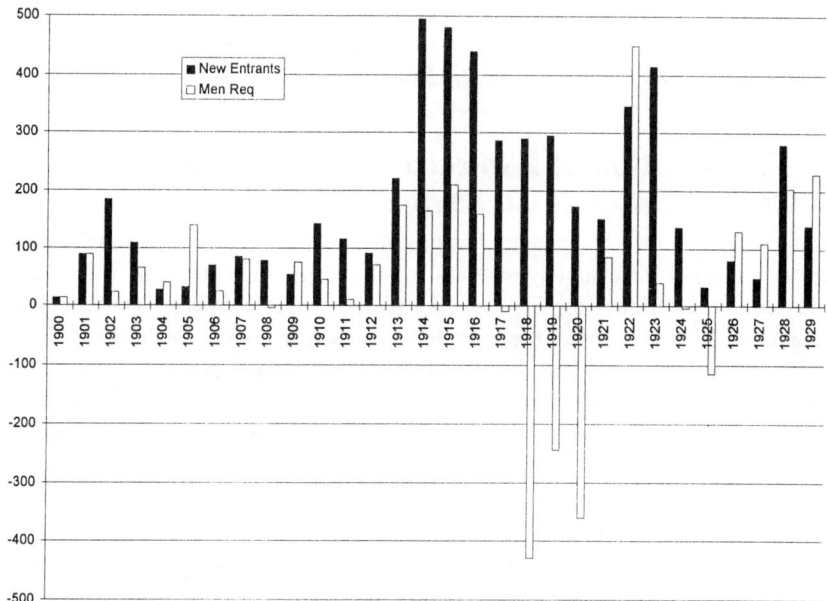

Figure 9 *A histogram to illustrate the great turnover of refinery workers each year because of the unpleasant nature of refinery conditions. X axis: calendar years, 1900 to 1929; Y axis; numbers of hourly paid workers. The filled columns represent the number of workers recruited each year, and the unfilled columns represent the number actually required. The difference between the columns represents the number of workers who left in that year because they were not prepared to put up with the conditions. Thus in 1914, for example, in order to maintain the establishment requirement of an additional 160 men, nearly 500 were eventually recruited. During WW I the refinery was maintained largely by casual labour and in the following depression, large numbers of workers were laid off. A more usual balance of recruitment was not seen until 1922.*

hundreds of people in fact must have been employed for short periods. Thus, during these very difficult years when, because of the appalling casualties on the Western Front, there were severe labour shortages, the Clydach refinery was virtually run with casual labour, including many female workers. This was probably inefficient but feasible, because in these early decades the production of nickel by the Mond process was particularly labour intensive, because of the manual labour needed for the physical transport of larger quantities of material by barrow around the complex of buildings making up each nickel plant. It is interesting to note that in the work records of the subjects about 25% of them had from two to five years of service in World War I.

The dip in the number of cases, particularly in the number of nasal cancers in men recruited from 1916 to 1921, as can be seen in Figures 4 and 5, can now be understood in that there was virtually no recruitment during these years. However, it remains surprising that in the years 1911 to 1916 when, as can be seen from Figure 8, actual output, as indicated by feedstock input, increased until 1918; there was no corresponding increase in case recruitment. It was not until 1921 that clear recovery from the immediate postwar economic decline became apparent and net recruitment increased. Correspondingly, there was a sharp increase in the number of lung cancer cases recruited in 1921/22 (*vide* Figure 5).

4 The Duration of Exposures of the Respiratory Cancer Cases at Clydach

Nasal cancer is a rare malignancy. In the early years of the century nasal cancers were not classified separately in national mortality statistics, and it has been estimated that the figure for England and Wales would have been about one case in eight years. In a study of the records of death held by the medical officers of health in the districts surrounding the refinery, eight cases of nasal cancer (unspecified) were found in 15 247 industrial workers with no known nickel refinery work, for the years 1948 to 1956.[2] However, it is likely that the particular variety that predominated among those exposed at Clydach may constitute an even rarer category of nasal cancer. Thus, there is little doubt that the nasal cancers that occurred among the workers at the Clydach nickel refinery arose from exposure to some agent or agents in the environment at the refinery. The same cannot be said for the lung cancer cases because such cancers are relatively common in the general population, as well as constituting a recognised hazard in a number of industrial situations. In addition, there is the important ubiquitous factor of cigarette smoking to consider. Figure 10 is a modification of Figure 5 in that the lung cancer cases with a known history of smoking are indicated. From Figures 4 and 5 it is apparent that throughout the period under consideration there were more lung cancer than nasal cases among the recruits. The relationship between the two is more clearly seen if the cumulative recruitments are considered, set out in Figure 11. Here it can be seen that up until 1915 there was a steady increase in future cases with the lung cancer cases increasing at a slightly greater rate, 78 cases to 64 (ratio 1.22). Thereafter until 1922, the incremental increase in lung cancer recruitment was much greater than nasal cancer recruitment, 94 cases to 69 (ratio 1.36). There followed a dramatic growth in recruitment of lung cancer cases and by 1930 there were 164 cases to 82 (ratio 2.0), strongly suggestive of some new factor emerging in the situation. During the next 40 years, of all the respiratory cancers attributed to exposure to the Clydach refinery environment, 116 were lung cancers and 3 were nasal cancers. For reasons that are outwith the scope of this paper, it is highly probable that of the 280 lung cancer cases attributed to employment at some time at Clydach about 65 were 'non-Clydach' cases, and in addition there is evidence to suggest that a small group of workers whose exposure was predominantly to the environment in the large gas production plant were specifically at risk. Thus, the apparent much greater occurrence of lung cancers (280 to 85, ratio of 3.3) may turn out, when studies are complete, to be much nearer the 1921 ratio of 1.4 for those whose exposures were almost exclusively to the nickel refining processes.

In view of the complexity of the lung cancer situation, a greater depth of analysis is required than can be detailed in the present paper. Thus the remainder of this exposition will concentrate on the nasal cancer cases where there can be

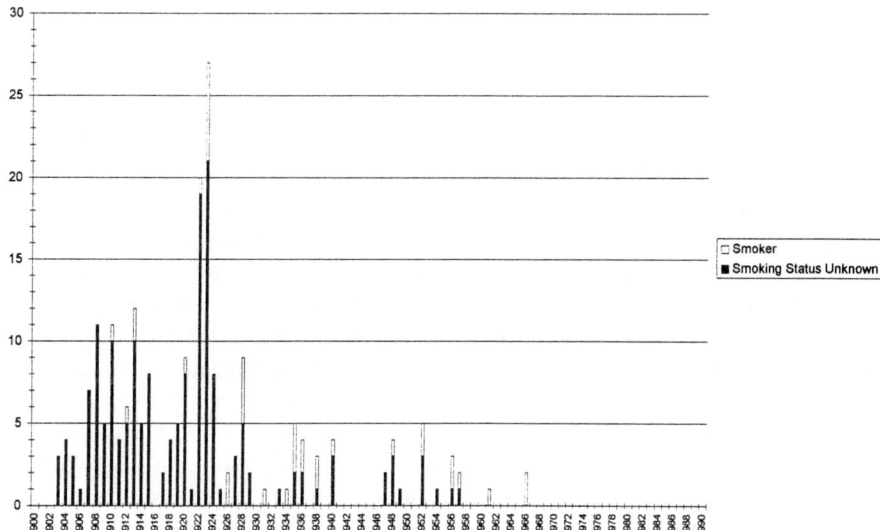

Figure 10 *A histogram illustrating what was known about the cigarette smoking habits of the lung cancer cases. x-axis: calendar years, 1900 to 1990; y-axis: numbers of hourly paid workers recruited each year. The filled portion of a column represent those workers whose smoking status were unknown, and the unfilled columns represent those who were known smokers. The workers plotted are those considered to be 'true' Clydach cases, i.e. the reduced cohort.*

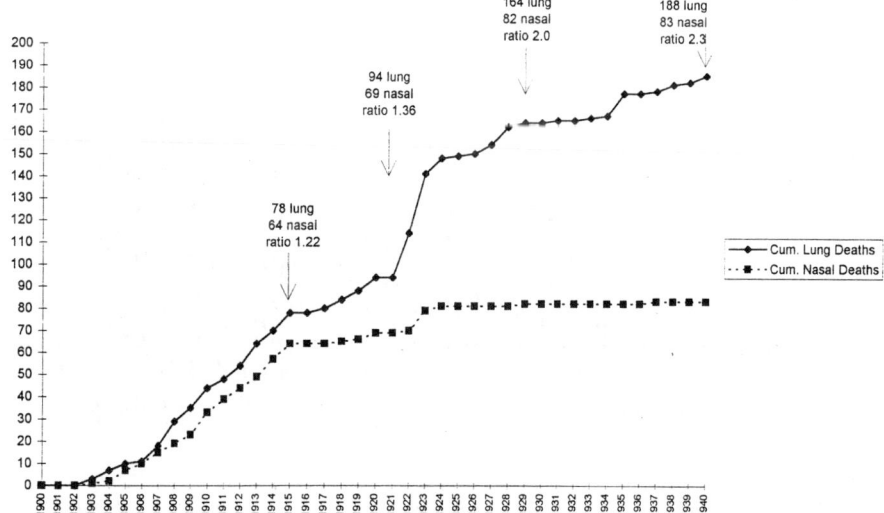

Figure 11 *Plot of the cumulative deaths of those recruited in the period 1900 to 1940. The continuous line and filled circles represent the lung cancer cases and the dotted line and filled squares represent the nasal cancer cases. x-axis: calendar years, 1900 to 1940; y-axis: numbers of hourly paid workers recruited each year who subsequently died of respiratory cancer. It can be seen that there is a marked divergence after 1921. The ratio of lung to nasal cancers indicated for various times in the figure can be seen to increase from 1.22 in 1915 to 2.26 in 1940. If all the lung cancer cases attributed to any exposures, however short, are included the final ratio becomes 3.29 (280/85). The interpretation of these ratio changes is complicated.*

little doubt that the source of the cancer induction was at Clydach. Figure 12 is a histogram of the employment durations (filled columns) and the time from cessation of work, usually retirement to pension, to death (blank columns), set out in order of employment duration. Recorded employment, particularly for those with many years of service, was complicated in that there were from 1914 to the late 1930s many logged periods of absence from the works. Some of these were voluntary, for example service in the armed forces; many were involuntary, brought about by temporary plant closures caused by the difficult economic times referred to above. However, in the works pension records such employees were 'kept on the books' and the bare record of date of start and date of going on pension thus give no indication of such absences. Working with the individual wage records that list year by year each successive job reveals these discrepancies, and Figure 13 is a histogram set out as in Figure 12 with the filled columns representing the actual time spent at work in the refinery, and the blank columns, as before, being the time from cessation of work at Clydach to death. The effect of the correction is clearly evident in the profile up to about 30 years of exposure. While the overall effect is small, for example the midpoint duration changes from

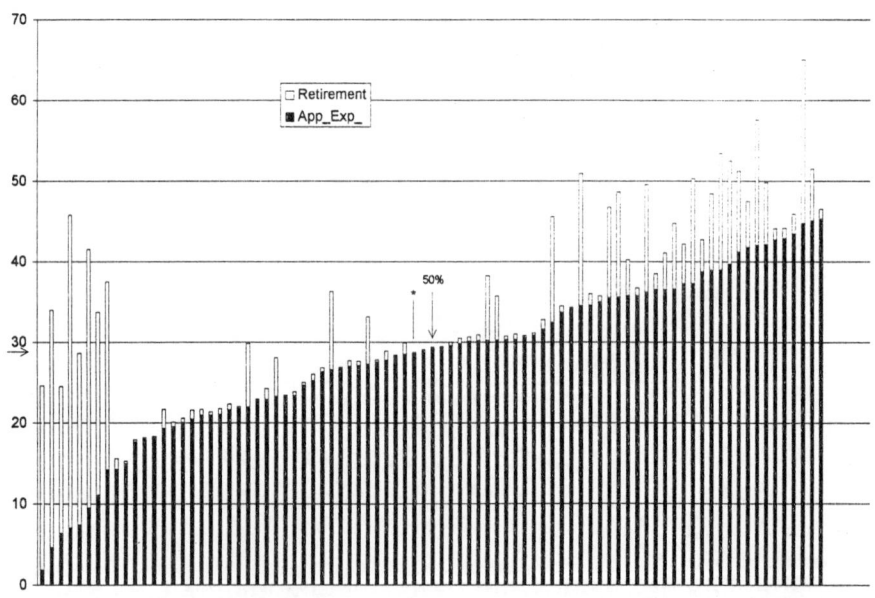

Figure 12 *A histogram setting out the 85 nasal cancer cases in order of their apparent exposure at the Clydach refinery. Apparent exposure is calculated by subtracting the date of recruitment from the date of going on pension or termination of employment for other reasons, because this figure could include long absences from the refinery, for example, war service and thus could be misleading. The asterisked case in the centre of the figure is an extreme example to be indicated in Figure 13. x-axis: cases set out in order of length of apparent service. The lower filled portion of a column indicates the calculated work period and the clear portion the period from cessation of work at Clydach to death. y-axis: years.*

29 to 26 years and the number of cases with 25 years or less of actual exposure increases from 29 to 38, the effect on the interpretation of an individual's particular exposure history can be important, particularly in those recruited in the years immediately before World War I.

A number of important general points can be seen in Figure 12. Consideration of the overall profile shows that there are three kinds of history. Firstly, there is a small group of cases with apparent exposures of 14 years or less (11 years true exposure from Figure 13) who nevertheless survived for two to three decades after leaving employment at Clydach before developing their nasal cancer. Secondly, there is a major group who developed their cancers and died 'in harness' as it were, after 15 to 30 or so years of exposure. Finally there is a group who retired, as will be discussed, at the usual age of 65 and who succumbed to nasal cancer some 10 or even 20 years later. It can also be seen from Figure 12 that from the commencement of employment to death could be as little as 15 years and that from cessation of work, effectively from diagnosis, in a large number of cases was a matter of months. Thus, it seems clear that whatever brought about these nasal cancers was a remarkably potent agent. Indeed, as will be seen, only five years of exposure could be sufficient to induce nasal malignancy.

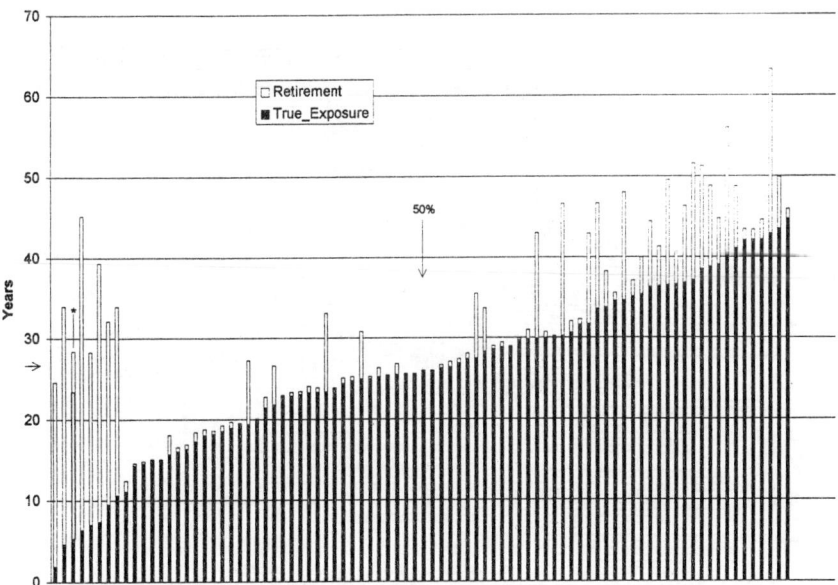

Figure 13 *This histogram is the same as in Figure 12, but the actual exposures at the refinery are plotted. The asterisked case shows that an apparent exposure of some 30 years in Figure 12 was in fact an exposure of about 5 years followed by an absence of some 20 years and then a return to work of about 4 years. Inspection of Figures 12 and 13 indicates that there are three groups of cases. There are those who worked for a short time, left and died many years later. Then there are the majority of 'characteristic' cases, 58 in all, who worked from 15 to 35 years and then died more or less in harness, usually well before the age of 60 years. The final group were those who worked for 30 to 45 years or so, retired and yet lived to an advanced age.*

Important features about the cluster of cases can be visualized by considering their occurrence in real time; firstly, in order of recruitment and then in order of death. Figure 14 sets out the 85 nasal cancer cases in real time in order of recruitment. Each horizontal rectangle represents a case, starting with a cross hatched section for his prework period, then a clear section for the period at Clydach and finally a filled section for the period from leaving Clydach to death. Reviewing the profile in the figure along the commencement of work points it can be seen that about 75% of the cases had five years or more exposure before 1920. It is also apparent that the earliest deaths tended to occur among the earliest recruits, and that there is a tendency for the later recruits to have longer lifespans. A final general observation to be made is that at all stages there were cases that survived for at least an expected lifespan.

Figure 15 sets out the cases in order of deaths. This arrangement demonstrates again that the earliest deaths (1923 to 1940) occurred predominantly among the earliest recruits. Furthermore, it is clear that, with the exception of four cases, all subjects had substantial exposure before 1930. This is in agreement with the view expressed by other authors that after 1930 there was a considerable reduction in risk, falling in later years to insignificant levels.[1–5] The implications of this are clear. A potent hazard existed in the Clydach environment with a maximum effect

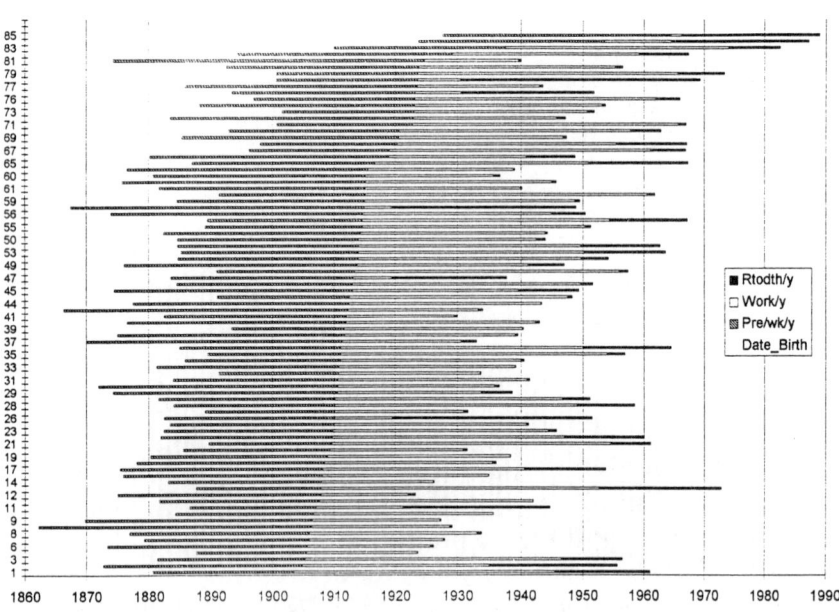

Figure 14 *This sets out in histogram format, in real time from 1860 to 1990, for each of the 85 nasal cancer cases: the time from birth to recruitment at the Clydach refinery (cross hatched portion of column); the apparent duration of service (unfilled portion) and finally the period from cessation of work to death (filled portion). The cases are arranged from below upwards in the order of the date of recruitment. It can be seen that some 75% of the cases had at least 5 years of exposure before 1920 and, apart from 5 cases, all had some 7 years of exposure before 1930, in fact 8 had died before 1930.*

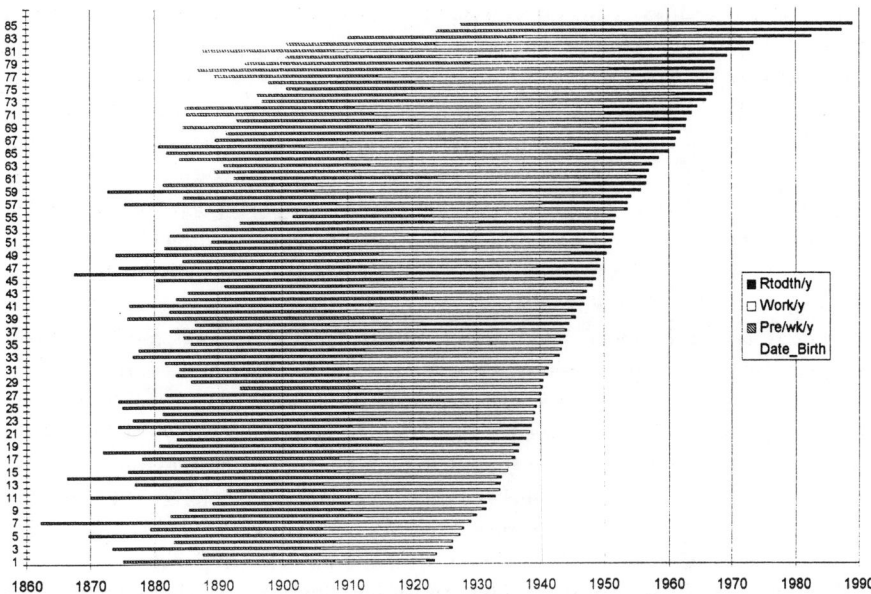

Figure 15 *As in Figure 14, this sets out in histogram format, in real time from 1860 to 1990, for each of the 85 nasal cancer cases: the time from birth to recruitment at the Clydach refinery (cross hatched portion of column); the apparent duration of service (unfilled portion) and finally the period from cessation of work to death (filled portion). The cases are arranged from below upwards in the order of the date of death. It can be seen that in the first 50% of cases most died shortly after cessation of work as was apparent in earlier figures. Furthermore, progressing from the earliest deaths it can be seen again (vide Figure 14) that there are increasing numbers of cases who were recruited later and who survive longer, many with long periods in retirement.*

only for some three decades from the commencement of refining. Furthermore, reference to Figures 6 and 8 shows that the incidence of cancers was not related to the mass flow of bessemer matte through the nickel plants, suggesting an origin from a relatively minor part of the process.

5 The Mond Nickel Refining Process

Before analysing the individual job sequences of the cases in order to identify specific parts of the operation that posed hazards to the workforce, it is necessary to establish an outline of the unique Mond nickel refining operation as set up at Clydach. Dr Langer was a chemical engineer who designed on a grand scale, and the refinery that he designed, and whose construction and operation he supervised, was greatly admired throughout the refining industry. A particular aspect of his ingenuity, of considerable relevance to the carcinogenicity problem, was the integration of all the processes so that the only waste products from the entire refinery operations were sulfur as sulfur dioxide and the mainly iron-containing siliceous slag from the furnaces. Thus, all the valuable metallic constituents of the original bessemer matte were eventually processed to a commercially useful end.

This was achieved by a complex series of recycling operations, and, as will be shown, one of the major recycling steps resulted in the inadvertent production and concentration of a complex arsenical nickel compound.

The chemistry underlying the Mond process is straightforward. The nickel and copper sulfides in the bessemer matte were burnt in air at 780°C to produce nickel and copper oxides. A considerable fraction of the copper was removed by treatment with dilute sulfuric acid, the copper sulfate solution was filtered off and the residue dried and transferred to reduction towers where the nickel oxide was reduced to nickel metal by the hydrogen in the water gas. The metallic nickel was then combined with carbon monoxide to produce nickel carbonyl. This was decomposed to nickel metal and carbon monoxide. The residue remaining still contained significant quantities of copper and nickel, because in the early years the processing was only about 80% efficient at best. However, there was a valuable gain in that there were increased concentrations of cobalt, silver, platinum and palladium and, unfortunately, in the early years, arsenic from the added sulfuric acid. In order to understand the possible exposure situations confronting the workforce, it is necessary to understand the physical aspects of the refinery plants in which these chemical reactions were carried out on a massive scale.

Figure 16 shows a plan of the Clydach refinery complex as it was in 1932. By that time there were six nickel plants on the main site and a seventh had just been commissioned on an adjacent area. The early plants consisted of a calciner shed, a copper shed and a nickel plant proper. Later, one calciner fed two copper shed/ nickel plant complexes. One such unit is shown in bold outline in the top RHS of the figure. The buildings immediately adjacent to each nickel plant housed the units producing copper and nickel sulfate, cobalt salts and the final precious metals-containing residues. The processing began in a calciner shed where crushing and grinding machinery reduced the imported bessemer matte to a powder of 60 mesh fineness (60 µm particle size) and six to eight calciner units. Basically, a calciner was until 1911, a 15 m long, 2.5 m diameter brick tunnel with a series of shuttered ports and a flat floor 1 m wide along which a thin layer of ignited, powdered bessemer matte was slowly dragged by a series of plough shares attached to an endless chain. The operating temperature in the tunnel was 780°C and because of the rather primitive method of progressing the material through the tunnel, frequent manual raking through the ports was necessary. Conditions in the calciner sheds were described as from awful to appalling, because of the high dust levels and the sulfur dioxide in the atmosphere. In 1911 the calciner design was improved, the length was increased to 30 m and the ignition bed was encased in an iron sheet tunnel with shuttered access apertures, which were often left open, and the movement of the burning material through the tunnel was mechanically much improved. In 1924 the calciner design was once again greatly modified to a 30 m long double-decker system. The calcining process is an obvious area to place high on the list of potentially hazardous environments. At this stage it should be noted that there were more modifications in this department in the early decades, than in any other. This implies that the oxidation of the copper and nickel sulfides was a bottleneck step. A careful scrutiny of the rates of annual feedstock imports shown in Figure 8 supports this view.

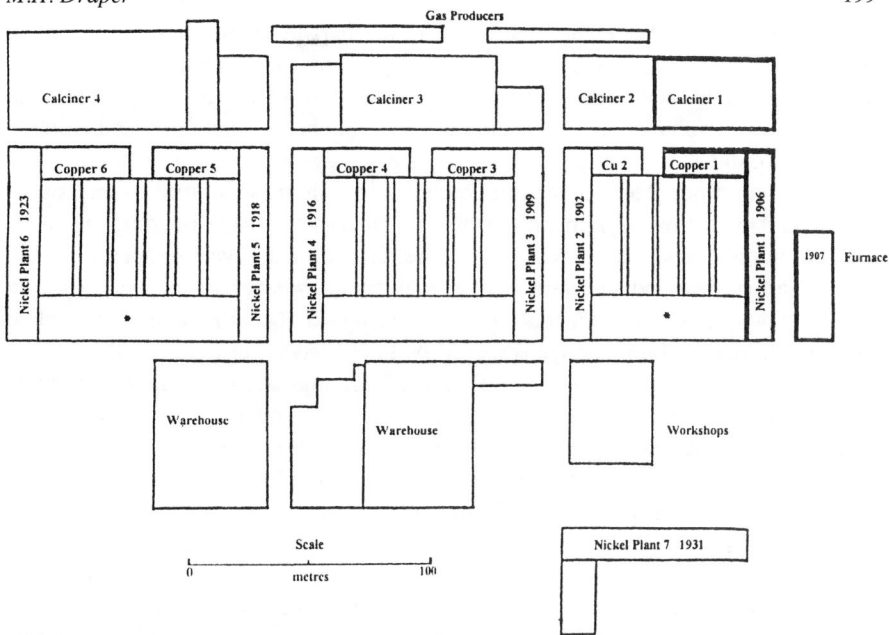

Figure 16 *Plan of the INCO Europe nickel complex at Clydach, Wales UK, as it was in 1932. It consisted of six nickel refinery units or plants set up in parallel, with a final seventh plant commissioned in 1931. A nickel plant consisted of a separate calciner unit (CALC) where bessemer matte was calcined. This calcine was transported manually in barrows to an adjacent copper shed (COSH) that was set at right angles and directly linked to the building that housed the nickel carbonyl extraction process, referred to as the nickel shed (NISH). From 1916 somewhat larger calciners fed a pair of nickel plants as can be seen in the figure. Nickel plant number 1 (historically no 2, 1906) is indicated in bold on the RHS of the figure. The product, nickel metal pellets, was transferred to a warehouse, and the residue, a concentrate or conc (CONC 1, 2 or 3), was transferred manually in barrows to the furnaces (FURN). In the large area between the nickel plants there were a series of long buildings that housed the ancilliary operations, mainly crystallization processes. The transverse building (*) that closed off a two nickel plant block housed 'wet processes', such as nickel sulfate production (NISU) and storage, cobalt salts production and final precious metals concentrations. The approximate dimensions of plant number 1 were: calciner shed, 48 × 28 m; copper shed, 23 × 12 m and the nickel shed, 74 × 12 m.*

The next process step was the leaching out the copper from the calcine with 10% sulfuric acid. In view of the controversy concerning the carcinogenic role of arsenic proposed by the medical officers at the refinery during these early years,[3] but not supported by later investigators,[1] the physical aspects of this operation need careful consideration. This operation was carried out in the copper shed which was set immediately adjacent to, but at right angles to the nickel plant (labelled copper 1, 2, *etc.* in Figure 16). This was a building with three floors some 11 m high × 23 m long × 12 m wide. The calcine was delivered in manually operated, supposedly covered (but often not), barrows from the calciner to a vertical, endless belt bucket elevator. Each ascending bucket was filled from the barrow by manual shovelling. The buckets discharged at the third floor onto a rotating horizontal worm that in turn discharged into a lead-lined tank filled with 10% sulfuric acid and agitated by piped steam. After about 15 minutes, the tank contents were siphoned off to a similar tank on the second floor and, after a further 15 minutes, this tank was discharged into a system of rotary vacuum filters on the ground floor. The copper sulfate-containing filtrate was piped to the copper sulfate crystallization department that was situated in two large buildings in the L formed by the copper and nickel sheds. The residual 'cake' was scraped from the filter drum surface and transferred to special dryers, whence as 'copper extracted matte' it was ready for manual barrow transfer to the adjoining nickel plant building through a short passage screened by thick canvas curtains.

The nickel plant occupied a large building 11 m high × 74 m long × 12 m wide and it contained some 70 cast iron reaction towers, some being 9m high. The calcine was sequentially progressed through a complex of 8 reducing, 12 volatilizing and 47 decomposing towers to produce nickel pellets and a residue or concentrate, universally referred to in the industry as a 'conc', at this stage a 'conc 1'. From the exposure point of view it is important to note that from the entrance of the calcine to a reducer through to the final production of the nickel deposited on seed pellets and the accompanying concentrate, the processes are totally enclosed and gas tight because of the toxic nature of nickel carbonyl and carbon monoxide, and the explosive nature of hydrogen gas. Thus, the exposure possibilities should be at the beginning and end of the nickel plant operations. There are, however, some feedback steps in the latter stages that could have posed risks. The possibility is mentioned here, but the details are too detailed to be presented in this text.

A critical aspect is the fate of the residue or, at this stage in the refining operation, the conc 1 remaining after the final volatilizing stage. The processing so far has removed about 70% of the copper and about 80% of the nickel. Thus, a conc 1 still contains valuable quantities of metal and, furthermore, because the processing removes specifically copper and nickel, all remaining metals, such as cobalt, silver, palladium and platinum will have been concentrated by a factor of about three. Unfortunately, this concentrating also applied to an element that was not desirable, namely arsenic. This arsenic came from two sources. The first from bessemer matte itself, in which the concentration of arsenic (as As), probably present as an arsenide, was about 0.2%, and the second, a much more serious matter, was from the chamber sulfuric acid which could contain as much as 3%, probably in some acid form analogous to the phosphorus acids. Although the chemistry is not understood, it is clear that the majority of this arsenic travels with the solids, not with the liquid copper sulfate. The refinery analytical records show that copper-extracted calcined mattes from 1908 to 1924 had arsenic levels ranging from 1 to 2.5%. As will be discussed, once this arsenic entered the system the majority of it persisted through the many high temperature treatment cycles of the refining process. It is contended that it is highly probable that herein lies the key to the Clydach respiratory cancer problem.

From the above it can be appreciated that a conc 1 was a valuable residue and as such was up until 1907 simply collected until an economic quantity was accumulated, probably a hundred tonnes or so, and passed back to a calciner to be processed as a slug to a conc 2. This recycling in these early days could be repeated up to seven times. However, after 1907 the concs were further processed by a furnacing operation in a separate building. Here the concs were heated in a reverberatory furnace with the addition of gypsum, sand and coke. The object of this step was to convert the oxides back to sulfides and to remove iron as a siliceous slag. A second treatment of the slag in a cupular furnace conserved any remaining economically important metals, leaving the final iron-containing slag to be discarded, the only waste discarded, as mentioned, in the refining operation (apart from sulfur dioxide). After the passage through the furnaces the material was referred to as a matte 1 or 2, *etc*. This matte was transferred, when a sufficient quantity had accumulated, to a calciner to go through the refining cycle as before. The formation of mattes from concs improved efficiency so that only three to four cycles were needed to win the required nickel and copper and have the final conc 3 or 4 with workable concentrations of cobalt and precious metals. However, in addition the arsenic levels had also increased, despite all the calcining and furnacing, to as much as 10% in some final concs.

The presence of arsenic in the process materials was well-known and some concern about the medical implications was expressed by the medical staff, because there was some evidence of arsenicism among the process workers. On the management side, there was the problem of arsenic interfering with the extraction of nickel itself, particularly in the last fractions. Throughout the first three decades there was a continuous effort to increase the efficiency of the nickel extraction. There was the furnacing operation to eliminate the iron, the three major modifications of the calcining process, and finally in 1920, incidentally when production was low, it was decided to change to arsenic-free sulfuric acid. The probable reason why this obvious step was deferred was cost. As mentioned, sulfuric acid was a major feedstock and clearly the cheapest source was sought, namely chamber acid made from iron pyrites, of which there were immense deposits in Spain. Arsenic-free acid is made using the contact process and is appreciably more expensive. Although steps were taken in 1920 to eliminate the contaminated acid, the analytical records indicate that some such acid was still in use as late as 1924, and it is likely that in some of the processes arsenic contamination lingered on for many years. By 1930 it was decided to eliminate the entire copper leaching process and, during the years 1932 to 1936, the linear calcining and leaching processes were eliminated and, as can be seen from Figure 7, the rate of nickel production greatly increased.

Thus far, the major central refining procedures have been sketched out for the years 1903 to 1930. However, linked to these processes were a number of sub-production cycles that drew their feedstock in various mixtures from calcined bessemer matte, copper-extracted calcined mattes, or concs 1, 2 or 3, and nickel-rich liquors from the recycled sulfuric acid copper leachates. These operations, which were conducted in the sheds between the nickel plants, produced nickel sulfate crystals and other nickel salts, cobalt chloride and the final precious metal-rich residues remaining after all the metal values of nickel, copper and cobalt had been extracted. Important points to note about these processes are that there was a considerable movement of materials not only along the principal refining chain of the calciners, the copper sheds, the nickel plants and finally the furnaces, but from all of these units to the subsidiary production units. Finally, mention must be made of the furnace building. As has been mentioned the various concs were barrowed to the furnace building from the nickel shed, actually from a discharge point outside the shed. Because the residue from the volatilizers was rather unpleasant, it was heated in an enclosed conveyer to eliminate the toxic gases and transferred mechanically out of the building. The conc from each cycle was stored separately in barrels in the furnace area until a sufficient tonnage had accumulated to make it economic to return it to a calciner. In addition, all of the various final residual sludges from the ancillary operations were also returned to the furnaces to be eventually incorporated in the recycling process.

6 The Job Sequences of the Nasal Cancer Cases

The above presentation describes the general background of the nickel refinery at Clydach, the complex evolution of the refining process itself, and the relationship of the lives of those workers that contracted respiratory cancers to these events, during the first three decades of operations at the refinery. Furthermore, there has been a recognition since the earliest investigations that the only workers at risk were those intimately involved with the main refinery processes. The other half of the workforce, the blacksmiths, the coopers, the carpenters, the machine shop mechanics, the packers, the yardmen, the drivers and the many other categories of worker necessary to support the core operations of the refinery did not appear to have been at significant risk.[3] The current study of the job sequences of these cases supports these views.

Figure 17 sets out in chart form in real time from 1905 to 1939 the job sequences of the first 20 deaths from nasal cancer. Also indicated for each are the case code number, the age at recruitment and the age at death. It seems reasonable to accept that these should reflect the response of the most vulnerable to the most hazardous situations. The first death was in 1923 and the last of the twenty 14 years later. The average age at starting work was 29.6 years (SD 7.1) and the average at death was 52.2 years (SD 8.1). The mean duration from recruitment to death was 22.6 years (SD 3.6) and the mean year of recruitment was 1909 (SD 2.9). By any standards these are remarkable figures. Table 2 sets out a comparison of these statistics with those of the next two successive groups of characteristic, that is essentially those cases depicted in the central portion of Figure 12.

Table 2 shows some interesting features, for example the later a worker was recruited the longer he would survive and furthermore the average age of commencing work was essentially the same for the three groups. Even though these are rather simple comparisons the figures can be interpreted to indicate the presence of a dose–response, in that, as 1924 was approached, there was a reduction of the available time to be exposed to the potent carcinogenic agent(s), because they were disappearing from the system.

Consideration of Figure 17 shows at first an indigestible mass of data, a considerable number of features of which can only be interpreted in the context of the 2000 years of data from all the cases. This cannot be undertaken in the present

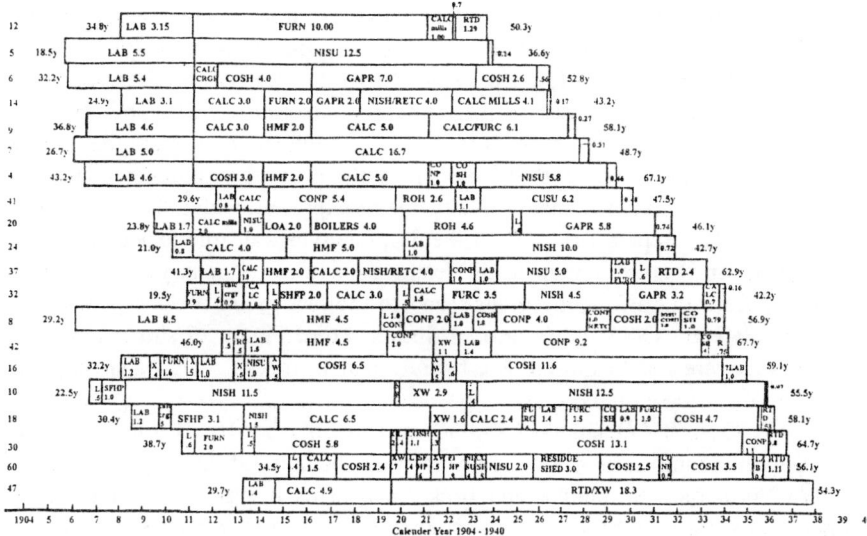

Key: CALC = Calciner shed; COSH = copper shed; NISH = nickel shed; CUSU = copper sulfate crystallization sheds; NISU = nickel sulfate production shed; FURN = furnaces; LAB = labouring work; XW = ex-works, absent from the refinery; HMF = service in the army; CONP = concentration processes

Figure 17 *This depicts the job sequences in real time of the first 20 Clydach refinery workers who died from nasal cancer in the period 1923 to 1937. Each row sets out the case identity number, the age when recruited; there then follows the specific jobs allocated, or absences from the refinery, e.g. for military service, and their duration followed by the time from cessation of work to death and finally the age at death. The averages of some of these events are set out in Table 2.*

Table 2 *Nasal cancer cases at Clydach: the 58 'characteristic' cases considered in 3 groups; first 20 deaths, second 20 deaths and the last 18 deaths*

	Mean age (years) at start ± SD	Mean calendar year of start ± SD	Mean age (years) at death ± SD	Mean duration in years of true exposure ± SD
First 20 deaths	29.6 ± 7.1	1909.0 ± 2.9	52.2 ± 8.1	19.1 ± 4.9
Second 20 deaths	31.7 ± 7.4	1913.0 ± 4.5	60.6 ± 5.0	24.2 ± 6.4
Third 18 deaths	28.2 ± 6.0	1918.0 ± 4.8	65.2 ± 5.0	32.2 ± 7.3

circumstances. However, some general features are clearly evident even in this small sample, which is typical in overall pattern of all the case histories. It can be seen that, in general, workers started with a stint as a labourer for periods of from half a year to some five or more years. Short periods of labouring were often interspersed between other jobs in later years. This was typical and was part of a familiarization process before being allocated to a specific job in one of the main processes. Before 1907 when the furnaces were introduced, there was a need for much labouring because of the transport of material around the refinery for the six or seven cycles in operation at that time, and hence labouring stints were longer than usual. Thus, labouring, if associated with the principal processing operations, would seem to be potentially hazardous.

Another important feature apparent in Figure 17 is that, with few exceptions, exposures were in the core refining processes, namely the calciner (CALC), copper (COSH), nickel and furnace sheds (NISH & FURN), together with their two main subsidiary processes, nickel sulfate production (NISU) and the final precious metal concentration process (CONP). It is interesting to note that copper sulfate (CUSU) only appears once (case 41) in these first cases and as a final 6.2 years period at the end of a total of 14.9 years of exposure, with 2.6 years absence (1919–1921). This represents 1.6% of the 382 man-years of true exposures of the whole group. In the next group of 20 deaths there were 60 man-years of copper sulfate exposure (12.6% of the group), and in the last group of 18 characteristic cases 25 man-years (3.6% of the group). Inspection of Figure 17 shows that three (9, 7 & 47) were only exposed as labourers and in the calciners, and that four others were also essentially single exposure cases; no. 12, furnaces no. 5, nickel sulfate; no. 10, in the nickel shed and no. 30 (possibly no. 16 also) in the copper shed. The remaining cases had mixed exposures, the majority moving from one of the above processes to another and back again. This is typical for most of the cases with a significant number working some 80% or more of their service in one job. An important category of worker that does not fit the pattern of Figure 17 is the class of worker, labelled 'all sites' (ALLS); these are the skilled and semi-skilled fitters and riggers that maintained the highly mechanized transport machinery that moved the calcines and mattes through the various processes.

Case 47 is particularly of note because his entire exposure was 1.4 years as a labourer and then 4.9 years in the calciner shed when he left the refinery to unknown other work, dying 18 years later at the age of 54 years. All the others died shortly after ceasing work, which was in fact shortly after diagnosis as it was the practice of the company to place a worker on pension as soon as a diagnosis of cancer was apparent. As was noted in Figure 13, there were nine cases where exposures were short, and all the six major process sites were represented in the histories of these cases. Although exposures in the calciner area are prominent in the histories set out in Figure 17, particularly after 1911 when the calciners were modified, this predominance is not apparent in the subsequent case histories considered in the same sequence of year of death as in Figure 14. What is observed is an increase in the occurrence of long periods of exposure in the copper and nickel sulfate areas, the furnace area and the appearance of the workers involved in maintenance. This is consistent with a contamination spreading through the processes of the refinery and its likely association with arsenic. This in turn raises the question of the importance of the one class of process material that does cycle repeatedly around the processes that appear consistently in the histories of the respiratory cancer cases, namely the concs 1 to 3 or more.

7 Chemicals in the Environment at Clydach

Through the history of the refinery at Clydach, management have analysed the various process materials for their element content, namely Ni, Cu, Co, Fe and S as these were important for the control of the refining process. In addition, because of the possible deleterious effect on the efficiency of the extractive procedures, arsenic content was frequently monitored. However, because there were no indications of serious health effects until the mid 1920s, no quantitative information about the environment was obtained then or for some decades later, and even then no chemical species determinations were attempted. In the course of the present investigation some important samples of process material from the 1920s were discovered. These were a sample of a conc 4 from 1920, a conc 3 from 1929, a sample of bessemer matte, undated, but almost certainly from the 1920s and a similar sample of a copper-extracted matte from this period. These samples, particularly the two concentrates, have been subjected to a wide range of physical and chemical analyses during the present investigation.

The analytical studies of the two concs are of most interest in the present context. The results of an inductively coupled plasma emission spectroscopical analysis are set out in Table 3.

These results confirm the analyses carried out when the samples were first obtained and are in line with other analyses carried out at Clydach on similar material. The striking difference between the two concentrates is the arsenic content. The 1920 sample has 9.6% and the 1929 1.0% arsenic. Of interest for later considerations the respective Fe contents were 4.4% and 0.75%. Possibly the most surprising result was that the mean particle size was 3.0 μm for the 1920 sample and 1.5 μm for the 1929 sample, remarkably small particles for an industrial refinery process material. Electron microscopy showed that the particles consisted of clusters of ultramicroscopic crystals. Studies with laser beam ionization microprobe analysis (LIMA) showed that the particles had a uniform composition and that arsenic was a prominent surface feature among the expected

Table 3 *Quantitative analysis of refinery samples by inductively coupled plasma emission spectroscopy*

Element	Molecular weight	1920 Conc 4		1929 Conc 3	
		$g\ 100g^{-1}$	$moles\ kg^{-1}$	$g\ 100g^{-1}$	$moles\ kg^{-1}$
Ni	58.7	37	6.4	27	4.5
Cu	65.6	21	3.3	29	4.5
Fe	55.8	4.4	0.8	0.75	0.1
Ca	40.1	0.03	0.01	0.27	0.07
Co	58.9	0.86	0.15	1.53	0.26
S	32.1	4.1	1.3	4.5	1.4
As	74.9	9.6	1.3	1.0	0.1
Ag	107.9	0.29	0.03	0.12	0.01
Sb	121.8	0.30	0.03	0.09	0.001
Pb	207.2	3.6	0.2	1.0	0.05
Pd	106.7	0.15	0.003	0.03	0.003
Pt	195.2	0.12	0.001	0.03	0.001
		81.5		**65.3**	

Ni, Cu, Co, Fe, S and precious metals. However, the most significant results were obtained from powder X-ray spectroscopy, because this showed that a prominent feature of the 1920 conc, but not the 1929 conc, was the presence of a molecule with the crystalline structure of a complex nickel arsenide mineral known as orcelite.[6] The composition of orcelite is believed to be of the form $(NiFeCu)_{4.2}(AsS)_2$ with the 'defect' elements present maximally at concentrations of (w/w) Fe, 2.4%; Cu^0, 3%; S, 1.4% (Caillere, 1961). With such a composition a particle from a conc 4 would consist of some 25% orcelite-like material.

A pecularity of the nasal cancer cases at Clydach is that the origin of so many of them was in the ethmoidal region with a typically eventual invasion through the cribriform plate into the orbit and brain. This suggests that very fine particles from the dusty environments at Clydach were carried to the olfactory region by the small back eddies that move postero-anteriorly from the inspired air at the back of the tongue to the ethmoid region. In this way odorous molecules at very low concentrations can be presented to the extremely sensitive olfactory sensory cells. Because these sensory cells can respond to just a few molecules, functionally, the air flow from the eddies must be small and gentle to avoid saturation of the receptors. It thus follows that only very fine particles would travel with the eddy currents and reach the ethmoidal surface epithelium. Thus, the picture emerges of a 1 µm diameter particle sited on the 100 $µm^2$ surface of an epithelial cell, and that each particle presents arrays of ultramicrocrystals to a few of the tens of thousands of growth factor receptors that extend through the cell surface.[7,8] Furthermore, in each microcrystal there are molecular matrices that could present, at the atomic level, at least two potentially biologically reactive atoms, namely arsenic, with the possibility of interfering with vital receptor phosphorylation reactions, and Fe, with the possibility of free radical generation.

8 Concluding Comments

The great puzzle about the Clydach nickel refinery was that there could be no doubt that, for a time, an extremely potent carcinogenic agent existed in the plant, but the situations in the industrial world, where nickel and other products were extensively used showed little, if any, evidence of carcinogenicity, especially nasal cancer. Thus attention has to focus on the processes at Clydach. Here, as has been discussed, the more detailed the examination, the less likely the conventional explanations appear. The study of the job sequences in particular has provided convincing evidence as to the reason why the 'epidemic' was confined to the first three decades of operation. Indeed, even after two decades, there was evidence of a decline in the danger. The lack of any correlation between the enormous throughput of bessemer matte and cancer incidence not only casts doubts on the involvement of sulfides, but also lessens the case for the oxides, which is weak in any event, because oxidic feedstocks continued in use long after the carcinogenicity issue disappeared. The job sequence studies show that the hazard spread through the central processing units, but was essentially confined to those units that utilized calcines and the various sequential concentrates that were derived from them. When the arsenic from the sulfuric acid was eliminated, coincidentally the 'epidemic' subsided. The identification of a complex nickel arsenide that resulted from the processes themselves is not only convincingly associated with the induction of the respiratory cancers at Clydach, but a causal hypothesis is testable particularly in the context of recent molecular understanding of carcinogenesis.

When the investigation at the Clydach refinery complex began, the objective was to obtain a clearer view of the process sequence than was available in the epidemiological literature and in so doing to establish some more facts about the presence of the reported high levels of arsenic. The first visit revealed a surprising wealth of new information that was obviously highly relevant to the arsenic problem. This raised the question as to why such readily available information had been overlooked. Upon reflection it became clear that what had happened was that in the 1960s the horrendous figures of the cancer rates, particularly of the rare nasal cancers, were so compelling that the obvious conclusion had to be drawn that it was 'nickel', a term used to include any inorganic nickel compound, that was the carcinogen. Thus there was little point in investigating any further, particularly as there were big problems looming up on the organic chemical front. It also became clear that the investigation at Clydach would need to be on a much larger scale than had been envisaged. This included not only a detailed history of the evolution of refinery and its many individual processes and the detailed compilation of the job sequences of the 360 respiratory cancer cases attributed to exposures at the refinery, but also a many-faceted analysis of a number of samples of process materials from the 1920s that were discovered in laboratory cupboards. There were also many curious fluctuations in the incidence of cases that were finally resolved from considerations of the complex industrial history of South Wales through the prosperity of the early decades, the First World War and the subsequent depression, and then the Second World War. It was demographic factors rather than changes in the refining processes that had influenced the fluctuations in case incidence. This aspect was far from clear at first and it was not until the significance of the very large turnover of workers each year was understood that an explanation emerged.

The resources that eventually were needed to be mobilized were well beyond those of any single conventional academic department, which is usually by tradition based on a single discipline such as epidemiology, occupational health or toxicology. It was fortunate that the investigators were able to draw upon the expertise and goodwill of colleagues at two universities, one with a medical school, as well as the collaboration of the INCO management at Clydach and in Canada. In this unique situation it proved possible to co-ordinate a multi- disciplinary approach to a most complex industrial environmental problem with a considerable degree of success, so much so that it suggests that this pattern of investigation could serve as a model for similar studies in the smelting and refining industries.

The basic concept is the establishment of a central co-ordinating group with the facility to organize and finance a number of strands of enquiry all directed to one end, for example in this case: what was the real cause, at the molecular level, of the cancers at Clydach? With this *modus operandi*, clearly a re-examination of the lung and nasal cancers at other nickel refineries could reveal a similar misinterpretation of the evidence and confirm that here also neither nickel nor simple inorganic derivatives are the cause of the cancers. An obvious candidate is the electrolytic nickel refinery at Kristiansand in Norway where there is a recycling and furnacing of process material containing up to 10% of arsenic of unknown species. An important conclusion to be drawn from the present studies is that mere knowledge of concentrations of elements in an environment is of little help. What is needed is identification of the species and this also needs qualification, in that it is the physical nature of the molecule and the kind of lattice that is presented to the target cell surfaces that really matters.

In conclusion, it has to be remembered that the suggested, co-ordinated multidisciplinary approach evolved as new aspects of the problem at Clydach were uncovered. Thus this is not a hypothetical approach, but one that actually worked. However, such an approach does not sit well with the way current academic disciplines are compartmentalized or funded. Research projects are usually required to be sharply focused by grant giving bodies. Thus at present there is no descriptive collective name such as epidemiology, or toxicology, or occupational health that quite describes the activities considered above. Because of the necessity to call upon the assistance of many scientific disciplines and the need for on-site field work, the name metademography is suggested to designate this type of endeavour in the area of industrial health studies. Metademography is concerned with the co-ordination of a multidisciplinary investigation with the defined objective of the elucidation of the ultimate cause or causes of specific diseases such as cancer that occur in sub-populations of people working in industrial situations. It concentrates on human biological reactions to these situations and asks questions such as: why did the epidemic stop, or why was the ethmoidal region so vulnerable and what was the reagent responsible, or why were there apparently three categories of latency? Some answers are appearing to these particular questions but many questions still remain regarding nickel refining. Other metal refining operations still await a metademograhic analysis.

Acknowledgements

These studies could not have been conducted without the extensive collaboration made available by the Management of INCO Europe Limited at Clydach, Wales, in particular by Dr L.G. Morgan and Dr L.M. Metcalfe. Dr J.H. Duffus and Dr M.V. Park, Edinburgh Centre for Toxicology (EDINTOX), Heriot-Watt University, Edinburgh, who gave invaluable advice on technical and chemicals aspects, particular thanks are also due to Dr Park for his identification and discussions concerning orcelite. Finally we are all indebted to Dr P. Johns, for his invaluable contribution to the LIMA studies carried out in his department (Edinburgh Surface Analysis Technology (ESAT), Heriot-Watt University, Edinburgh) and to Professor Ian Parsons and Dr Martin Lee, Department of Geology, University of Edinburgh for their X-ray crystallographic studies.

The funding of much of the analytical studies was provided by a grant from the Nickel Producers Environmental Research Association, Inc. (NiPERA). A grant towards the production costs of the illustrations was received from the Ethel and

Gwyne Morgan Charitable Trust. Grateful thanks are due to both these bodies. Finally, throughout all this work the computer and other skills of Mrs Faye Craig have been invaluable.

References

1. ICNCM (1990) Report of the International Committee on Nickel Carcinogenesis in Man, *Scand. J. Work Environ. Health*, **16**, 1–82.
2. Doll, R. (1958) Cancer of the lung and nose in nickel workers, *Br. J. Ind. Med.*, **15**, 217-223.
3. Morgan, J.G. (1958) Some observations on the incidence of respiratory cancers in nickel workers, *Br. J. Ind. Med.*, **15**, 225–234.
4. Doll, R., Morgan, L.G. and Speizer, F.E. (1970) Cancers of the lung and nasal sinuses in nickel workers, *Br. J. Cancer*, **24**, 623-632.
5. Easton, D.F., Peto, J., Morgan, L.G., Metcalfe, L.P., Usher, V. and Doll, R. (1992) Respiratory cancer mortality in Welsh nickel refiners: which nickel compounds are responsible. In, Nieboer, E. and Nriagu, O., eds. 'Nickel & Human Health: Current Perspectives', Wiley, New York, pp. 603–619.
6. Oen, I.S., Keift, C., Burke, A.J. and Westerhof, A.B. (1980) Orcelite and associated minerals in the Ni–Fe–As–S system in chromites and orthopyroxenites of Nebral Malaga, Spain, *Bull. Mineral.*, **103**, 198–208.
7. Sporn, M.B. and Roberts, A.B. (1992) Autocrine secretion – 10 years later, *An. Inter. Med.*, **117**, 408–414.
8. Zimmerman, G.A., Laurant, D.E., McIntyre, T.M. and Prescott, S.M. (1993) Juxtacrine intercellular signalling – another way to do it, *Am. J. Respir. Cell Mol. Biol.*, **9**, 573–577.

Assessment of Carcinogenic Risk: International Activities, Regulations and the Industrial Situation

Representatives of international organizations, regulatory bodies and industry were asked to describe the basic philosophy and current practice of their organizations in relation to carcinogenic risk assessment with particular reference to inorganic substances. These presentations provided a basis for subsequent panel discussion which addressed the following points:

- Risk assessment starts from hazard identification. Are current procedures of hazard identification, and further classification systems, appropriate for inorganic substances? What changes might be made to improve and harmonise the current procedures?
- Identification of hazard is followed by assessment of possible exposure to the hazard. Are current procedures for exposure monitoring adequate for risk assessment purposes?
- Assessment of exposure can only be interpreted in the light of good dose–response data. How should we cope with the inadequacies of the data available at present? How can we obtain valid dose–response data for inorganic substances and use them in the assessement of carcinogenic risk?
- Is the current risk assessment methodology adequate? What information and computational support are necessary for this purpose and how should they be provided?

The Role of the International Agency for Research on Cancer (IARC)

P. BOFFETTA

INTERNATIONAL AGENCY FOR RESEARCH ON CANCER (IARC), LYON, FRANCE

The International Agency for Research on Cancer is involved in two aspects of the assessment of carcinogenic risk of inorganic substances. On the one hand, it participates in the accumulation of scientific data on carcinogenic risk following exposure to inorganic substances. This is done mainly through the co-ordination of epidemiological studies of workers employed in specific industries: these studies are often of multicentric nature (for example, a study of welders,[1] and a study of gold miners[2]); other research activities are less direct and deal with aspects such as the discussion of methodological problems of epidemiological studies on workers exposed to cadmium[3] or a meta-analysis of epidemiological studies of workers exposed to inorganic lead.[4] On the other hand, the International Agency for Research on Cancer has evaluated the evidence of carcinogenicity of several inorganic substances within its Monographs programme. A number of metals or their compounds are classified as human carcinogens. The evidence is based mainly on the results of epidemiological studies on occupationally-exposed populations. Arsenic and arsenic compounds, nickel compounds, and chromium(VI) compounds have been recognised as human carcinogens (Group I) several years ago. More recently, the data on the carcinogenicity of beryllium and its compounds and of cadmium and its compounds have been reviewed and both metals have been classified as human carcinogens. The examples of beryllium and cadmium are illustrative of the difficulties of modern occupational epidemiological studies in reaching conclusions as for the carcinogenicity of metals. Three inorganic compounds are classified as possible human carcinogens (Group IIB): cobalt and its compounds, inorganic lead compounds, and metallic nickel. A number of other metals and salts are not classifiable as for carcinogenicity to humans (Group III): chromium(III) compounds, metallic chromium, ferric oxide, inorganic mercury compounds, metallic mercury, selenium and its compounds, and titanium dioxide. The IARC Monographs programme deals essentially with hazard identification, although quantitative aspects of the association between exposure and cancer are taken into consideration, when available, to contribute to the elucidation of the mechanism of carcinogenicity.

References

1. Simonato, L., Fletcher, A.C., Andersen, A., Anderson, K., Becker, N., Chang-Claude, J., Ferro, G., Gerin, M., Gray, C.N. and Hansen K.S., *et. al.* (1991) Historical prospective study of European Stainless Steel, mild steel, and shipyard welders, *Br. J. Ind. Med.*, **48**, 145–154.
2. Simonato, L., Moulin, J.J., Javelaud, B., Ferro, G., Wild, P., Winkelmann, R. and Saracci, R. (1994) A retrospective mortality study of workers exposed to arsenic in a gold mine and refinery in France, *Am. J. Ind. Med.*, **25**, 625–633.
3. Boffetta, P. (1992) Methodological aspects of the epidemiological association between cadmium and cancer in humans, *IARC-Sci.-Publ.*, **118**, 425–434.
4. Fu, H. and Boffetta, P. (1995) Cancer and occupational exposure to inorganic lead compounds: a meta-analysis of published data, *Occup. Environ. Med.*, **52**, 73–81.

IPCS Activities on Risk Assessment of Inorganic Compounds with Reference to Potency of Carcinogenicity

B.H. CHEN AND E. SMITH

INTERNATIONAL PROGRAMME ON CHEMICAL SAFETY, WORLD HEALTH ORGANIZATION, GENEVA, SWITZERLAND

1 Introduction

The International Programme on Chemical Safety (IPCS), established in 1980, is a joint venture of three United Nations agencies: the United Nations Environment Programme (UNEP), the International Labour Organization (ILO) and the World Health Organization (WHO). The WHO Programme for the Promotion of Chemical Safety is the Central Unit of the IPCS; it is responsible for the overall management and the implementation of many of the IPCS activities on behalf of the three co-operating organizations.

IPCS responds to the needs of Member States for information and guidance on the safe use of chemicals. Chemical safety in this context refers to the prevention and management of adverse effects, short-term and long-term, on humans and the environment of natural and synthetic chemicals, and of production, storage, transport, use and disposal of synthetic chemicals.

The objectives of IPCS are to catalyse and co-ordinate activities in relation to chemical safety and in particular to: (1) carry out and disseminate evaluations of risk to human health and the environment from exposure to chemicals, mixtures of chemicals or combinations of chemicals, physical and biological agents; (2) promote the development, improvement, validation and use of methods for laboratory testing and ecological and epidemiological studies and other methods suitable for the evaluation of health and environmental risks as well as hazards from chemicals; (3) promote technical co-operation with Member States; (4) promote effective international co-operation with respect to emergencies and accidents involving chemicals; (5) support national programmes for prevention and treatment of poisoning involving chemicals; and (6) promote training of the required manpower.

One of the major outputs of IPCS is the publication of Environmental Health

Criteria (EHC) documents, which include EHC documents on specific chemicals and EHC methodology documents on risk assessment. Up to now, 170 EHC documents have been published.

The Environmental Health Criteria programme was initiated with the following objectives:

- to assess information on the relationship between exposure to environmental pollutants and human health and to provide guidelines for setting exposure limits;
- to identify new or potential pollutants;
- to identify gaps in knowledge concerning the health effects of pollutants;
- to promote the harmonisation of toxicological and epidemiological methods in order to have internationally comparable results.

EHC documents on specific chemicals contain a comprehensive review and evaluation of available information on the biological effects of the selected chemicals that can influence human health and the environment. Whenever possible, the evaluations contain numerical values (criteria) such as 'Guidance Value (GV)', to provide advice and assist regulatory authorities in setting exposure limits or standards.

EHC methodology documents are devoted to evaluating toxicological methodology (genetic, neurotoxic, teratogenic and nephrotoxic effects), environmental epidemiology, biomarkers, effects of chemicals on high-risk groups, *e.g.* the elderly, and evaluation of short-term tests for carcinogens, *etc.*

2 IPCS Principles and General Approach on Risk Assessment for Carcinogens

Risk assessment comprises hazard identification, exposure assessment (based upon environmental monitoring and biological monitoring data, history of exposure and changes over time), dose–response relationship for the agent under consideration and estimation of the likely risk in the population involved.

Within the context of risk assessment of environmental carcinogens, the principles followed by IPCS are as follows:

- IPCS is guided by the basic philosophy of the International Agency on Cancer (IARC) and its classification of chemical carcinogenic risk;
- IPCS has developed a general approach for deriving health-based guidance values for chemicals including potential carcinogens; the values are incorporated in more recent EHC documents where applicable;
- distinction is made between genotoxic and nongenotoxic carcinogens in risk assessment;
- IPCS tends to be conservative in deriving health-based guidance values for chemicals of potential carcinogenicity with the aim of protecting the health and well-being of the human population;
- IPCS endeavours to provide to the best of its ability environmental criteria

recommendations (guidelines, guidance values) for chemicals. The setting of national or local standards for chemical exposures, however, is the sole responsibility of the government concerned.

IPCS has developed a general approach for the derivation of health-based guidance values for environmental chemicals and this has been published in 1994 in EHC No. 170 entitled 'Assessing Human Health Risks of Chemicals: Derivation of Guidance Values for Health-based Exposure Limits'.[1]

The aim of the guidance value is to provide quantitative information from risk assessment to risk managers to enable them to make decisions in protecting human health. The term 'guidance value' is defined as a value such as concentration of a chemical in a medium (air, water or food) which is derived from toxicological and epidemiological studies after appropriate allocation of 'tolerable intake' through different media of exposure. Combined exposure from all possible media at the guidance value level over a lifetime span is expected to be free of appreciable risk or at least with a risk that is considered to be acceptable. It was felt that guidance values (criteria) for exposure to chemicals in environmental media should be developed in all IPCS EHCs to the extent possible and that these values can be used as required by national or local authorities in their development of exposure limits or standards.

Concerning the derivation of health-based guidance values for potentially carcinogenic chemicals, IPCS considered the following approaches, which are cited in EHC No. 170 but left the decision on their use to individual Task Groups:

- unit risk assessment approach, *i.e.* quantitative extrapolation by mathematical modelling of the dose–response curve to estimate the risk at likely human intakes or exposures (low-dose risk extrapolation);
- relative ranking of potencies in the experimental range;
- division of effective levels by an uncertainty factor.

In 1993 the WHO European Regional Office/European Centre for Environmental Health (WHO/EURO/ECEH) in Bilthoven initiated the revision and updating of WHO Air Quality Guidelines for Europe in collaboration with IPCS. The Working Group on Methodology and Format for updating Air Quality Guidelines discussed the derivation of air quality guidelines for carcinogens.[2,3]

In deriving guidance values for potentially carcinogenic chemicals, the IARC classification for the specific chemical is taken into consideration by IPCS Task Groups and Working Groups for updating Air Quality Guidelines for Europe. In general, for Group I and IIA chemicals, the unit risk assessment approach is used; for Group IIB, III and IV chemicals, the uncertainty factor approach is used. For chemicals classified as IIB, a separate factor for the possibility of a carcinogenic effect in humans may be incorporated. Exceptions to the above are that, for Group I or IIA chemicals, the uncertainty factor approach can also be used provided that there is strong evidence from exposed human beings that the mechanism of carcinogenicity is a threshold phenomenon, *i.e.* that it can be established with certainty that an increase in exposure level to the chemical is associated with an increase in cancer incidence above a certain level of exposure; for Group IIB chemicals the unit risk assessment approach can also be used, provided that the mechanism of carcinogenesis in animals is likely to be a non-threshold phenomenon.

The unit risk assessment approach is used for Group I and IIA chemicals, though the imprecision of the estimates has been noted. Therefore, all sources of error and bias should be addressed.

In all cases, a distinction may be made between genotoxic and nongenotoxic carcinogens. The role of biomarkers of exposure and early pre-carcinogenic effects should be considered.

For pesticide residues in food, the WHO/FAO Joint Meeting on Pesticide Residues (JMPR) approach has been the use of safety factors for carcinogens.[4] JMPR considered that:

- when evidence of neoplasia has been identified, the safety factor may be increased, depending on the available auxiliary (ancillary) data and the establishment of an NOAEL;
- most of the mechanisms of chemical carcinogenesis are not fully understood. In view of the uncertainty surrounding the use of various mathematical models for carcinogenicity assessment, the 1983 JMPR meeting decided that the use of safety factors remained a reasonable approach. This pragmatic approach is used by JMPR in the absence of satisfactory alternatives;
- where there is the need for a very high safety factor due to concern about the safety of the use of the pesticide in question, it may be prudent to recommend that the pesticide should not be used where residues in food may occur. However, a pesticide for which there is limited evidence of carcinogenicity in animal studies should not necessarily be prohibited for use.

The evaluation of carcinogenic responses from exposure to chemicals is a universal concern. Therefore, the use of a consistent approach is of considerable interest. There is an ongoing IPCS project on the harmonisation of risk assessment for carcinogenicity and mutagenicity. An IPCS Workshop on the Harmonization of Risk Assessment for Carcinogenicity and Mutagenicity (Germ Cells) – A Scoping Meeting was held in February 1995 in the United Kingdom and aimed to define and describe the process of carcinogenicity and mutagenicity risk assessment, in order to establish how future harmonisation of these might be attained.

3 Products of IPCS and Joint IPCS Activities on Risk Assessment for Inorganic Carcinogens

The risk evaluation of inorganic carcinogenic chemicals in the IPCS EHC series is summarized in Table 1.[5-16]

Table 1 *Cancer risk evaluation of inorganic chemicals*

Chemical	EHC No. & year	Cancer risk evaluation
Arsenic	EHC 18 1981[5]	Lung cancer (inorganic arsenic) skin cancer (through drinking water).
Barium	EHC 107 1990[6]	Barium chromate is the only barium compounds that is a human carcinogen.
Berylium	EHC 106 1990[7]	Significantly elevated risks of lung cancer from occupational exposures.
Cadmium	EHC 134 1992[8]	Lung cancer; prostatic cancer inconclusive.
Chromium	EHC 61 1998[9]	Elevated risk for bronchial carcinoma in workers exposed to chromate salts and pigments. Evidence is insufficient to implicate chromium as a causative agent of cancer in any other organs than lung.
Manganese	EHC 17 1981[10]	No data available on its carcinogencity.
Nickel	EHC 108 1991[11]	Very high risks of lung and nasal cancer in nickel refinery workers and those exposed to soluble nickel, often combined with nickel oxide exposure. Slightly increased risk of lung cancer for nickel–cadmium battery workers. Soluble nickel is more dangerous than metallic nickel.
Platinum	EHC 125 1991[12]	No data available on its carcinogenicity.
Selenium	EHC 58 1987[13]	Possible association between low levels of selenium and a high incidence of cancer. However, it is still difficult to draw a firm conclusion.
Tin and organotin compounds	EHC 15 1980[14]	Carcinogenicity studies on animals were negative.
Titanium	EHC 24 1982[15]	Low carcinogenicity in animal studies.
Vanadium	EHC 81 1988[16]	Animal studies did not provide any indication of carcinogenic effect.

In the 2nd edition of the WHO Air Quality Guidelines for Europe a number of inorganic chemicals was reviewed (Table 2).[2,3]

In summary, IPCS prepares evaluations of risk to human health and the environment from exposure to chemicals, including carcinogens, and provides environmental criteria recommendations for chemicals. The setting of national or local standards for chemical exposure is the responsibility of the government concerned.

Table 2 *Risk assessment and unit risks estimate of inorganic air pollutants as recommended by WHO in the Updating and Revision of the WHO Air Quality Guidelines for Europe* [2,3]

Chemical	Risk assessment [a]	Unit risk estimate	Concentration associated with a life time cancer risk of $1/10^5$
Arsenic[b]	Human carcinogen. Increased lung cancer incidence through occupational exposure to inorganic arsenic compounds. Moderately elevated risk of lung cancer among people near emission sources.	1.5×10^{-3}	7 ng m^{-3}
Chromium[b,c]	Human carcinogen. Lung cancer.	4×10^{-2}	0.25 ng m^{-3}
Nickel[b]	Human carcinogen by inhalation exposure. Strongly allergic.	3.8×10^{-4}	25 ng m^{-3}
Cadmium[b]	Human carcinogen (Lung cancer). Renal effects.	None[d]	

[a] Additional lifetime cancer risk occurring in a hypothetical population in which all individuals are exposed continuously from birth throughout their lifetime to a concentration of 1 μg M^{-3} of the agent in the air they breathe. These risk estimates should not be regarded as being equivalent to true cancer risk.
[b] No safe level of exposure can be recommended assuming a linear dose–response relation.
[c] Cr(VI)
[d] Threshold for increased risk of renal dysfunction from a lifetime exposure extrapolated to 0.3 μg m^{-3} (current ambient air levels are around 1/50 of this value). In order to prevent any further increase of cadmium in agricultural soils likely to increase the dietary intake of future generations, a guideline of 5 ng m^{-3} is established.

References

1. WHO (1994) IPCS Environmental Health Criteria **170**: Assessing Human Health Risk of Chemicals: Derivation of Guidance Values for Health-based Exposure Limits, World Health Organization, Geneva, pp. 18–38.
2. WHO Regional Office for Europe (1994) Methodology and Format for Updating and Revising the Air Quality Guidelines for Europe. Report on a WHO Working Group, World Health Organization, Copenhagen, Regional Office for Europe. pp. 5–6, 16–17.
3. WHO Regional Office for Europe (in press) Update and Revision of the Air Quality Guidelines for Europe: Inorganic Air Pollutants. Report of the Final Consultation on Updating and Revision of the Air Quality Guidelines for Europe, World Health Organization, Copenhagen, Regional Office for Europe.
4. WHO (1990) IPCS Environmental Health Criteria No. **104**: Principles for the toxicological assessment of pesticide residues in food, World Health Organization, Geneva, pp. 47–53.
5. WHO (1981) IPCS Environmental Health Criteria No. **18**: Arsenic, World Health Organization, Geneva.
6. WHO (1990) IPCS Environmental Health Criteria No. **107**: Barium, World Health Organization, Geneva.

7. WHO (1990) IPCS Environmental Health Criteria No. **106**: Beryllium, World Health Organization, Geneva.
8. WHO (1991) IPCS Environmental Health Criteria No. **134**: Cadmium, World Health Organization, Geneva.
9. WHO (1988) IPCS Environmental Health Criteria No. **61**: Chromium, World Health Organization, Geneva.
10. WHO (1981) IPCS Environmental Health Criteria No. **17**: Manganese, World Health Organization, Geneva.
11. WHO (1991) IPCS Environmental Health Criteria No. **108**: Nickel, World Health Organization, Geneva.
12. WHO (1991) IPCS Environmental Health Criteria No. **125**: Platinum, World Health Organization, Geneva.
13. WHO (1987) IPCS Environmental Health Criteria No. **58**: Selenium, World Health Organization, Geneva.
14. WHO (1980) IPCS Environmental Health Criteria No. **15**: Tin and Organotin Compounds, World Health Organization, Geneva.
15. WHO (1982) IPCS Environmental Health Criteria No. **24**: Titanium, World Health Organization, Geneva.
16. WHO (1988) IPCS Environmental Health Criteria No. **81**: Vanadium, World Health Organization, Geneva.

The Role of the US Occupational Safety and Health Administration

W. PERRY

OCCUPATIONAL SAFETY AND HEALTH ADMINISTRATION (OSHA), WASHINGTON DC, USA

Abstract

The Occupational Safety and Health Act, passed by the US Congress in 1970, establishes the authority and basic principles by which health standards are promulgated by the Occupational Safety and Health Administration. These principles have been interpreted and refined by 25 years of case law that has developed as a direct result of rule-making activity. The Act requires OSHA to develop health standards that "ensure to the extent feasible, on the basis of the best available evidence, that no employee will suffer material impairment of health or functional capacity even if such employee has regular exposure to the hazard ... for the period of his working life." Thus, OSHA has a fundamental responsibility under the Act to establish health standards that are based on health evidence and that are designed to minimize risk to each individual who may be exposed to a particular hazard for a working lifetime. The Act and accompanying case law also explicitly require OSHA to consider the technological and economic feasibility in developing new standards. In practice, when promulgating new standards dealing with potential carcinogens, the stringency of the standard is more constrained by feasibility considerations than by the weight of health evidence or magnitude of risk.

In one of the more significant legal rulings relevant to standards development, the US Supreme Court has ruled that OSHA, before issuing a new standard, must demonstrate that a significant risk of harm exists among workers exposed to the hazard in question, and that the new standard will substantially reduce that risk. For carcinogenic hazards, this necessitates conducting quantitative risk assessments to meet this legal test. Although OSHA has not defined any risk level as representing a *de minimis* risk, the Agency has determined as a policy matter that lifetime risks of a magnitude of 1 cancer case per 1,000 workers constitute a significant risk that warrant further reducing exposures, provided that such reduction is both technologically and economically feasible.

OSHA evaluates both human epidemiological and experimental animal data in determining the potential carcinogenicity of workplace contaminants, and often relies on scientific organizations such as IARC for evaluating the overall weight of evidence. Quantitative risk assessments are conducted both in-house and by outside experts. During the public comment periods and hearings conducted for each rule-making, OSHA also receives quantitative risk assessments conducted by interested parties, and must evaluate those assessments in reaching regulatory decisions. Because each new standard must be developed under this formal public process, OSHA's risk analysis is subjected to a high degree of scientific scrutiny.

Specific to the subject of inorganic carcinogens, OSHA has promulgated comprehensive standards covering three hazardous materials (inorganic arsenic, asbestos, and cadmium), and is developing a proposal to issue a new standard on another substance (hexavalent chromium). Comprehensive health standards include requirements to: comply with permissible exposure limits; conduct exposure monitoring, medical surveillance, and employee training; and develop programs for using respirators and other protective equipment. The permissible exposure limit for a new standard is developed directly from quantitative risk assessment results and feasibility considerations. Epidemiologic studies have provided the data that the Agency has relied upon for quantitative risk assessment for these standards; however, in the cadmium rule-making, animal data also played an important role in evaluating the relative potency of various cadmium compounds. It has been OSHA's experience in conducting risk assessments for both inorganic and organic compounds that risk estimates based on animal data are consistent with, and often lower than, estimates based on epidemiological studies, that is, use of animal data in risk assessment has not been shown in OSHA's experience to systematically overestimate risks to workers. OSHA has been collecting and reviewing scientific information on two other materials, silica and man-made mineral fibres, as part of an ongoing prioritization process to identify future candidates for either rule-making or information dissemination activities.

The Role of the European Commission, Directorate General V

G. ARESINI

EUROPEAN COMMISSION, DG V, PUBLIC HEALTH & SAFETY AT WORK DIRECTORATE, LUXEMBOURG

In order to tackle the problem of workers' protection against exposure to carcinogenic agents in greater depth, the Commission's intention has been to depart from Directives on individual agents, such as vinyl chloride monomer and asbestos, and move towards the production of a more general Directive.

Nowadays the legal framework for the protection of workers is represented by the Council Directive 90/394/EEC, which has as its aim the protection of workers from risks to their health and safety, including the prevention of such risks arising or likely to arise, from exposure to carcinogens at work. In order to achieve this aim, the Directive lays down the following main principles:

(a) *Replacement and organizational control of exposure.* The employer is obliged to minimize exposure of workers to carcinogens by applying safety measures, such as the replacement of the carcinogenic agent with another which is not dangerous or is less dangerous to the health and safety of workers, or by the implementation of closed systems (where appropriate) or by other measures, such as restriction of access to certain areas, suitable working procedures, hygiene provisions, *etc.* in order to ensure that the level of exposure of workers is reduced to as low a level as is technically possible.

(b) *Limit values.* Limit values are to be set on the basis of the information available, including scientific and technical data, in respect of all those carcinogens for which this is possible. These limit values shall be set out by the Council in accordance with the procedure laid down in Article 118A of the Treaty.

In carrying out this activity the Commission is advised by a Committee of scientific experts, whose mandate has recently been made official by a Commission Decision. For certain specific carcinogens, for which a threshold can be identified, the Committee will try to make a recommendation for a limit value, on a case-by-case basis, if the data are adequate.

Fortunately, the cancer records which have been kept in certain industrialized countries for nearly 50 years, as well as the dermatological studies carried out in all parts of the world, have shown that only a few hundred chemical substances out of the 100 000 or so listed in the European inventory of chemical substances are able, in certain circumstances, to induce cancers in humans.

At the time of the adoption of the above Directive, the Council decided on an automatic mechanism for adapting the Directive to progress, the Directives concerned being Council Directives 67/548/EEC and 88/379/EEC on the approximation of the laws, regulations and administrative provisions of the Member States relating, respectively, to the classification, packaging and labelling of dangerous substances and to dangerous preparations. This means that any substance or preparation that is classified as carcinogenic, in the framework of the activities carried out under the above Directives, is automatically covered by the Directive on the protection of workers.

Even if this procedure simplifies the implementation of this Directive, because the employers are able to make an easier hazard identification (substances classified at the Community level with the risk phrase R45 'may cause cancer' or R49 'may cause cancer by inhalation'), some limitations are included in this system; among those, the obligation that one substance must be classified at the Community level to be covered by the Directive (and this means that, for instance, pharmaceutical or cosmetics will not be covered) and the assumption that the provisions of a Directive intended to deal with the introduction of substances into the market could be applied *sic et simpliciter* to the matter of the workers' protection. The latter is even more important if we consider the alterations that substances usually undergo during working processes, often with the creation of different molecules with different biological action.

Further to several scientific debates on this matter (most of them in the course of the preparation of some monographs on carcinogens published by the Services of the Directorate 'Public Health and Safety at Work' in Luxembourg) and to political discussion with delegations of the Member States (during consultations on the implementation of the 90/394/EEC Directive), it became evident that the protection of workers exposed to carcinogens would be improved through an amendment of this Directive, namely of its Article 2 (Definitions).

This amendment, which has recently been adopted by the Commission and will undergo the examination by the Council and the European Parliament probably later this year, is now proposing to use the criteria for the definition of carcinogens in accordance with the Annexes of the Directives 67/548/EEC and 88/379/EEC, and not to consider substances only classified R45 and R49 as the subject of the Directive on the protection of workers.

This hazard identification, which represents the first step of the procedure intended to carry out the assessment of the risks to the workers' health and safety, is particularly important for substances whose carcinogenic potency may depend on their chemical or physical characterisation.

It is expected that the scientific outcome of this seminar may help not only scientists and epidemiologists towards a better interpretation of the results of their studies but also help regulators in a better definition of the targeted population for their legislation.

Chemical Substances – Classification and Labelling, Risk Assessment and Protection of Workers

J.M. COSTA-DAVID

EUROPEAN COMMISSION, DIRECTORATE GENERAL XI, BRUSSELS, BELGIUM

In the European Commission, Directorate General XI is responsible for a significant part of European Union legislation on chemicals. The essential piece of legislation, particularly as far as carcinogens are concerned, is Directive 67/548/EEC on the Classification, Packaging and Labelling of Dangerous Chemical Substances (the activities involved at European Commission level under the provisions of the directive will be described below in more detail).

It is an important piece of legislation for several reasons but, as far as this Seminar is concerned, perhaps mainly because it triggers or conditions other pieces of European Union legislation. One good example is the Protection of Workers Directive 90/394/EEC, since the classification of a particular chemical as a carcinogen automatically implies the taking into consideration of that property and the need for measures that have to be taken in the workplace to protect workers from exposure to it.

Another related Directive is 88/379/EEC, on the approximation of the laws, regulations and administrative provisions of the Member States relating to the classification, packaging and labelling of dangerous preparations. For obvious reasons the classification of a particular substance as a carcinogen implies the need for taking that into consideration whenever the classification of a preparation that contains it is being considered. In terms of the criteria that both the substances and the preparations directives resort to in the analysis process, they have obviously to be interrelated. The substances directive being an older one (1967) it is only natural that it can claim the right of precedence in this respect. Proof of this are the several references to Directive 67/548/EEC in the text of the various adaptations to technical progress of the preparations directive.

Still, other pieces of European Union legislation draw on Directive 67/549/EEC, such as the Council Regulation no. 793/93 on the evaluation and control of the risks of existing substances and Commission Regulation no. 1488/94 laying down the principles for the assessment of risks to man and the environment of existing substances in accordance with Council Regulation 793/93. Commission Directive 93/67/EEC addresses the same issues in respect of new substances.

The risk assessment regulations in particular make a very specific reference to Directive 67/548/EEC. In its Article 3 (c), Regulation 793/93 mentions the classification of chemicals under Directive 67/548/EEC as necessary to the data reporting on high volume production or import of existing substances. The regulation also makes other references to the Classification and Labelling Directive and it is worth emphasizing that the classification data are also important for the establishment of the list of priority substances foreseen under the above mentioned regulation.

The link between classification and labelling and risk assessment becomes more obvious if we bear in mind that the latter contains two main elements: firstly the identification of a hazard, based on the intrinsic properties of a substance, and secondly the consideration of the exposure to the chemical. The identification of a hazard is carried out using the criteria for classification. The risk assessment directive is based almost entirely on the hazardous properties as described by the classification criteria.

Risk management is the next logical step and here is where the so-called 'downstream legislation' comes into play. The workers protection directive (90/394/EEC) is one of those.

Thus, Directive 67/548/EEC is unavoidable in considering problems related to carcinogenic risks (as well as others).

At this stage it is perhaps worth emphasizing some of the features of the organizational structure governing the classification and labelling activities of the European Commssion. Here are a few features:

- The European Commission legal instruments concerned with the classification and labelling of chemicals are as follows:

    Directive 67/548/EEC on the approximation of laws, regulations and administrative provisions relating to the classification, packaging and labelling of dangerous substances – responsibilities of the European Commission under the provisions of this directive
    Directorate General XI (Environment, Nuclear Safety and Civil Protection) – Unit XI/E/2 (Chemical substances and Biotechnology), formerly XI/A/2 and XI/C/4.

Unit XI/E/2 is responsible for the administrative and political dimensions of the work on Classification and Labelling of chemicals. In practice, this means that it drafts directives on the basis of the elements furnished by the European Chemicals Bureau whenever those directives are mainly of a technical nature. This is certainly the case with adaptation to technical progress directives where for instance the classification of chemicals is addressed. Changes to Annex I of the Directive 67/548/EEC are mainly either about the inclusion of chemical substances that have been classified for the first time under the provisions of the directive or about changes to an already existing classification.

Directives of a less technical nature are drafted by DG XI. However, the European Chemicals Bureau is always involved in the form of the Working Group on Classification and Labelling.

- The European Chemicals Bureau at Ispra hosts the following meetings:

    Working group meetings for discussions on the health effects of existing substances.
    Working group meetings for discussions on the environmental effects of existing substances.
    Working group meetings for discussions on new substances.
    Working group meetings for discussions on pesticides.
    Special *ad hoc* meetings.

The Working Group meetings serve the purpose of discussing the wide range of issues, which are of relevance for keeping Directive 67/548/EEC up to date. For the most part, this means discussing the proposals for classification of chemical substances that are submitted by Member States or Industry. Each different Working Group addresses the classification of chemicals from their own perspective, *i.e.* the Environment Group assesses chemicals from the point of view of environmental effects and so on. Not too long ago this would have been a much simpler process than is the case today. With 15 Member States the needed consultation is certainly more weighty and time consuming than when there were only six. What was lost in terms of simplicity though was certainly gained in terms of bringing in a much wider range of points of view and contributions to debates.

Chemical substances before environmental criteria were first set up as part of the provisions of Directive 67/548/EEC were classified basically in terms of their health effects alone. Today, chemical substances are passed on for the vote of the Technical Progress Committee only when both Working groups, dealing with the environmental effects and health effects, have had the opportunity of expressing their views.

Special *ad hoc* meetings serve the purpose of addressing a specific problem that deserves a thorough analysis that would probably take an inordinate amount of time from a regular Working Group were it to be analysed in one of its regular sessions. Sometimes the subject has already been addressed in a regular Working Group meeting where it was realised that the matter deserved a different level of analysis. To these *ad hoc* meetings, experts on the specific topic are normally invited, their presence being very much needed for classification, along with representatives from the Member States. In 1996, one of the *ad hoc* meetings was devoted to discussing the issue of metal alloys classification. This decision stemmed from the expressed need by the metal industry (mainly through EUROMETAUX) to address the specific case of alloys since the classification of these, according to the views of the metal industry, is not properly taken care of under the provisions of either the dangerous substances or the dangerous preparations directive.

- The Technical Progress Committee

    The Technical Progress Committee, which includes a representative from each member State of the European Union, is the body responsible for voting adaptations to the Technical Progress of Directive 67/548/EEC, which is the same as saying that they vote on the changes on the classification of chemicals and also on the wording of new risk and safety phrases and other changes to Directive 67/548/EEC. The vote is regarded as favourable whenever there is a qualified majority of the representatives of Member States that make up the Technical Progress Committee.

Related EU Legislation in Respect of Directive 67/548/EEC – see above

As was already said, the implications of Directive 67/548/EEC span beyond the classification of chemicals, that since there is a considerable body of European Union legislation which is intimately linked to or triggered by it. In practice this means that changes in Directive 67/548/EEC can bring about changes in other bits of European Union legislation. This is also one of the reasons why changing the directive on Classification and Labelling of chemical substances is a slow and relatively cumbersome process. It implies a lot of consultation of the competent Directorates General of the Commission to make sure that the proposed changes to Directive 67/548/EEC are compatible with the related bits of European Union legislation and that such changes are acceptable from different points of view, be they economic, logistic or political, *etc*.

Directorates General normally involved in such a consultation process are DG III (Industry), V (Social affairs), VI (Agriculture), VII (Transport), XII (Research), XXIV (Consumer policy) and the Legal Service.

One important element to emphasize is that the European Commission does not take part in the vote of the Technical Progress Committee, which means that, contrary to what some people believe, it does not unilaterally decide on the classification of chemicals (the European Commission does no more than ensure that the provisions of Directive 67/548/EEC are respected) and neither can it change those same provisions unilaterally. Ultimately, Directive 67/548/EEC is made by Member States for Member States. The Commission could be described as refereeing an activity whose rules have to be respected.

It is true to say that there are a lot more organic substances classified under the provisions of the classfication directive than there are inorganic ones. This does not reflect, of course, a specific intention on the part of the European Commission to neglect inorganic substances. The reason is an altogether different one.

Having the right of initiative, the European Commission uses it as a reflection of the needs it perceives at the level of Member States or of groups such as Industry. As time passes such needs reflect themselves in the changes introduced to Directive 67/548/EEC. This directive is not static as its twenty-one adaptations to technical progress and seven amendments imply. The previous reference to the 1996 *ad hoc* meeting to address the classification of alloys is one such example. It cannot be denied that this meeting reflects the capacity of the European Commission to react positively to a 'perceived need' and eventually such an *ad hoc* meeting and the activities that will follow will clarify the situation.

These characteristics surely mean that the directive is 'condemned' to becoming more rich in the future and meetings such as this Luxembourg Seminar certainly have had a positive contribution to make to the debate.

Industry Views from a North American Perspective

I.M. ARNOLD

NORANDA INC., TORONTO, CANADA[1]

1 Introduction

This paper reflects a North American and an Occupational Health perspective on the assessment of carcinogenic risk, particularly as it applies to inorganic substances. Protection of the workforce and communities from untoward risk is integral to this discussion. In preparation for this meeting (Luxembourg, Assessment of Carcinogenic Risk, Session 7) four questions were posed for consideration by the Session 7 workshop. In the broad perspective, they are:

(1) Are current hazard identification procedures and classification systems appropriate for inorganic substances; what changes are needed to improve these systems, and, how/should they be harmonised?
(2) Are current systems for assessment of exposure adequate for risk assessment models?
(3) Are there data inadequacies with respect to knowledge on dose–response; how should they be addressed; how should data be used?
(4) Is the current risk assessment methodology adequate?

These are very important questions that need to be answered. However, they only reflect some of the areas that must be considered in the assessment of carcinogenic risk for inorganic substances. The following discussion also explores some of the other factors that impact the risk assessment process and suggests methods to improve the current approaches to hazard/risk assessment.

2 Discussion

To ensure that the definitions being used in this presentation are clear, the following explanations are given to describe 'hazard' and 'risk'.

---

1 Current address: Alcan International Ltd., Montreal, Canada

2.1 Hazard

For the purposes of this discussion, hazard is defined as 'The (inherent) property of a substance to cause harm'. Therefore, hazard is an attribute. It does not denote risk. Certain conditions must be met for risk to occur. All too often the term 'hazard' is misused as a surrogate for risk and as a justification for the development of standards and regulations. The public consistently misinterprets hazard as risk and adds to their interpretation, the factor of public outrage. In scientific risk communication terms, Peter Sandman[1] has defined the public interpretation of risk as 'Risk = Hazard + Outrage'.

2.2 Risk

Risk is a significantly different term than hazard. The definitions of risk are numerous but they consistently reflect exposure to a hazard. The adverse health impact (toxic effect) of that exposure is a function of concentration of the substance and the duration of exposure. Add to that the function of probability of exposure and the definition of risk starts to become clearer. Risk is, therefore, the probability that a hazard may be expressed. Speciation must also be considered as it will affect the hazard associated with a particular substance. Not all species are born equal. They possess different hazards and consequently different risks. It is no more acceptable to put nickel and all its compounds in the same hazard category than it is to say that all carbon-containing compounds have the same hazard profile.

Much of the modern world has skewed the true meaning of risk as it applies to health end points. Risk must be placed in the proper perspective. Unquestionably, life, itself, may be considered as a sexually transmitted condition, or even 'disease', with 100% risk of mortality. Living is a hazard. It carries with it a high risk factor. An interesting concept of safety – in itself a perspective on risk – is that cited by Hrudey and Krewski in a paper recently submitted for publication.[2] In their paper they quote Malcolm Dawson, a Yukon First Nation Leader, as saying 'A safe level is one that you do not need to worry about'. This is, at first glance, a rather simplistic concept. In reality, it is very perceptive. It reflects, again, the fact that perception of risk is a very subjective one and begs the question as to whether it is really feasible to find a level of risk that is acceptable to all stakeholders. In the same paper the authors advance a theory that there are scientific methods available to calculate the lower boundary of concern using known potency factors. This may represent a starting point for a mathematical risk determination. Their calculations, 'applying conservative assumptions underlying the linear, no-threshold hypothesis for carcinogenesis', suggest that 'under any realistic definition of safety, there exists a safe level of exposure to cancer-causing chemicals'.

Many inorganics, such as metals, have consistently shown up as a source of concern for regulators because they fit three criteria originally applied to organic compounds and to environmental situations. That is, they are toxic, persistent and bioaccumulative. The draft NAFTA Trilateral Framework for the Sound Management of Chemicals[3] states, 'Persistent and bioaccumulative toxic chemicals merit special attention because of their linkages to human health and ecosystem integrity'. Is it, in fact, reasonable to apply these criteria to metals with respect to health-related end points such as cancer? There is considerable debate as to the wisdom of using the latter two criteria, persistent and bioaccumulative, for placing metals in the target area of regulatory concern. Some organisms require the uptake and bioaccumulation of metals to support their physiological processes. They are truly essential micronutrients. As elements, metals are constituents of a multitude of naturally occurring substances and therefore are inherently persistent. The toxic effect of metals and their compounds is dependent upon their concentration, duration of exposure and speciation. Toxicity, therefore may be a reasonable starting point to consider regulation of some substances. There are, however, problems associated with this approach. Bruce Ames, in papers presented at international meetings held in Ottawa and Washington in 1993,[4,5] and in other published papers,[6,7] noted several salient points to consider in determining risk. Although the major causes of carcinogenicity are known, a disproportionate amount of attention is paid to pollution. This stems, in part, from the use of animal test data in risk assessment. These data are frequently generated through the use of MTD (maximum tolerated dose) testing and misinterpreted to mean that low dose industrial pollutants are relevant to the development of human cancer. Dr Ames and his colleagues have also noted that:

- there are no epidemics of human cancer caused by synthetic industrial chemicals. In fact, life expectancy continues to rise;
- zero exposure to carcinogens can not be achieved, carcinogens are naturally ubiquitous. The human body has defence mechanisms that are equally effective against both natural and synthetic chemicals;
- regulating trivial risks of exposure to substances erroneously inferred to cause cancer at low doses can (and probably does) harm health by diverting resources from programmes that could be more effective in protecting the health of the public.

In areas not related to cancer end points, other authorities have also expressed concern about risk communication and priority setting. Professor Sergio Piomelli[8] of the College of Physicians and Surgeons of Columbia University has decried the use of misleading terminology for the communication of risk. In a letter referring to the draft CDC document on childhood lead poisoning he requests that 'lead poisoning' be redefined as blood leads ≥ 20 µg dl^{-1} and that levels between 14 and 19 µg dl^{-1} be classified 'by a more appropriate and not alarmist definition'. He also speaks out against the use of funds for 'screening children at no risk' and carrying out 'wasteful and unnecessary universal screening'.

There are numerous factors to promote regulatory development and reflect the risk assessment process. Only some that are felt to be of a critical nature are discussed below. There is a genuine professional concern about health impacts of some metals and their compounds. There is also considerable professional discord about the interpretation of many health end points. Workplace health and industrial hygiene data are frequent sources of information. The data used to make decisions on carcinogenicity, however, reflect workplace conditions that are at least 20–30 years old. These data were further confounded by the potential synergistic interactions of materials no longer found in modern processes and workplaces. In those areas where stringent regulations are just now being applied, exposures have already significantly decreased as a result of improved plant processes, engineering changes and the use of personal protective equipment. The impacts of a major lifestyle confounder, smoking, is also diminishing as smoking rates decrease in many sectors of society. Health end points such as cancers, are, for many reasons, less likely to occur. For hazard determination, older workplace data have value from an historical perspective. Relating them to the control of risk is misplaced.

Untoward fear of unwelcome health end points, such as cancer, in many sectors of the population, is driven by hazard data misinterpreted as risk data. This misperception results in a political response to the fear, pushing action on regulatory activity to an even higher level.

The ongoing quest for better scientific understanding means that the art of hazard identificiation can be used to determine impacts at minute levels of exposure. Many of these impacts may not even reflect a hazard, but rather a normal response to a biological challenge. Once again, however, these changes are interpreted as posing a significant health risk. Some data gaps do, indeed, exist in our understanding of the hazards of inorganic substances. As a result, even larger data gaps exist in risk determinations. The former gaps need to be identified, and appropriate and co-ordinated research undertaken, to provide the basis for relevant risk assessment.

The current regulatory approach in many jurisdictions is often based on hazard classification systems. Unfortunately, these classifications frequently fail to introduce the parameters necessary to manage risk. Risk management of carcinogens must not be based solely on the decision that there is a hazard associated with a substance. There are at least five steps in a risk management system. The initial step is to simply identify the hazard. Many hazards are well-recognised, but even the use of hazard data is fraught with difficulties, as discussed earlier. The second level of approach, once the hazard is identified, is to determine the risks that it poses. This step is frequently overlooked and risk is equated with hazard. This clearly introduces problems in many spheres – risk communication, regulation, unrealistic expectations, misplaced use of public funds and, potentially, economic hardship. Other steps in the risk management process include risk control, risk communication and system evaluation. The latter is used to measure the effectiveness of the process and recommend ways to better manage the risks – in short, continuous improvement. This multistep approach appears to be consitent with the establishment of risk reduction programmes as advocated in Programme Area D by the Intergovernmental Forum on Chemical Safety (IFCS).[9]

3 Opportunities for Improvement in Assessment of Carcinogenic Risks

There are several significant opportunities for improvement in the assessment of carcinogenic risks. The following points reflect some of those opportunities and are meant to serve as a stimulus to further discussion of this topic.

There are a number of data gaps in our knowledge of hazards related to inorganic materials. These require better definition through an appropriate needs analysis. As a result of such an exercise, research activities can be directed towards narrowing the gaps. As an example, criteria for hazard and risk classification systems may be found to be inadequate, inconsistently applied or subject to considerable global variability. This would suggest a need for more consistent criteria. A co-operative approach by government, industry, academia and other relevant non-governmental organizations is more likely to result in a beneficial result.

A second area for improvement relates to the previous point. Consistent criteria for the interpretation of data will also enhace co-operation and reduce discord within the scientific community.

Uncertainties in the risk determination process must also be clarified. This will result in better understanding of risks and a decreased need to use approaches such as that exemplified by the precautionary principle for minimizing risks to human health. The precautionary principle indicates that where there are threats of serious or irreversible damage, the lack of full scientific certainty is not a reason to postpone cost-effective measures to prevent environmental degradation. Diminishing the uncertainty surrounding the assessment of carcinogenic risk is of vital importance to the ongoing protection of the human component of the global environment.

The last opportunity for improvement is, perhaps, the most important one to consider. The current hazard-based classification system causes confusion and is open to misinterpretation and misuse. A risk management system that recognises all factors that impact risk, offers the opportunity to deal fairly with those problems that are truly important to society.

4 Benefits of Improved Assessment of Carcinogenic Risk

There are several benefits to be accrued from the improvement in the assessement of carcinogenic risks. These are:

- improved carcinogenic risk management in the workplace and in the community;
- use of a consistent global approach to develop an appropriate, cost-effective, regulatory strategy;
- improved understanding of carcinogenic risks through the use of scientifically-based risk information and the application of sound risk communication principles;
- promotion of appropriate priority setting to reduce health risk.

5 Conclusions

Industry continues to participate actively with other agencies in the sponsoring of pertinent health effects and exposure assessment research. Organizations such as the International Council on Metals and the Environment, through their Health Advisory Panel, are actively pursuing the improvement of health and industrial hygiene data gathering and management systems.

Industry believes that greater attention must be paid to speciation, risk communication issues, and the development of classification systems that reflect risk concepts such as carcinogenic potency. Current methods for the assessment of carcinogenic risk are confounded by scientific data gaps, misperceptions and conflicting priorities. It is imperative that a co-operative, scientifically and socially responsible approach be instituted for the management of (carcinogenic) health risks from inorganic materials.

References

1. Sandman, P. (1991) Risk = Hazard + Outrage – A formula for Effective Communication, *AIHA*.
2. Hrudey, S.E. and Krewski, D. (1995) Is there a safe level of exposure to a carcinogen? Submitted for publication in *Environ. Sci. Technol.*, Revised Edition, May 2, 1995.
3. NAFTA (1995) A Trilateral Framework Statement for the Sound Management of Chemicals, Draft Document, North American Committee on Environmental Co-operation, September 1995.
4. Ames, B.N. (1993) Does Current "Risk Assessment" Harm Health?, American Society for Cell Biology, Congressional Biomedical Research Caucus, Washingon, D.C., November 5, 1993.
5. Ames, B.N. and Gold, L.S. (1993) Comparing Synthetic to Natural Chemicals is Essential for Perspective in "Risk Assessment", Managing Risks to Life and Health, Ottawa, Canada, October 18, 1993 (also submitted to *Risk Analysis*).
6. Ames, B.N. and Gold, L.S. (1993) Environmental Pollution and Cancer: Some Misconceptions. In, 'Phantom Risk: Scientific Interference and the Law', Foster, K.R., Bernstein, D.E. and Huber, P.W., eds., MIT Press, Cambridge, MA, pp. 153–18.
7. Gold, L.S., Slone, T.H., Stern, B.R., Manley, N.B. and Ames, B.N. (1992) Rodent Carcinogens: Setting Priorities, *Science*, **258**, 261–265.
8. Piomelli, S. (1995) Letter to Henry Falk, M.D.; Environmetal Hazard and Health Effects (NCEH) Re: CDC Draft Document on Childhood Lead Poisoning; Sept. 7, 1995.
9. Intergovernmental Forum on Chemical Safety – Background, purpose, functions, WHO, September, 1994, pp. 38–43.

Industry Views from a European Perspective

L.G. MORGAN

CONSULTANT IN OCCUPATIONAL HEALTH, SWANSEA, UK

1 Introduction

Firstly, I propose to discuss the procedures for decision making by the Authorities, and the need for more openness as well as the establishment of an appeal procedure.

Then through the topic of **Hazard Identification**, I will look at classification and labelling, including:

- the continuing need for *speciation* (and the avoidance of generic classification);
- the re-evaluation of *epidemiological* evidence;
- the problem of *classification of alloys*;
- and hopefully stimulate some discussion about *potency*.

In respect of **Risk Assessment**, I will raise the issues of its proper use on all relevant directives. I will point out that it is wrong just to take a classification table, such as the carcinogens classification, and use it in other directives without assessing whether it is relevant.

In the area of **Risk Management**, I will mention some of our concerns, including,

- the setting of exposure limits and encourage the use of the *de minimis* concept;
- the difficulties of the *small and medium enterprises* and how they need rules that are simple to understand and that are easy to apply and monitor;

and a little about *atmospheric sampling* and the need for relevant instrumentation.

In my **conclusion**, I will mention the fact that steps are now in hand to develop the same systems of classification and labelling on a world-wide basis.

My remit is to explain the non-ferrous metals industry position in respect of hazard and risk assessment and risk control of occupational carcinogens, and to do this from a European point of view. My colleague Dr Arnold has talked in very general terms. I have homed in on the European legislation and make no apology for this. What is interesting is that we have both independently felt it necessary to define hazard and risk. We clearly share a feeling that this is necessary, so that not only are we all talking about the same thing but also that we know what we are talking about!

To represent industry is no easy task as, not surprisingly, even within industry there are differences of opinion and emphasis, and clearly I have not been able to talk to all concerned. I hope, however, to have steered a middle course and, anyway, my objective is to stimulate debate in the subsequent discussion.

Those of you who have other ideas for the future or even problems that you want to raise will have the opportunity to discuss them later, but I would remind you that the purpose of the whole meeting, and this session is no exception, is to be constructive. We want ideas for the future, not criticism of the past.

I am very grateful to all those of my colleagues who have helped me in the preparation of this paper.

2 Discussion

Among the inorganic substances, the metals of interest in respect of carcinogenicity include arsenic, antimony, beryllium, cadmium, chrome, cobalt, lead and nickel and we are also interested in sulfuric acid mists and welding fumes.

It is now generally recognised that Health, Safety and Environmental issues should have a prominent position in the business plan of any company. Indeed, many major organizations employ experts who are very knowledgeable about these issues in their particular industry. In this context, while accepting and indeed welcoming any rule that protects individuals and the environment that they live and work in, industry would wish to be consulted in the development of those rules. Using our expertise, we can help to ensure that the rules are relevant and do not place unnecessary burdens upon those who have to apply them.

How does that basic philosophy of running a healthy and safe workplace fit in with the aims and objectives of this conference, namely to review current procedures of assessment of carcinogenic risk and see whether any changes are needed and if so, what?

It must be said at the outset that a lot has been achieved by the current system over the years and there are now many laws in member states that result from the EU legislation we are discussing this week. It is to be hoped that they are reducing cancer risks to individuals. Clearly the rule-making was 'State of the Art' at the time it was established, originally in 1967, and science has moved along since that date. The European Commission does update directives by means of amendments and adaptations 'as a result of technical progress' but we have felt for some time that a complete rethink was needed. We were therefore pleased to assist when this conference was proposed.

My first point will come as no surprise; it is that industry frequently does not agree with every one of the assessments that are made, particularly when they are at odds with evaluations published elsewhere. In this respect we would wish for complete openness in decision making, especially in respect of carcinogenic classification and of permitted exposures and limit values. These decisions are of great importance for everybody. So, we believe, the data and arguments on which they are based should be published for all to see. Finally, there should be an appeal procedure where these decisions can be disputed in front of a neutral, disinterested body by the people they concern. The composition of such a body would be a matter of debate.

Since there is frequently confusion about what is meant by risk and hazard, I trust you will bear with me in establishing what is meant by these terms before we talk about them.

According to Council Regulation 793/93/EEC on the Assessment of Risks from Existing Substances:

> *Hazard* is defined as 'the intrinsic property of a chemical agent with the potential to cause harm'
> *Risk* is defined as 'the likehood that the potential for harm will be attained under the conditions of use and/or exposure'

2.1 Hazard Identification

2.1.1 Classification and Labelling

2.1.1.1 Introduction Back in 1967 the Commission decided upon a system of classification of dangerous chemicals based on hazard. Classification by risk was not used because it involved bringing in external factors of exposure. However, the EU Classification Directive was coupled with labelling and in that respect it now fails to comply with the modern concepts of risk management The label should refer to the risks associated with first handling of the product held within the container, while the safety data sheet should describe all the hazards that could occur in handling and use. Instead, it refers to the hazards that have been identified during classification. Thus what is called a *risk phrase* in the official European literature is in fact a *hazard phrase*.

2.1.1.2 Speciation My next point in respect of hazard classification is that of speciation, which is one of the areas that causes us considerable concern

In organic chemistry there are large groups of materials with similar structures, so it is possible to predict that since one member of the group has certain chemical or toxicological properties many or all other members will also have the similar properties, to a lesser or greater degree.

It worries us that some official bodies and legislators do not recognise that the same is not true in respect of metals and other inorganic chemicals. They frequently consider that if a metal in one chemical form is shown to have a particular property such as being carcinogenic, then all forms, metals or metal compounds must, have the same property. Such a position does not allow for the widely different properties, including valency and solubility, of the metal compounds, of which perhaps the most important is bioavailability to the target cell, or even the specific intracellular site of action.

Several bodies concerned with classification of carcinogens have recognised that species differences do occur, and for example, most have recognised the differences in respect of chromium(III) and chromium(VI).

On the other hand, IARC failed to recognise that there might be differences between the carcinogenic potential of the less soluble and more soluble nickel compounds and failed completely to recognise any difference in carcinogenic potential between beryllium and its compounds, cadmium and its compounds and cobalt and its compounds.

The ACGIH, although given sound scientific evidence showing that the toxic properties of pure cobalt powder are different from those of hard metal powder (a mixture of carbide with only a small percentage of cobalt), has never taken this into account in its assessments on cobalt. On the other hand, recognition of the biological differences in the speciation of nickel and its compounds are being recognised by the 'Dusts and Inorganic Committee'. In addition, it is interesting to stress that the recent National Toxicology Programme in the US has shown that some forms of nickel are carcinogenic by inhalation but others (nickel sulfate) are not.

Industry would wish that during both hazard and risk assessment, speciation be recognised as being fundamental. For the sake of simplicity some grouping may be required, or appropriate, but the physical and chemical properties affect the toxicology enormously and it would therefore be quite wrong to lump a metal and its compounds together just because they may under certain conditions release a common cation. The EU did at least make a start in respect of nickel compounds by classifying some as category 1 and the metal and certain other compounds as category 3 and has recognised that some cadmium compounds are category 3 and some category 2 carcinogens but that the metal is not. We need more of this type of speciation, where it is appropriate. Both the chrome-producing and -using industries are particularly concerned to see implementation of the recommendation by the Specialised Experts advising DG XI that all chromium(VI) not yet classified should be made category 2.

2.1.1.3 Epidemiology My third point is in respect of epidemiology, the use of which needs careful evaluation and a possible rethink. Dr Draper has already stressed how in the nickel industry at Clydach it was a combination of 'within process' factors that led to the problems, and that classification of individual substances may in some cases be incorrect. DG V has indeed recognised this point and classified the relevant process. The beryllium industry has frequently pointed out that classification of beryllium, resulting from the tiny (and often disputed) excess of cancer in the refining industry, does not take into account the possibility of overload, nor does it take account of potency.

Secondly, we want a guide from epidemiologists to assist the uninitiated in the intepretation of their findings. They frequently give a figure of a threshold below which the relative risk should not be considered significant. We ask the Commission to consider producing a general guide that takes into account the important variables such as the size of the population and the effect of smoking and other lifestyle factors. This could be of value to all concerned, but particularly 'non-experts' from small industries, in interpreting epidemiological data for classification and standard setting.

Epidemiologists have a problem in assessing exposure from historical data for the purpose of both hazard and risk assessment. Frequently there are some data available but they may be sparse and of doubtful quality, or they has been collected in a form that is not relevant to today's needs. However, the tools are often available for a rough guess. The semi-quantitative nature of this rough guess should be clearly explained and, preferably, a sensitivity analysis performed on the various guesses to decide how large an uncertainty factor should be included. This needs more research.

Epidemiologists have another set of problems which are going to render the science less effective a tool in the industrial setting in the future These are that the smaller numbers of the workforce, the greater its mobility within a particular operation resulting in mixed exposure, and the improved environmental conditions compared with the past will mean that the ability to study large populations will not be possible.

No mention has been made of the use of biological monitoring as an epidemiological tool. In the case of some metals it is possible to measure the excretion of the metal, or its level in the blood relatively easily. The data can provide a good index of recent exposure to some chemical forms, taking all routes of such exposure into account. Can these data have a place in epidemiological studies, or indeed do they have a more useful role in the risk management field, which we will be discussing later?

It is also possible to measure metabolites and markers of genetic damage in blood and/or urine, but at the moment these are very nonspecific and as much affected by lifestyle as industrial exposure. Is this a field for development in the future?

2.1.1.4 Alloys No presentation by industry on the problems of carcinogenicity of metals and related inorganic substances would be complete without some mention of alloys. The current position is that the Preparations Directive specifies default values for the administrative concentration limits of a substance in a preparation that activates classification of that preparation. Thus if a preparation contains more than 0.1% of a category 1 or 2 carcinogen, or more than 1.0% of a category 3 carcinogen, then the preparation has to be treated as if it were a carcinogen of that category. For example, stainless steel (FeCrNi), and copper–beryllium alloys are thus category 3 and category 2 carcinogens, respectively This classification assumes that alloys are simple mixtures, which is chemically incorrect. This is where the problem arises. In many alloys the dissolution of certain metals in others renders the alloy insoluble and so it is believed that no harm should result. But how can this negatively be proved? The philosopher's stone that we need is a screening test, or batch of tests, to determine the carcinogenicity of metals and metal alloys that is acceptable to all concerned. However, this is for the future and will probably result from advances in molecular science and genetics. What is needed now is special consideration for alloys taking into account their physical, chemical, metallurgical and toxicological properties. We are pleased that talks on this issue are now to take place.

2.1.1.5 Potency This subject has been raised at DG XI discussions under the debates on the dangerous substances and preparations directives, so I am considering it here under hazard, but some may feel it should better be considered under risk.

The DG XI group concluded that for extremely potent carcinogens the concentration in a preparation that triggers classification should be lowered from the default value. Furthermore, that for certain nongenotoxic carcinogens (*i.e.* those that work by irritation or some other method not directly on the gene) can be raised from 1 to 5%. Considering that the default values were arbitrary in the first place, one wonders why, when there is information available and clearly there are such vast differences in potency even amongst the genotoxic carcinogens, that the figures cannot be raised higher than 5%. These are questions that remain open and could well be the subject of discussion this afternoon.

2.2. Risk

EU hazard classification is not directly concerned with the workplace or with protection of the worker. It was originally devised for the purpose of labelling so that the warning on a container should be the same in each country of the community. The hazard does not vary, it being an intrinsic property of the substance, but the risk it poses varies greatly depending upon the exposure. The EU relies on other directives for workplace legislation, notably the Carcinogens at Work and the proposed Chemical Agents Directives.

These two directives require a risk assessment to be made when dangerous substances are used in the workplace either as such or as preparations. However, there is concern within industry that a hazard classification, designed for one specific purpose, may be used for another without appropriate risk assessment. This may lead to unnecessary substitution. There is no argument that substitution is quite correctly one of the main planks of the carcinogenic substances regulation, but the risk has to be recognised and evaluated and substitution deemed appropriate and necessary. Other directives, in order to avoid re-classification, have used the Classification and Labelling Directive 67/548 as a yardstick and have not included a relevant risk assessment. The clearest example was the original, 1982 version of the Seveso Directive. In this version, implemented before nickel had been classified, any facility with more than 100 kg of nickel metal on the premises was required to register as a major accident hazard site! Clearly no assessment of the risks involved had been undertaken! I should mention that this has since been revised but nickel monoxide, for instance, still remains on the list. The simplest risk assessment would show that this is unjustified.

Another similar problem could arise from the fact, already mentioned, that all chromium(VI) compounds are to be classified as category 2 carcinogens. We are concerned that there is a possibility this could lead to indirect classification of certain metallurgical processes such as stainless steel production or welding, in spite of the fact that epidemiological surveys have not shown any significant excess cancer risks.

While no directive can legislate for all industrial use, it should at least map out the relevant risk assessment procedures that are needed to evaluate risk so that the user can understand that the material can be used safely. In this way we might be able to reduce the ever increasing problem of commercial concerns banning the use of certain materials within their organizations to the detriment of their products (and our markets) with no recognisable gain in safety. We are advised that it is industry's problem to educate customers how their materials can be used safely. However, the natural suspicion of the truthfulness of a supplier, coupled with the lawyer's natural caution against anything that might result in a law suit, makes this a very difficult task. We do need all the help we can get from the Commission here.

2.3 Risk Management

The legislators have an important role in recommending and then rule-making, in respect of safe working practices and deciding upon *exposure limits*. To do this they have to assess the risks involved from exposure to the substance.

We are exposed to carcinogens all day and every day and yet many of us can expect to live to a ripe old age and not die of cancer. In this context, a *de minimis* level of exposure to carcinogens is now recognised to exist. At this level of exposure the risk is so small that it may not be detectable, and anyway is acceptable to the community at large. This applies both inside and outside the workplace. In deciding on occupational exposure limits for carcinogens for the workplace, legislators should determine an acceptable level of risk and then recommend exposure limits taking into account both the *de minimis* exposure value and the state of control technology in the using industries to ensure feasibility. Thus, any limits should not be set so low that they unnecessarily put people out of work, causing other problems.

The actual mechanics of risk management depend upon a number of factors, many of which are relevant only to specific industrial operations, the equipment and procedures being used, and the materials being handled. Each employer must therefore organize and run his business in such a way that the risks are contained and that exposure levels comply with the appropriate standards. The rules that are made should place an obligation on the user to have a safe operation.

We stress again that this does not necessarily mean banning hazardous materials, but it does mean that, in certain situations, substitution might be considered and where not appropriate, the original material be used with due diligence so that the relevant risks are contained. This is therefore more of an industry problem than one for the Authorities, as to prescribe how an operation should be carried out would be to stifle all innovation and be counter-productive.

However, there is one point that requires consideration and that is in respect of *monitoring*. The science of atmospheric monitoring is now well-developed, although further advances are on the way and legislation must be able to allow for them. One can get surprises, thus it has recently been shown that an air sampler specifically designed to pick up only the inhalable fraction of dust gives a higher result than previous models believed to be collecting the total airborne dust! New samplers and sampling methods are currently being introduced and considerable work is being performed to facilitate extrapolation from the old to the new. The Commission should support this research, which will be a vital adjunct to risk assessment and management in the future.

Biological monitoring has a limited place in the control of exposure to occupational carcinogens at the present time because, in respect of metals and related inorganic substances, it has not been possible to equate levels in the biological matrix with the risk of developing cancer. This is a field where we can expect new developments.

I should mention here the problems of the *SMEs* (*small and medium enterprises*). They generally cannot afford the expertise employed by larger industries and need all the help and information they can get. Most importantly, while they should not be allowed to put their employees at risk, the rules that they have to work to should be simple and easy to apply and monitor.

Thus, in addressing you, I have:

- reminded you that we should be looking forward to the future, questioning existing practices and putting forward ideas for improved legislation;
- requested more openness in decision making with the possibility of an appeal procedure;
- discussed the need for speciation in hazard identification;
- mentioned the problem of potency and questioned whether it should play a greater part in classification;
- raised the problem of classification of alloys and suggested that the time has come for special legislation to meet the needs of these remarkable materials while also preserving the purpose of the legislation, namely to protect human health and the environment;
- suggested mapping out relevant risk assessment procedures for individual directives to reduce unnecessary 'banning';
- discussed the need for pragmatism in the setting of occupational exposure limits;
- stressed the need for help from the Commission in the education of users of hazardous materials, to make it clear to them that they need not be banned and can be used safely if proper risk assessments are undertaken and risk control measures applied.

Finally, I would remind you that work is already in hand to make the rules common to all OECD Member States. I would therefore ask whether the time is right for a complete rethink of the classification procedure and the risk assessment of metals and related inorganic substances?

Summary of the Panel and Plenary Discussion

M.H. DRAPER

EDINBURGH CENTRE FOR TOXICOLOGY, EDINBURGH, UK CHAIRMAN OF THE PANEL DISCUSSION

This session consisted of seven presentations from regulatory interests, international advisory bodies and industrial concerns on different aspects of the assessment of carcinogenic risk. Papers were given by representatives from:

(1) The International Agency for Research on Cancer of the World Health Organization (P. Boffetta);
(2) The International Programme on Chemical Safety of the United Nations Environment Programme, the International Labour Office and the World Health Organization (B.H. Chen);
(3) The Occupational Safety and Health Administration of the USA (W. Perry);
(4) The European Commission, DG V, Public Health and Safety at Work Directorate (G. Aresini);
(5) European Commission, DG XI – Environment, Industry and Environment Directorate (J. Costa-David);
(7) Industry views from a North American perspective (I.M. Arnold);
(8) Industry views from a European perspective (L.G. Morgan).

The representative invited from the Trade Unions was unable to attend. These presentations were followed by a panel discussion with the presenters which was moderated by M.H. Draper, Edinburgh Centre for Toxicology. The panel discussion was then extended to plenary discussion.

The question of carcinogenic potency was raised. This is clearly an important practical issue as far as industry is concerned, yet it is rarely addressed. In reply it was pointed out that IARC does not do risk assessment. IARC experts endeavour to evaluate the available relevant scientific literature and, if possible, to derive a dose–response relationship in so far as it helps with hazard classification. Thus, no classification is attempted in terms of potency.

A particular cause of concern to industry regarding metals and metallic compounds was the apparent lack of transparency in the way classification decisions were reached. In some organizations, despite the presence of delegates from interested parties at the meetings of experts, there was limited participatory scientific debate and sometimes the dossiers containing the data on which the decisions were based were not made available. There was also a strong feeling that in some cases there were political issues, rather than scientific issues, influencing the outcome. It was explained that, strictly, it was not DG XI that made classification decisions but a panel of national experts and that, if there was considerable controversy concerning a particular chemical, a decision to seek the opinion of a further panel of specialized experts could be taken. In addition, it was possible for delegates to be assisted by their own experts. It should also be remembered that classification is not a purely scientific matter and other factors may have to be considered.

With regard to the exact chemical speciation of exposures, it was suggested that this called for more experimental data, particularly from animal experiments. However, on the regulatory side, this is politically unacceptable because the official policy of the European Commission is to reduce as far as possible the use of animals in testing procedures. Alternatives to animal studies are being pursued actively, but it will be some time before adequate substitutes are available.

The plenary discussion was dominated by two linked issues, the immediate practical issue of the current rules for classifying alloys and the more general issue of the classification of inorganic substances.

Concerning alloys, it was pointed out that the properties of metal atoms in an alloy were quite different from those of the same metal atoms in the pure metals or other types of compound. There could be no scientific justification for extrapolating from pure metal to alloy, or for extrapolating from other chemically distinct compounds. To cite one example among many, the importance of beryllium–copper alloys used in computer circuits was raised. Here, the classification of beryllium and beryllium compounds into the IARC class 1 was greatly influenced by the mortality from lung cancers reported among men and women who were exposed to a complex beryllium phosphor of unknown composition during the early days of fluorescent lamp manufacture. This is clearly not relevant to exposure to alloys.

From the regulatory side it was recognised that the existing methodology was not appropriate for alloys and it was stated that this problem was being studied further. However, it had to be emphasized that the system had been developed by the Member States and to change it was difficult. Perhaps those who consider that the system has faults should propose a better system and then engage in dialogue at the appropriate level within the Member States.

Criticism was made of the way that inorganic entities were understood in toxicological and, subsequently, in regulatory contexts. It is scientifically unacceptable to assign toxicological properties simply to an element. Organic chemists would never talk about the carcinogenic properties of carbon, nitrogen, oxygen or hydrogen, although these may be the components of a carcinogenic substance. Likewise, we should not talk about the carcinogenicity of cobalt, chromium or nickel, *etc*. It is the combined form of elements into specific compounds that should concern us. It was agreed that the current toxicity classification systems for inorganic substances, such as metals and minerals, are unsatisfactory from a scientific point of view. The disciplines of inorganic chemistry and mineralogy have been developed to identify and describe inorganic substances and should provide the basis for a better system. A positive step that might be taken would be for expert inorganic chemists and mineralogists to collaborate in some way with the ongoing work of the Intergovernmental Forum on Chemical Safety (IFCS). In order to improve international harmonisation of classification systems, it was suggested that the proceedings of this seminar should be submitted by the European Commission to the IFCS.

Other more practical issues were also considered, for example, the problem of mixed exposures. There is also continuing concern about the adequacy of exposure measurements, particularly in relation to chemical species determination. There may be a need to develop more sophisticated sampling and analytical equipment. In any new developments, the importance of correlating old and new exposure data must be borne in mind. Regulatory authorities should encourage these developments.

Much discussion centred around the activities of the European Commission, which are guided by committees of national experts. Concern was expressed that, although Commission decisions reflected the advice of a committee of national experts, sometimes these experts might not be sufficiently knowledgeable about the inorganic substances under consideration. It was recognised that there are sensitive issues to be considered here and changes may be needed at the national level as well as at the international level if good science is to prevail.

Concern was expressed that inaccurate hazard identification must result in ineffective risk management. It was proposed that industry should co-operate with the Commission in producing an explanatory document showing clearly the relationship between hazard identification, risk assessment and risk management.

In conclusion, satisfaction was expressed at the opportunity that the seminar had afforded for regulators, research scientists from many disciplines, and representatives from a range of industries, covering smelting, refining and usage, to meet together and discuss mutual problems. It was hoped that this might be a starting point for further discussion and debate about the problems that had been identified.

Working Party Reports

Working Party 1: Physico-chemical Characterisation of Exposures

CHAIRMAN: A. LANGER

RAPPORTEUR: M.V. PARK

1 Remit

Working Party 1 reviewed the current methods used routinely to characterise exposures to inorganic substances and considered how they can be improved.

Questions under consideration included:

- Are current methods adequate for the assessment of carcinogenic hazard and risk?
- Can improvements be made in sampling strategies?
- Are there techniques now used only in research laboratories that could be applied with benefit to the work environment on a routine basis?
- Can we target monitoring on genuinely hazardous chemical species and move beyond simple elemental analysis?
- Can we make use of bio-markers as measures of exposure to inorganic substances?
- Should we routinely monitor for organic carcinogens such as polycyclic aromatic hydrocarbons, which may contaminate inorganic substances and confound carcinogenicity assessment?

2 Report

2.1 Assessment of Workplace

2.1.1 Identification and Characterisation of the Materials in the Workplace and Local Environment to which People are Exposed

This would include and focus on their speciation, as far as possible (elemental analysis alone is insufficient), their structural composition and physical characterisation, (including particle-size distribution and solubility in biological fluids) and surface properties of solids. This would be carried out in existing industries, in plants where feedstocks are from multiple or changing geological sources, in new processes, or where a health risk is indicated in a current process. The possibility of contamination by potentially harmful organic compounds at levels known to be hazardous should also be examined.

2.1.2 Analytical Techniques

Standard techniques currently used for elemental analysis are inductively-coupled plasma atomic emission spectrometry (ICP-AES) and X-ray fluorescence (XRF). Other techniques of characterisation and/or speciation are scanning and transmission electron microscopy with energy-dispersive X-ray spectrometry (SEM/EDX and TEM/EDX) and X-ray powder diffraction spectroscopy (X-ray PDS), with, for aerosol analysis, aerosol spectrometry. Consideration could also be given to the application of more advanced techniques such as X-ray PDS using synchrotron radiation (to increase sensitivity), laser ionization mass spectrometry (LIMS), electron spectroscopy for chemical analysis (ESCA) and Auger electron spectroscopy (AES).

In many cases the specificity of solid phases could best be indicated by reference to their X-ray powder diffraction spectrum. Limitations are recognised in the analysis of complex mixtures, mixtures with trace crystalline phases, the presence of amorphous phases and crystalline members with structures not yet described. Single particle heterogeneity is recognised to occur in some industrial environments, and these pose special problems in identification and characterisation.

2.2 Monitoring

2.2.1 Following this periodic characterisation of the plant and process, routine area and personal samples could be examined with less critical detail, with the possible use of 'surrogate' marker compounds, *i.e.* the monitoring of the level of a particular species as being indicative of the overall level of a single toxic substance, or of a group of potential toxicants.

2.2.2 Monitoring samples collected over regular intervals should be archived to permit retrospective examination. Special storage environments and conditions may be required, and archiving will be necessary over a long period, decades rather than years.

2.2.3 Records of this monitoring must be maintained to allow the results to be related to individuals and their work locations. These data may be pertinent for future study.

2.2.4 Ideally, procedures should be standardised as far as possible to allow inter-plant comparisons (external auditing and quality control will be required for this).

2.2.5   The purpose of this monitoring may be to ensure compliance, but has other uses such as exposure assessment and reduction on a continuing basis, and the testing of the effectiveness of control systems.

2.3 Biological Assessment

2.3.1   Workplace air monitoring may form part of exposure assessment, but it requires full validation of its effectiveness for that assessment in the workplace, and the conditions confirmed.

2.3.2   A more direct procedure is monitoring for bio-markers by a non-invasive technique such as urine analysis, and could range from elemental analysis, monitoring of specific bio-marker compounds, to examination of the structures of cells shed in the urine. Sputum, blood, hair or other materials may also be appropriate.

2.3.3   The use of immunological markers is also possible, although such methods are presently in the developmental stage.

2.3.4   There are other techniques slightly more invasive, such as pulmonary lavage, which have been used to bio-assess toxic agents. Occasionally, tissue analysis on biopsy and autopsy materials have to be used as well. The latter are powerful research tools.

2.4 General Comments

2.4.1   It is important that scientists testing substances in animals appreciate the significance of their work for the Regulatory Community. They should publish their data and conclusions with an appropriate degree of rigour and with interpretative caveats, where appropriate.

2.4.2   Communication between Industry, Labour, Academia and the Regulatory Authorities should be encouraged. It is recommended that procedures be set up to bring about regular meetings among these interested parties, and that new data be evaluated on a periodic basis.

2.4.3   It is imperative that Management at the highest level is made aware of the importance of these recommendations and that resources are made available accordingly.

2.4.4 In order to assist those involved in hazard assessment, publicity should be given to substances which are potentially biologically harmful, *e.g.* arsenic, ozone, and also to chemical processess such as those producing valence changes, *e.g.* Cr(III) to Cr(VI), which may result in increased hazard.

2.4.5 Archiving of appropriate tissues, particularly lung tissues, obtained at autopsy appropriate for further studies is highly recommended.

3 Working Party 1 – List of Participants

Core members

Dr A. LANGER (Chairman)
Dr M.V. PARK (Rapporteur)
Dr R.P. NOLAN

Dr A. ROBERTSON
Prof A. CHURG

Participants

Mr N. ANDRE
Dr E.G. ASTRUP
Dr S.R. BERGE
Dr R. BROWN
Dr R. CORNELIS
Dr G. CRAWFORD
Dr B. CONARD
Mr G. ETHIER
Dr K. ETZLER
Dr C. FERNICOLA
Prof B. FUBINI
Prof P.J. HEWITT
Mr H.W. HOEKSEMA
Ms M. JAROSZEWSKI
Mr T. JEPSEN
Dr A. JONES
Dr M. KOPONEN

Dr A. LAMBERTY
Mr J.C. MAHIEU
Dr B. MANTOUT
Mr S. MARASCHIN
Dr R. MONTAIGNE
Dr R. MURRAY
Mr L. OKSANEN
Dr F. PELLET
Dr S. SHCHERBAKOV
Mr M. STENTIFORD
Mr E. SUNDQUIST
Dr C.F. TAYLOR
Dr D. TOLLERUD
Prof R.J.P. WILLIAMS
Mr F. WILMOTTE
Dr M. WYART-REMY

Working Party 2: Interpretation of In Vitro and In Vivo Experimental Studies

CHAIRMAN: U. SAFFIOTTI

RAPPORTEUR: K. SCHUMANN

1 Remit

Working Party 2 reviewed the current *in vitro* and *in vivo* test systems to identify their limitations as bases for predicting human carcinogenicity of inorganic substances, and to recommend ways for increasing their relevance.

Questions under consideration included:

- How can clearly identified limitations of such studies be minimized or discounted?
- Can the existing data be better utilized for assessment of carcinogenicity?
- How relevant are carcinogenic reactions caused by administration routes inappropriate with regard to human exposure (*e.g.* injection)?
- Is it possible to make exposures in experimental studies relate better to normal human exposure patterns?
- Are the most appropriate species and strains of animals being studied?
- Are there any new test systems which offer clear advantages over the methods currently in use?
- Is it possible to provide an experimental data base for assessment of mixed exposure situations?

2 Report

The following conclusions and recommendations were agreed upon.

2.1 Standardisation and Speciation of Inorganic Test Material

(1) Standardised and well-defined test materials are needed to compare results from different *in vivo* and *in vitro* methods and between different laboratories. The materials should include samples representative of real life exposure. Alteration of the samples in individual laboratories, prior to testing, should be avoided.

(2) In the case of particles, such as crystalline silica, the quantitative

measurement of exposure for both *in vivo* and *in vitro* tests should be expressed not only as weight, but also in relation to surface area and to particle size distribution and number. These parameters are important for the evaluation of dose-response relationships. In some cases, it may also be appropriate to consider the surface area of cells involved in the interaction with particles.
(3) Chemical speciation and physical characteristics need to be considered in the design of toxicological tests. Oxidative states may vary though the different compartments of the organisms. Methods should be developed to characterise such changes. It may be possible, however, to quantify exposure to certain metal species during the production process or in the use of the finished product.

2.2 Toxicokinetics, Bioavailability and Dose–Response Relationships

(1) Further studies are needed on the role of mixed exposures to different inorganic materials. In the case of metal alloys, it is recommended that the bioavailability and tissue deposition of critical components be determined. Differences in the nature of the chemical and electrochemical reactions of metals and alloys with body fluids should be considered.
(2) Research efforts should be pursued to understand better the importance of the physico-chemical features of metal compounds and mineral particles and fibres (*e.g.* surface properties, morphology) in the sequence of events leading to genetic damage and cancer initiation. For cancer risk assessment, it is crucial to determine whether physico-chemical features of metal compounds affect mainly the bioavailability and distribution of the metal ion or are directly or indirectly involved in DNA damage, and whether they influence the release of mediators by the metal.
(3) Both *in vivo* and *in vitro* experimental results can serve to identify carcinogenic hazards. They can also be used to study dose–response relationships and, thus, to assist in risk extrapolation. Experiments should be critically designed, including negative and positive controls.
(4) Comparison between *in vivo* and *in vitro* systems is difficult. An increasing variety of cellular and cell-free tests are becoming available and can provide relevant data for evaluation of carcinogenesis mechanisms. *In vitro* data may not relate to *in vivo* bioavailability, nor to *in vivo* latency of the carcinogenic effect. However, determination of dose–response relationships in *in vitro* systems may be of value.
(5) There are differences among animal species regarding susceptibility to inorganic carcinogens. The choice of the most appropriate species, for long-term studies and for hazard assessment, can be optimised on the basis of short-term experiments comparing the toxicokinetics of the compound(s) or the response of relevant cellular and molecular targets. This approach is also useful to give a rationale for the extrapolation to humans. The use of human tissues and cells may be particularly useful in this context.

(6) Different routes of exposure may lead to different reaction patterns in the body, which may be due to differences in distribution, to the overwhelming of the local defence systems (*e.g.* oxygen scavengers, DNA-repair mechanisms) or to other factors, especially at excessive exposure levels. Investigation of early and intermediate reaction end points can be helpful in assessing exposure conditions for long-term bio-assays.

2.3 End points, Test Systems and Preservation of Biological Materials

(1) Short-term and intermediate end points can be effectively used on the basis of underlying mechanisms of carcinogenesis. Such end points can be based on early tissue reactions, induction of neoplastic transformation, quantification of cell surface receptors, growth factors and gene expression, production of free radicals, as well as on modification of defence mechanisms, such as macrophage recruitment and persistent inflammation.
(2) The exposure levels used in the different systems need to be judged in relation to the sensitivity of the system. The choice and sequence of different *in vitro* and *in vivo* assay systems to study a substance under consideration depends on the goal of the study and on the available knowledge about the substance. Batteries of test systems have been described in detail, for example, in the Organization for Economic Co-operation and Development (OECD) 'Guidelines for Testing Chemicals' (1982 onward). However, these guidelines are mostly applicable to the testing of organic substances. Guidelines and databanks should be developed for inorganic compounds and properly updated.
(3) Biological materials derived from experimental animals and from occupationally-exposed humans with corresponding diseases should be sampled and preserved for future molecular and genetic analysis.

3 Working Party 2 – List of Participants

Core members
Dr U. SAFFIOTTI (Chairman)
Dr K. SCHUMANN (Rapporteur)
Dr M.-C. JAURAND

Prof H. MUHLE
Dr C. KENNEDY
Prof P. MORROW

Participants
Dr L. ARINGER
Ms A.C. BOUTIN
Mr C. BOZEC
Prof J. BRUCH
Dr B. CHEN
Ms R. CÔTÉ
Dr L. CURCIO
Dr B. DERO

Dr A. GIOVANETTI
Prof U.F. GRUBER
Dr S. JACOBI
Dr D. JANS
Mr C. LAMBRÉ
Dr H. LINDEMANN
Dr M. MACLAINE PONT
Prof P. MORROW

Participants

Mr J. DESCAMPS
Prof S.G. DOMNIN
Dr J.H. DUFFUS
Prof J. DUNNIGAN
Dr Z. ELIAS
Mr R. FAJARDO
Dr U. FÖST

Mr J.M. PUJADE-RENAUD
Dr B. REHN
Dr G. REYES
Dr M. ROLLER
Mr F. SEILER
Dr G. SCHOETERS
Dr W. VON DER HUDE

Working Party 3: Conduct and Interpretation of Epidemiological Studies

CHAIRMAN: L.G. MORGAN

RAPPORTEUR: S. PORRU

1 Remit

Working Party 3 reviewed the limits to the effectiveness of current epidemiological practice to identify how they may be overcome, and how methods must develop in order to provide clear identification of human carcinogens.

Questions under consideration included:

- Can the available database be improved in practice, for example, by improved death certification, cancer registries with linkage to employment history, taking account of confidentiality rules, *etc*.?
- What information would epidemiologists like to have from exposure monitoring and how can data be best recorded and made available when required?
- Can retrospective epidemiological studies be improved?
- Is there scope for further application of metademographic techniques in epidemiology?
- Are there ways of obtaining relevant historical data, particularly relating to exposure assessment, that have not been sufficiently exploited?
- Can the statistical techniques for analysis of epidemiological data be improved?

2 Report

The remit was (1) to review the limits of current epidemiological practice in respect of the identification of human carcinogens (of inorganic substances); and (2) to consider how methodology needs to be developed so that these limits may be overcome.

In considering the limits it was agreed that, except for physico-chemical characterisation and speciation, there were no specific problems in relation to inorganic substances that did not concern all occupational epidemiology.

The topics discussed were (1) available databases on inorganic substances; (2) the problems of exposure monitoring; (3) the role of historical data and retrospective epidemiology; (4) the statistical techniques; (5) quality assessment.

2.1 The Database

(1) *Use of registries.* In respect of the studies based on routinely collected data (such as cancer registers, death certificates), it was considered that the quality of the data is such that their main value is for hypothesis generating, although it was recognised that some agencies are of such a high quality that their material can also be used for hypothesis testing. It is certainly feasible to improve the sometimes poor quality of the data, for example, by record linking of variables such as occupation and life habits (such as smoking history).

Recommendation (i)   Better liaison between data collectors is recommended.

Recommendation (ii) Routine data collection should be made according to standardised procedures or classification systems, for example, use of international codes for occupations, economic activities of the employers, diagnosis of illnesses and death.

Recommendation (iii) Eurostat recommended the use of the Community Data Collection system.

(2) *Registration laws.* In the EU there is an obligation placed upon employers to register workers who are exposed to classified carcinogens on the basis of hazard assessment and risk evaluation carried out at a workplace.

Recommendation (i)   Make this rule universal.

(3) *New industries.* New industries pose a problem because they may be unwittingly working with hazardous substances.

Recommendation (i)   They should be made aware of the need to collect data in a standardised format, in order to enable good quality epidemiological studies in the future.

Recommendation (ii) Biomonitoring. Studies on biomonitoring in new industries could be useful in giving advance warning. This needs further research.

(4) *Information.* There is a need to improve skills in collection and interpretation of data.

Recommendation (i)   Regulatory agencies/authorities should provide information on how to record data and how to perform and interpret Good Epidemiology Practices (GEP). A good example is 'the Guidelines for good epidemiological practice for occupational and environmental epidemiologic research' published by the Chemical Manufacturers Association in 1991 (*J. Occup. Med.*, Vol. 33, No. 12, 1221–1229).

Recommendation (ii)   The updating of the DG XII booklet 'Searching for Causes of Work-Related Diseases' (J. Olsen *et al.*, Oxford Medical Publications, Oxford University Press, 1991).

(5) *Confidentiality.*

Recommendation (i)   The issues of confidentiality both in relation to individuals and the workplace have to be addressed.

2.2 Exposure Monitoring

(1) *Meeting epidemiological needs.*

Recommendation (i)   Monitoring of exposure should be made by means of Good Occupational Hygiene Practices, taking into account especially the need for physico-chemical characterisation of inorganic carcinogens, (for example, solubility, valency, chemical status, particle size and density) and the variables that are well-known to influence the exposure profile, such as the use of protective equipment, work practice variability and life habits, such as tobacco smoking.

(2) *Sampling.* In respect of how sampling should be performed it was considered that while personal monitoring has the greatest value, in its absence static monitoring can provide useful information.

Recommendation (i)   Personal monitoring is the preferred option.

(3) *Compliance sampling.*

Recommendation (i)   This should be designed and set up to provide at least basic minimum of data for future research.

(4) *Storage.* Correct storage of sampling is essential to help elucidate future problems.

Recommendation (i)   Standardised procedures should be recommended.

(5) *Bio-markers.* There is a need to know more about the correlation, if any, between biomarkers of cause and effect with actual cancer risk. This is especially true of inorganic substances. The generally non-specific nature of 'molecular' biomarkers of early carcinogenic effect and the current non-availability/applicability of such markers for practical health surveillance should be taken into account.

Recommendation (i)   Research in this field is essential.

(6) *Dose–response.* In respect of dose–response correlation. There are major uncertainties in respect of (1) the presence or absence of a threshold dose for inorganic carcinogens, and (2) the extrapolation of the risk to low level exposure. It was considered that it is unlikely that the available data or the use of classic epidemiological techniques will give insight as regard the threshold level but that molecular markers perhaps may give some new information. The working party was also aware of the ethical aspects involved in the use and interpretation of such markers.

Recommendation (i)   Need for future work in this field is recognised.
(7) *Small–medium enterprises*. The needs of small–medium enterprises were noted.

2.3. Historical Data

There is a need for improving retrospective exposure assessment. Useful data can be obtained using information collected from historical sources and older workers, or performing nested case control studies. In community based studies the assessment of occupational histories collected by questionnaires, by an expert team, could improve the quality of the data.

Recommendation (i)   The working party recommends the joint handling of data by industry and academic institutes in order to enhance the quality of retrospective epidemiology.

2.4 Multidisciplinarity

The working party noted that the multidisciplinary investigation adds a valuable dimension to the evaluation of the carcinogenicity of inorganic substances.

2.5 Statistics

(1) The working party agreed that most of the statistical issues in epidemiology studies are well understood and catered for.
(2) The working party also addressed the use of meta-analysis in occupational cancer epidemiology. Publication bias, namely the non-publication of negative or 'non-positive' studies, has to be addressed in the interpretation of the results of such an analysis.

   The issue of data sharing is also important when pooling raw data from multiple studies. It was noted that using this technique data can be reclassified and reprocessed, providing more direct analysis for subcohorts and for estimation of dose–response relationship.

Recommendation (i)   Meta-analysis is a cost-effective and useful tool and its use is to be recommended provided the problems are recognised.

2.6 Interpretation

The working party feels that the interpretation of epidemiological studies is of paramount importance. The main issue is that epidemiological studies on occupational inorganic substances are observational, and are subject to potential bias and confounding.

   Care should be taken not to overemphasize a positive study and to address properly all possible limitations, but try to discuss (1) evidence of bias or confounding; (2) the implication of bias, such as its direction and magnitude.

The problem of confounding from smoking is important, especially when evaluating studies on lung and bladder cancer.

Recommendation (i)   The working party considers that an accurate collection of data on possible confounders is necessary for good epidemiological research.

2.7 Quality Assessment

In respect of the quality assessment of epidemiological studies, the following were required:

(a) a clear statement of a research hypothesis;
(b) accurate epidemiological data;
(c) proper exposure measurements;
(d) the discussion and analysis of possible bias and confounders.

Recommendation (i)   The availability of negative or non-positive findings and the data sharing after publication may be used to enhance the quality of epidemiology studies.

2.8 Good Epidemiological Practice

The working party considered that the epidemiological study of occupational exposure to (carcinogenic) inorganic substances cannot be fully undertaken without a thorough appreciation and implementation of Good Epidemiology Practices. Although GEP will not guarantee good epidemiology, it should provide an essential framework for performing epidemiological studies, promoting sound epidemiological research and encouraging quality data collection and analysis.

The challenges for the future of occupational epidemiology have been listed recently by Hernberg. The working party recognised these and considered them essential prerequisites for good research (Hernberg S., Occupational Epidemiology: Developments and Perspectives, *Med. Lav.*, 1995, **86**, No. 2, pp. 95–105).

3 Working Party 3 – List of Participants

Core members
Dr L. MORGAN (Chairman)
Dr S. PORRU (Rapporteur)
Prof A. BERNARD

Prof J.M. HARRINGTON
Dr M.H. DRAPER

Participants
Dr G. ARESINI
Dr I.M. ARNOLD
Dr S.R. BERGE
Dr P. BOFFETTA
Dr B. BUCLEZ

Dr P. LINNETT
Mr J.J. MOULIN
Mr W. PERRY
Dr E. PIRA
Ms M. PON

Participants

Mr E. CHANTELOT
Dr C. CHATZIS
Dr B. DAVISON
Mr N. FLINT
Prof J.B.L. GEE
Dr W. HANKE
Dr J. HOSKINS
Mr F. LECHENET
Dr C. LESNE

Dr S. PUGH-WILLIAMS
Dr T. RANTANEN
Dr M. REFREGIER
Dr T. SORAHAN
Dr B. SWENNEN
Mr W. TAYLOR
Dr L. VAN DER BERG
Dr R.A. WALK
Ms K. ZIEGLER-SKYLAKAKIS

Conclusion

Looking to the Future

J.H. DUFFUS

EDINBURGH CENTRE FOR TOXICOLOGY, EDINBURGH, UK
CHAIRMAN OF THE SCIENTIFIC COMMITTEE

Uniquely, this seminar has concentrated on the assessment of risk associated with workplace exposure to inorganic substances. It has brought together scientists and physicians working at the forefront of science with regulators and managers who must devise and implement regulations based on their discoveries. It has also provided a forum in which doubts could be aired in specialist working parties, debated, and possible solutions identified.

We have been very lucky in having contributions from world leaders with a lifetime's experience who have provided an incomparable base of knowledge and understanding of their fields to support our deliberations.

Many people worked very hard to make the seminar a success but perhaps none more than our chairmen and rapporteurs. The results of their efforts in guiding and summarizing discussion may be read elsewhere in these proceedings in the consensus working party reports. These reports provide much food for thought for the reader. In making recommendations, the working parties did not have to consider costs of implementation or to quantify the benefits resulting, but these are matters relating to specific situations. Each reader will have a different idea of what is most important. However, in spite of some repetition, I think it may be useful to pick out what I consider to be the main recommendations for future action.

1 Working Party Recommendations

1.1 General Recommendations

It is striking that some common themes are to be found within the Working Party Reports. Firstly, all three working parties call for the availability of defined materials for referencing test methods. Secondly, they all call for improved and extended archiving of samples and records. Thirdly, they all call for improved communication between the various groups involved in risk assessment. This last recommendation may be the most important and in some ways the most difficult. Looking at the papers presented in this seminar, it is clear that there is such rapid progress in fundamental science that it is very difficult for non-specialists to keep abreast of it. The only solution may be to refer all risk assessment to a regular meeting of a representative multidisciplinary team.

The current approaches to assessment of carcinogenicity, based as they are on assumptions appropriate to organic compounds, should be redefined to take into account the characteristics of inorganic substances and our knowledge of bio-inorganic chemistry. In particular, the ability of living cells to exclude partially or to concentrate certain ions must be carefully considered in determining exposure thresholds for possible carcinogens. For example, as pointed out by Professor Williams, under normal circumstances, nickel(II) ions have difficulty entering animal cells because of competition with magnesium(II) ions for uptake by the selective transport mechanism. Nickel(II) ions in the cytoplasm are pumped into vesicles. DNA itself in the normal cell is protected by bound magnesium(II) ions from any association with nickel ions. Further, magnesium(II) ions occupy the only binding sites in RNA, and many in proteins and other molecules, to which nickel(II) ions can bind.

Since magnesium(II) is normally present in living cells at 10^{-3} M concentrations and nickel(II) ions at 10^{-10} M concentrations, only massive exposure to nickel(II) ions can overcome the natural defences to any kind of biochemical interaction and toxic effect. The recent failure of US National Toxicology Program (NTP) carcinogenicity tests to demonstrate any significant tumour induction by nickel(II) sulfate bear out this analysis from first principles. This does not, of course, imply that other nickel compounds are not carcinogenic.

1.2 Physico-chemical Characterisation of Exposure

Identification and characterisation of potentially hazardous inorganic materials to which people are exposed in the workplace and local environment should focus on speciation as far as this is possible (elemental analysis alone is insufficient). Speciation should include structural composition and physical characterisation, (including particle-size distribution and solubility in biological fluids), and, in certain cases, surface properties of solids. More use should be made of X-ray powder diffraction spectrometry (PDS), taking account of limitations in analysis of complex mixtures, mixtures with trace crystalline phases, presence of amorphous phases and crystalline members with structures not yet described. The possibility of contamination by potentially harmful organic compounds, at levels known to be hazardous, should also be examined.

Following regular periodic characterisation of the industrial plant and process, area and personal samples taken routinely should be archived, to permit retrospective examination. Satisfactory records of this monitoring must be maintained to allow the results to be related to individuals and their work locations and to permit future study. Procedures should be standardised as far as possible to allow inter-plant comparisons with external auditing and quality control required for this.

To meet epidemiological needs, monitoring of exposure should be made by means of good occupational hygiene practice, taking especially into account the variables that are well-known to influence the exposure profile such as the use of protective equipment and work practice variability.

Biological materials derived from occupationally-exposed humans with corresponding diseases should be sampled and preserved for future molecular and genetic analysis. The issues of confidentiality both in relation to individuals and the workplace have still to be properly addressed.

1.3 Conduct and Interpretation of *In Vitro* and *In Vivo* Experimental Studies

Studies are needed on the role of mixed exposures to different inorganic materials. Research is needed to understand better the importance of the physico-chemical features of metal compounds (*e.g.* surface properties, morphology, *etc.*) in the sequence of events leading to genotoxicity and cancer initiation. For cancer risk assessment, it is crucial to determine whether physico-chemical features of metal compounds affect mainly the bioavailability of the metal ion or else act independently of any metal ion release to cause DNA damage or other carcinogenic changes.

In the case of metal alloys, the bioavailability and tissue deposition of critical components should be determined. The electrochemistry of the reaction of metals and alloys with body fluids must be carefully assessed.

Both *in vivo* and *in vitro* experimental studies must be carefully designed and the relationship between *in vivo* and *in vitro* results needs to be better established, especially in the context of dose–response relationships. In addition, differences between animal species must be further characterised to permit better extrapolation in predicting possible human effects. As part of this characterisation, the differing consequences of different exposure routes and excessive exposures must be considered.

1.4 Conduct and Interpretation of Epidemiological Studies

There are no recommendations of epidemiological practice that are specific to inorganic materials. Epidemiologists should always require precise qualitative and quantitative characterisation of exposures.

Where possible, the collection of data should be a collaborative process between those actually collecting the data, statisticians and those with an intimate knowledge of the industrial process. Routine data collection should be made according to standardised procedures or classification systems, which are now being implemented in the EU, for example, as regards international codes for occupations, economic activities of the employers, diagnosis of illnesses and death. However, such procedures should never be used as a substitute for a critical assessment of data requirements for specific workplace situations.

Regulatory agencies/authorities should ensure that all relevant bodies have full information on how to deal with recording, applying and interpreting good epidemiological practice (GEP). Issues of confidentiality both in relation to individuals and the workplace have still to be addressed if epidemiological studies are to be optimised.

The importance of occupational histories of workers for the retrospective assessment of exposure should be given emphasis and efforts made to obtain such histories and record them in carefully designed databanks for future use. Design of databases should pay attention to recording data on possible confounders.

2 General Overview

Apart from the deliberations of the Working Parties, a number of general conclusions were suggested in Plenary Sessions throughout the seminar and, although not formally agreed, appeared to have general support among the participants. Again there is some repetition of the working party conclusions but this may serve to indicate how strongly the participants felt about certain key issues.

The seminar was concerned broadly with the issues surrounding risk assessment for occupational exposure to inorganic substances and their potential carcinogenicity. Risk assessment may be divided into four phases:

(1) Hazard identification – does the agent really cause the adverse effect or is there merely a statistical association, however strong?
(2) Exposure assessment – what exposures are currently experienced or expected under prevailing conditions and what is the relationship between exposure and effective dose for the population at risk?
(3) Exposure–response assessment – what is the relation between exposure, severity of disease, and incidence in humans?
(4) Risk characterisation – what is the estimated incidence of the adverse effect in a given population as compared with (a) the general population and (b) the local population?

The conclusions listed in Table 1 should be considered in the context of the above risk assessment framework.

Hazard identification and subsequent risk assessment of inorganic substances has almost universally been based solely on elemental content. This has resulted in the condemnation as human carcinogens of whole groups of compounds containing a given element, often on poor epidemiological evidence relating at best to only one. Even this approach is not applied consistently since all chromium compounds are not condemned because Cr(VI) has been postulated to be a carcinogen. Living organisms are rarely exposed to pure elements as such but to their chemical species, and carcinogenicity, like other toxic properties, must depend upon the species of the element present at the target area. No one suggests that all carbon compounds should be banned because some carbon compounds are carcinogens. Following such considerations, it is clear that blanket identification of compounds of individual inorganic elements as human carcinogens is no longer tenable. There is a need for an expert meeting to consider how a new approach to toxicological evaluation of inorganic substances can be developed.

Table 1 *Conclusions*

(1) Current criteria for hazard identification of inorganic substances based simply on elemental composition are not scientifically tenable. Apart from this, it has often been assumed that the same criteria as for organic substances may be applied and this is not necessarily the case.

(2) Hazard identification and risk characterisation for organic substances should be based on a sound understanding of bio-inorganic chemistry. In particular, thee is a need to understand the biological speciation of metals as a result of valency changes in cells and binding to biological ligands.

(3) Inorganic hazards should be identified as the exact chemical species when referring to metals and by the exact name when referring to minerals, *e.g.* chrysotile, crocidolite, amosite, *etc.*, instead of 'asbestos', a commercial term which has confused published toxicological assessments.

(4) Hazard identification and risk characterisation for inorganic substances should not be confounded by results from overload observations in experimental animals. Many animal studies, especially on carcinogenicity, have been carried out with doses at near toxic levels, which can overwhelm local defence mechanisms such as oxygen scavengers or DNA repair mechanisms. Thus, false conclusions are reached when these results are extrapolated linearly to the much lower dose levels which are typical for humans.

(5) Hazard identification should be transparent. The evidence supporting the identification should be freely available and open to debate before decisions are made. Decisions once made must be open to change as scientific knowledge develops.

(6) Accurate use of terminology is essential; there may be a need for a new terminology to define mineral species more precisely.

(7) All mutagenicity and carcinogenicity tests should include dosimetry in order to provide a basis for potency and risk assessment.

(8) Epidemiological studies should be improved by applying good epidemiological practice (GEP) and by using all available information, including negative epidemiological studies.

(9) Law and regulation based on science should be open to regular revision as scientific understanding improves.

(10) Risk assessment is a multidisciplinary activity and all relevant parties should be involved. Further multidisciplinary meetings like this seminar should be planned for the future.

List of Seminar Participants

Mr N. André
Manager
European Tin Center
44 rue d'Arenberg Bte 33
B - 1000 Bruxelles
Belgium

Mr A. Angelidis
Administrator
Occupational Health & Hygiene Unit
European Commission DG V
Bâtiment Jean Monnet
L - 2920 Luxembourg
Luxembourg

Dr G. Aresini
Principal Administrator
Occupational Health and Hygiene Unit
European Commission DG V
Bâtiment Jean Monnet
L - 2920 Luxembourg
Luxembourg

Dr L. Aringer
Head of the Medical Unit
National Board of Occupational
  Safety and Health
S - 17184 Solna
Sweden

Dr I.M. Arnold
current address:-
Director, Occupational Health and Safety
Alcan Aluminium Ltd
(Vice-President, Alcan International Ltd.)
1188 Sherbrook Street West
Montreal, Québec
Canada H3A 3G2

Dr E.G. Astrup
Corporate Specialist: Ind & Env
  Toxicology
Elkem A/S
Corp HQ Dep HSE, PO Box 4282
N - 0401 Oslo
Norway

Dr S.R. Berge
Medical Officer
Falconbridge Nikkelverk AS
PO Box 457
N - 4601 Kristiansand S.
Norway

Prof A. Bernard
Université Catholique de Louvain
Unité de Toxicologie Industrielle
Clos Chapelle-aux-Champs 30 Bte
Bte 30–54
B - 1200 Bruxelles
Belgium

Dr P. Boffetta
International Agency for Research
  on Cancer
150 Court Albert Thomas
F - 69372 Lyon Cedex 08
France

Mr C. Bozec
EHS Department
Eramet
33 avenue du Maine
F - 75755 Paris Cedex 15
France

Prof J. Bruch
Institute Leader
Institute for Hygiene and Occupational
  Medicine
University of Essen, Virschowstrasse
D - 45122 Essen
Germany

Mr E. Chantelot
Eurométaux
12 avenue de Broqueville
B - 1150 Bruxelles
Belgium

Prof A. Churg
University of British Columbia
Dept of Pathology
2211 Westbrook Mall
CDN - Vancouver BCV 6T 2B5
Canada

Dr R. Cornelis
Research Director National Fund for
  Scientific Research - IUPAC
University of Gent, Lab. Anal. Chem.
Proeftuinstraat 86
B - 9000 Gent
Belgium

Ms A.C. Boutin
Chercheur
Usinor Sacilor
Immeuble Pacific TSA 10001
F - 92070 Paris la Défense
France

Dr R. Brown
Toxicologist
Tox Services
4 Bramble Close
Uppingham/Rutland
GB - LE15 9PH
United Kingdom

Dr B. Buclez
Medical Advisor
Aluminium Pechiney
BP 62
F - 13541 Gardanne Cedex
France

Dr B. Chen
Toxicologist
IPCS, World Health Organisation
20 avenue Appia
CH - 1211 Geneva 27
Switzerland

Dr B. Conard
Vice-President Health
INCO Ltd
145 King St West, Suite 1500
CDN - M5H 4B7 Toronto, Ontario
Canada

Mr J.M. Costa-David
European Commission DG XI
Unit XI/E/2 - Industry and
  Environment
5 avenue de Beaulieu, office 2/25
B - 1049 Bruxelles
Belgium

List of Seminar Participants

Ms R. Côté
Health & Safety Evaluator
Health Canada (Canadian Government)
Main Bldg, Room 2701, Tunney's Pasture
CDN - K1A 0LZ Ottawa
Canada

Dr L. Curcio
Executive Director
NiPERA
100 Capitole Drive, Suite 104
USA - 27713 Durham
United States

Dr B. Dero
Eurométaux
12 avenue de Broqueville
B - 1150 Bruxelles
Belgium

Prof S.G. Domnin
Director
Medical Research Center for Prevention
  and Health Protection of Industrial Workers
30 Popov Street
RU - 620014 Eckaterinburg
Russia

Dr J.H. Duffus
Director
EDINTOX
Heriot-Watt University
GB - Edinburgh EH14 4AS
United Kingdom

Dr Z. Elias
Chief of Laboratory
INRS
avenue de Bourgogne
F - 54501 Vandoeuvre
France

Dr G. Crawford
Consultant on Environmental Affairs
Nickel Development Institute
214 King St West
CDN - Toronto, Ontario
Canada

Dr B. Davison
NiDI
22 Whitegates, Mayals
GB - Swansea
United Kingdom

Mr J. Descamps
President
AFIC/AIP
30 avenue de Messine
F - 75008 Paris
France

Dr M.H. Draper
Consultant
EDINTOX
10 West Mayfield
GB - Edinburgh EH9 1TQ
United Kingdom

Prof J. Dunnigan
Director/Master Prog. in Envir.
Sciences
University of Sherbrooke
Sherbrooke
CDN - J1K 2R1 Québec
Canada

Mr G. Ethier
Executive Director Health
ICME
294 Albert Street, Suite 506
CDN - Ottawa, Ontario K1P 6E6
Canada

Dr K. Etzler
Hütten- und Walzwerks-
 Berufsgenossenschaft
Kreuzstrasse 45
D - 40210 Düsseldorf
Germany

Mr N. Flint
Consultant
Nickel Development Institute
25 Blythe Way, Solihull
GB - W. Midlands B91 3EY
United Kingdom

Prof B. Fubini
Professor Chemistry
University of Turin, Dept Chimica
Via P. Ciuria 9
I - 10125 Torino
Italy

Dr A. Giovanetti
Researcher
ENEA AMB BIO INTO
Cre Casaccia, Via Anguillares 301 SPO 5
I - 00060 Roma
Italy

Mr R. Haigh
Head of Unit
Occupational Health and Hygiene Unit
European Commission DG V
Bâtiment Jean Monnet
L - 2920 Luxembourg
Luxembourg

Prof J.M. Harrington
University of Birmingham
Inst. of Occupational Health
PO Box 363, Edgbaston
GB - Birmingham B15 2TT
United Kingdom

Dr C. Fernicola
Occupational Health
USSL 18-TSLL-Occup. Health Sce
Via A. Cantore 20
I - 25128 Brescia
Italy

Dr U. Föst
Bundesanstalt für Arbeitsschutz
Friedrich-Henkel-Weg 1–25
D - 44149 Dortmund
Germany

Prof J.B.L. Gee
Pulmonary and Critical Care Section,
Department of Internal Medicine
Yale University School of Medicine
333 Cedar Street, PO Box 3333
USA - New Haven CT 06501
United States

Prof U.F. Gruber
Toxicology, University of Basel
Grosspeterstrasse 25
CH - 4052 Basel
Switzerland

Dr W. Hanke
Assistant Professor
Institute of Occupational Medicine
8 Teresy Street, PO Box 199
PL - 90-950 Lodz
Poland

Prof P.J. Hewitt
Bradford University Research
University of Bradford
GB - BD7 1DP Bradford, W. Yorkshire
United Kingdom

List of Seminar Participants

Mr H.W. Hoeksema
Engineer Environmental Techniques
n.v. EPON
PO Box 10087
NL - 8000 GB Zwolle
The Netherlands

Dr W.J. Hunter
Director
Public Health and Safety at Work
 Directorate
European Commission, DG V
Bâtiment Jean Monnet
L - 2920 Luxembourg
Luxembourg

Dr D. Jans
Scientific Officer
IMA-Europe
75 avenue de l'Indépendance Belge
B - 1080 Bruxelles
Belgium

Dr M.-C. Jaurand
Director of Research
I.N.S.E.R.M., Unité 139
Faculté de Médecine
8 rue du Général Sarrail
F - 94010 Créteil
France

Dr A. Jones
Deputy Director
Scottish Poisons Info Bureau
Royal Infirmary of Edinburgh
GB - Edinburgh EH3 9YW
United Kingdom

Dr J. Hoskins
Scientific Staff
MRC Toxicology Unit
Hodkin Building, University of
 Leicester
GB - LE1 9NV Leicester
United Kingdom

Dr S. Jacobi
Degussa AG ZN Wolfgang
Industrielle Toxikologie
D - 63403 Hanau, Main
Germany

Ms M. Jaroszewski
Principal Administrator
Danish Working Environmental
 Service
Landskronagade 33
DK - 2100 Copenhagen
Denmark

Mr T. Jepsen
Senior Adviser
Danish Employer's Confederation
Vester Voldgade 113
DK - 1790 Copenhagen
Denmark

Dr C. Kennedy
Inhalation Toxicology Research
 Institute
PO Box 5890
USA - Albuquerque NM 87185
United States

Dr M. Koponen
Chief Consultant
Env. Affairs
Outokumpu Oy
Corporate Management/Po Box 280
FIN - 02101 Espoo
Finland

Dr A. Langer
Director
Inst. of Applied Sciences, Env. Sciences
 Lab Brooklyn College,
City University of NY
Bedford Avenue and Avenue II
USA - Brooklyn NY 11210
United States

Dr C. Lesne
Ingénieur de Recherche
CNRS
Unité de Prévention du Risque Chimique
avenue de la Terrasse
F - 91198 Gif-Sur-Yvette
France

Dr P. Linnett
Group Occupational Physician
Johnson Matthey PLC
Orchard Road, Royston
GB - Herts SG8 5HE
United Kingdom

Mr J.C. Mahieu
Engineer
INRS
30 rue Olivier Noyer
F - 75680 Paris
France

Dr A. Lamberty
IRMM
Retieseweg
B - 2440 Geel
Belgium

Mr F. Lechenet
Director
AFIC/AIP
30 avenue de Messine
F - 75008 Paris
France

Dr H. Lindemann
Scientist
Bayer AG
Fachbereich Toxikologie,
PO Box 101709
D - 42096 Wuppertal
Germany

Dr M. Maclaine Pont
Writer of Criteria Documents for
 Occupational Standards
Dept. of Toxicology, Agricultural
 University
PO Box 8000
NL - 6700 EA Wageningen
The Netherlands

Ms M. Malygina
Kiev State University of Civil
 Engineering and Architecture
31 Vozderhoflotsky Avenue
UA - 2522037 Kiev
Ukraine

List of Seminar Participants

Dr B. Mantout
Occupational Health Physician
Usinor Sacilor
Stainless Steel Line, Ugine-Savoie
F - 73400 Ugine
France

Dr R. Montaigne
Counsellor-Technical Affairs Dept
CEFIC
avenue E. van Nieuwenhuyse, 4, Bte 1
B - 1160 Bruxelles
Belgium

Prof P.E. Morrow
Emeritus Professor of Toxicology
University of Rochester
Medical Ctr, Box EHSC
USA - 14642 Rochester, NY
United States

Prof H. Muhle
Fraunhofer Institute für Toxikologie
 und Aerosolforschung
Nikolai-Fuchs-Strasse 1
D - 30625 Hannover
Germany

Dr R.P. Nolan
Associate Director
Brooklyn College of the City
 University of NY
2900 Bedford Avenue - room 5135
USA - 11210 Brooklyn - NY
United States

Mr P. Papadopoulos
National Expert Detached
 Occupational Health & Hygiene Unit
European Commission DG V
Bâtiment Jean Monnet
L - 2920 Luxembourg
Luxembourg

Mr S. Maraschin
Co-ordinator Health, Safety &
 Environment
Billiton
PO Box 436
NL - 2260 AK Leidschendam
The Netherlands

Dr L.G. Morgan
Consultant
NiPERA
Glynteg, Park Road, Ynystawe
GB - SA6 5AP Swansea
United Kingdom

Mr J.J. Moulin
Epidemiologist
INRS
avenue de Bourgogne
F - 54500 Vandoeuvre
France

Dr R. Murray
South Hill, Newton Green
GB - CO10 02R Sudbury, Suffolk
United Kingdom

Mr L. Oksanen
Industrial Hygienist
Outokumpu Polarit Oy
FIN - 95400 Tornio
Finland

Dr M.V. Park
Deputy Director
EDINTOX
Heriot-Watt University, Riccarton
GB - Edinburgh EH14 4AS
United Kingdom

Dr F. Pellet
Occupational Health Consultant
Pechiney
Pechiney Balzac
F - 92048 Paris la Défense Cedex
France

Dr E. Pira
Medical Assistant
Department of Occupational Medicine
Via Zuretti 29
I - 10126 Torino
Italy

Dr S. Porru
Servizio e Cattedra di Medicina del
 Lavoro, Univ. di Brescia
P. le Spedali Civili 1
I - 25125 Brescia
Italy

Mr J.M. Pujade-Renaud
12 avenue Lulli
F - 78530 Buc
France

Dr M. Refregier
Medical Advisor
Sté Talc de Luzenac
BP 1162
F - 31036 Toulouse Cedex
France

Dr A. Robertson
Head
Institute of Occup. Medicine, Edinburgh,
Analytical Division
8 Roxburgh Place
GB - Edinburgh EH8 9SU
United Kingdom

Mr W. Perry
Health Scientist Industrial Hygienist
Occupational Safety and Health
 Administration
200 Constitution Avenue, NW Rm
N-3718
USA - Washington DC 20210
United States

Ms M. Pon
Senior Health Advisor
BHP Minerals
550 California Street
USA - 94104 San Fransisco CA
United States

Dr S. Pugh-Williams
Medical Adviser
Inco Europe Ltd
Clydach
GB - Clydach, Swansea SA65QR
United Kingdom

Dr T. Rantanen
Occupational Health Dr
Outokumpu Harjavalta Metals Oy
FIN - 29200 Harjavalta
Finland

Dr B. Rehn
Scientific Staff Leader
Institute for Hygiene and
 Occupational Medicine
University of Essen, Virschowstrasse
D - 45122 Essen
Germany

Dr M. Roller
Med. Institut für Umwelthygiene
Auf'm Hennekamp 50
D - 40225 Düsseldorf
Germany

List of Seminar Participants

Dr U. Saffiotti
National Cancer Institute, Cancer Etio. Div.,
Laboratory of Experimental Pathology
Bldg 41 Rm C-105
USA - Bethesda MD 20892-0041
United States

Dr K. Schumann
Walther-Straub-Institute für Pharmakologie
  und Toxicologie
Nüssbaumstrasse 26
D - 80336 München
Germany

Dr S. Shcherbakov
Medical Research Center for Prevention and
  Health Protection of Industrial Workers
30 Popov Street
RU - 620014 Eckaterinburg
Russia

Mr J. Spaas
Secretary-General
Eurométaux
12 avenue de Broqueville
B - 1150 Bruxelles
Belgium

Mr E. Sundquist
Chief Engineer
Ministry of Labour
PL 536
FIN - 33101 Tampere
Finland

Mr W. Taylor
Managing Director
Brush Wellman Ltd.
Unit 4, Ely Road, Theale
GB - Berkshire R47 4BQ
United Kingdom

Dr G. Schoeters
Programme Manager
VITO
Boeretang 200
B - 2400 Mol
Belgium

Mr F. Seiler
Scientific Staff Member
Institute for Hygiene and
  Occupational Medicine
University of Essen, Virschowstrasse
D - 45122 Essen
Germany

Dr T. Sorahan
Senior Lecturer in Epidemiology
University of Birmingham
Institute of Occupational Health,
Edgbaston
GB - Birmingham B15 2TT
United Kingdom

Mr M. Stentiford
Group Technical Support Managner
Watts Blake Bearne + Co Plc
Park House, Courtenay Park
GB - TQ12 4PS Newton Abbot
United Kingdom

Dr B. Swennen
Head Occupational Health
  Department
Union Minière
Leemanslaan 36
B - 2250 Olen
Belgium

Dr C.F. Taylor
Chief Medical Officer
British Steel plc
Trosteworks
GB - Llanelli SAI4 8YU
United Kingdom

Dr D. Tollerud
Director Occupational Environmental
  Medicine
University of Pittsburgh
130 Desoto Street, Room A-718
USA - 15261 Pittsburgh, PA
United States

Ms M.Th. Van Der Venne
Principal Administrator
Public Health Unit
European Commission, DG V
Bâtiment Jean Monnet
L - 2920 Luxembourg
Luxembourg

Dr W. Von Der Hude
National Expert Detached
Occupational Health & Hygiene Unit
European Commission, DG V
Bâtiment Jean Monnet
L - 2920 Luxembourg
Luxembourg

Prof R.J.P. Williams
Professor (Emeritus) Chemistry
Oxford University (Inorganic Chemistry)
South Parks Road
GB - Oxford OX1 3QR
United Kingdom

Dr M. Wyart-Remy
Secretary-General
IMA-Europe
75 avenue de l'Indépendance Belge
B - 1080 Bruxelles
Belgium

Dr L. Van Der Berg
Chief Medical Officer
Impala Platinum - Gencor
PO Box 222
SAF - 1630 Springs - Gauteng
South Africa

Dr J. Vancleemput
Occupational Health Physician
N.V. Eternit
B - 1880 Kapelle-op-den-Bos
Belgium

Dr R.A. Walk
INBIFO
Fuggerstrasse 3
D - 51149 Köln
Germany

Mr F. Wilmotte
Responsible for Occupational
  Hygiene
Metaleurop s.a.
58 rue Roger Salengro
F - 94126 Fontenay-ss-Bois
France

Ms K. Ziegler-Skylakakis
GSF-Institut für Toxikologie
Neuherberg, Postfach 11 29
D - 85758 Oberschleissheim
Germany

List of Seminar Participants

Secretariat

Ms L. Beaumont
Eurométaux
12 avenue de Broqueville
B - 1150 Bruxelles
Belgium

Ms F.M. Craig
EDINTOX
c/o Craig Publication Services
Torsonce House, Newside
Charmichael ML12 6NG
GB - South Lanarkshire
United Kingdom

Ms E. Heighton
European Commission DG V
Bâtiment Jean Monnet
L - 2920 Luxembourg
Luxembourg

Ms C. Martin
Eurométaux
12 avenue de Broqueville
B - 1150 Bruxelles
Belgium

Ms G. Panichi
European Commission DG V
Bâtiment Jean Monnet
L - 2920 Luxembourg
Luxembourg

Subject Index

ACGIH, 238
Acid, 17, 27, 30, 34, 37, 137, 146, 155, 166, 178, 187, 198–201, 236
Actinolite, 41, 48, 61
Adenocarcinoma, 68, 73, 88, 89, 92–94, 97, 98, 100, 101, 137, 138, 146, 159, 169, 182
Adenoma, 137
Adenomatoid lesions, 137
Adrenaline, 31
Aerosol, 8–10, 39, 42, 49, 50, 53, 103, 106, 108, 109, 113, 114, 117, 131–133, 247
Air quality guidelines, 215, 217, 218
Airway fibrosis, 64, 66
Alkaline phosphatase, 166
Allosteric control, 26
Alloys, 3, 147, 148, 154, 159, 166, 171, 173, 226, 235, 239, 242, 244, 251, 262
Aluminium, 29
 refining, 164
 silicate, 59, 127
Alveolar clearance, 14, 70, 105–108, 111, 112, 119, 124, 125, 129
Alveolar macrophage, 14, 15, 104, 105, 107, 111, 113, 114, 116, 118, 123, 127, 131, 132
Alveolar particle clearance, 130
Alveolar retention, 118, 119, 125, 127
Alveolar type II cell hyperplasia, 137
Alveolar type II epithelial cells, 138
Alveoli, 14, 107, 117, 123, 124, 129, 132

Amosite, 10, 41, 52, 53, 61–69, 72, 73, 80, 84, 115, 116, 119, 132, 264
Amphibole, 10, 48, 50, 51, 54–56, 61, 62, 64, 65, 67, 73, 85
Amplifications, 138
Analytical techniques, 247
Anatase, 41, 139
Aneuploidy, 79, 84, 88
Animal data, 16, 66, 152–154, 174, 221
Anthophyllite, 41, 52, 56, 61
Antimony, 15, 147, 156, 157, 236
 trioxide, 147, 156, 157
Apoptosis, 13, 82, 85, 90, 91, 98–100
Archiving, 247, 249, 260
Arsenic, 2, 15–17, 146–149, 151, 153, 155, 156, 159, 165, 168, 174, 181, 182, 187, 198–201, 204–208, 211, 212, 217, 218, 221, 236, 249
Asbestos, 2, 8, 9, 10, 11, 41, 45, 47, 49, 50–58, 61–65, 67–74, 77–81, 83–85, 116, 131, 136, 142, 143, 150, 221, 222, 264
 bodies, 68, 70
 disease, 11, 56, 57, 64
 fireproofing, 62
Asbestosis, 10, 64, 66–69, 73, 74
Aspect ratio, 43, 44, 66, 67
Asthma, 151
Ataxia-telangiectasia, 87
Athymic nude mice, 138
ATP production, 81
Autocrine growth factor, 92
Autocrine loop, 92–94

Autopsy material, 126, 248
Autoradiography, 78

Bacteria, 77, 78
BALB/3T3/a31-1-1 cells, 138, 139, 143, 144
Barium, 25, 217
  chromate, 217
Benzo[a]pyrene, 84, 136, 150
Berylliosis, 166, 167
Beryllium, 2, 15, 16, 25, 33, 34, 148, 157, 160, 165–168, 174, 178, 181, 211, 217, 219, 236, 238, 239, 244
Bessemer matte, 17, 186, 187, 189, 190, 197–201, 205, 206
Bioaccumulative, 230
Bioassays in small rodents, 135, 142
Bioavailability, 149, 165, 238, 251
Bio-inorganic chemistry, 6, 20, 35, 261, 264
Biological assessment, 72, 248
Biological monitoring, 214, 239, 242
Biomarkers, 214, 246, 256
Biomonitoring, 255
Bladder cancer, 258
Bloom's syndrome, 87
Bromine, 32, 34
Buffering, 35

c-*myc* Amplification, 89
Cadmium, 2, 7, 15, 16, 32, 90, 147–149, 155–157, 159, 160, 165, 168, 169, 174, 176, 178, 181, 211, 212, 217–219, 221, 236, 238
Caesium, 21
Calbindin, 28
Calcine, 171, 198–201, 204
Calcined bessemer matte, 201
Calcine mattes, 200, 201
Calciner, 187, 198, 199, 201, 202, 204
Calcium, 7, 23–25, 27, 28, 30–33, 37, 60, 149, 150, 165, 170, 205
  carbonate, 60

  pump, 32
Calmodulins, 26
Canada, 58, 62, 65, 72, 154, 171, 172, 178, 186, 187, 190, 207, 229, 234
Cancer susceptibility, 86, 95
Carbon black, 104, 105, 112, 114, 116, 124, 127
Carbon monoxide, 188, 198, 200
Carbon particle, 130
Carcinogenic potency, 4, 11, 48, 49, 53, 75, 76, 78, 80, 150, 223, 234, 243
Carcinoma, 12, 61, 71, 74, 87, 88, 91–94, 96, 98–102, 104, 136–140, 143, 158, 178, 182, 183, 217
Case control study, 161
Catalase, 140
Cell nuclear antigen, 90
Cell surface receptors, 252
Chemical speciation, 4, 19, 149, 244, 251
Chloralkali workers, 153, 158
Chlorine, 24, 32, 34
Chromate, 35, 146, 149, 150, 156, 157, 169, 170, 179, 217
  pigments, 150
Chrome
  pigment, 157, 169, 170
  plating, 149, 170, 179
Chromium, 2, 15, 16, 29, 34, 146, 148–151, 153, 156, 160, 163, 165, 169, 170, 172, 174, 175, 179, 181, 211, 217, 219, 221, 244, 263
  compounds, 211, 238
  plating, 150
  trioxide, 149, 150, 156
Chromosomal alterations, 13, 91, 138
Chromosome
  aberrations, 79–81
  abnormalities, 76, 79, 81, 96
Chronic inhalation studies, 110
Chrysotile, 9, 10, 41, 43, 45, 50, 51, 53, 54, 56, 57, 61–69, 71–73, 83, 84, 264

Cigarette smoke, 12, 55, 60, 61, 67–71, 74, 87, 145, 148
Cigarette smoking, 10, 11, 17, 69, 70, 74, 96, 148, 190, 192, 193
Ciliated clearance, 103
Cirrhosis, 140, 144
cis-Platin, 34
Classification and labelling, 224–227, 235, 237, 240
Classification of carcinogens, 163, 173, 238
Classification, packaging and labelling, 1, 223–225
Clastogenicity, 79
Clearance rate, 14, 105, 108–110, 112, 117, 119, 121, 122, 124, 172
Clydach, 3, 15, 17, 18, 171, 172, 181–185, 187–200, 202–208, 238
Coal
 mine dust, 104, 113
 miners, 15, 126
 power plants, 104
Cobalt, 7, 15, 17–19, 28–30, 34, 36, 127, 133, 151, 158, 173, 181, 198–201, 204–206, 211, 236, 238, 244
 metal powder, 151
 salts, 198, 199
Co-carcinogens, 86
Cockayne's syndrome, 87
Cohort study, 149, 152, 161, 167, 178, 179
Coke, 104, 210
Communication, 165, 175, 230–234, 248, 260
Conc, 17, 199–202, 205, 206
Confidentiality, 254, 256, 262
Confounders, vii, 126, 153, 171, 258, 263
Connective tissue, 31
Conniston, 186
Contact inhibition, 80
Continuous scan X-ray diffraction, 42
Copper, 7, 15, 17, 18, 27–37, 147, 148, 151, 154, 158, 177, 179, 180, 187–189, 198–201, 204–206, 239, 244

beryllium alloys, 239
extracted matte, 200, 205
leaching process, 201
oxides, 198
pump, 32
shed, 198, 199, 202, 204
sulfate, 151, 187, 188, 198, 200, 204
sulfide, 17, 154, 189, 198
Cristobalite, 41
Crocidolite, 9, 10, 41, 43, 47–49, 52, 53, 55, 61–66, 73, 84, 115, 264
Crystalline silica, 59, 60, 111, 112, 116, 134, 136–138, 140, 141, 143–145
Cyclases, 7, 23, 27
Cyclin D, 90, 99
Cytochrome P-450, 12, 86, 95
Cytochrome P-450IIE1, 86, 95
Cytogenetic changes, 80
Cytokines, 124, 139
Cytotoxicity, 81, 84, 85, 111, 137–139, 143

Data, vii, 1–3, 10, 11, 15, 16, 42, 46, 49–51, 55, 58, 59, 61, 62, 64–66, 68, 69, 76, 78, 83, 94, 108, 110, 112, 122, 123, 127, 135, 137, 138, 147, 148, 150, 152–156, 158, 161–163, 165, 167–175, 179, 181, 189, 202, 210–212, 214, 216, 217, 221, 222, 225, 229, 231–234, 237, 239, 244, 245, 247, 248, 250, 251, 254–258, 262, 263
Data bases, 173, 174, 255, 263
Debrisoquine, 12, 86
de mimimis risk, 220
Dentists, 153, 159
Dephosphorylation, 7, 23, 25
Diesel particle, 108, 114, 124, 130, 131
Diesel soot, 105, 108, 112, 113
Directive, 1, 222–228, 235–237, 239–242
DNA, 11, 19–21, 23, 25, 27, 29, 31–38, 57, 76–79, 81, 82, 84, 85, 90–92, 98, 101, 137, 140, 141, 144, 145, 162, 166, 170, 175, 251, 252, 261, 262, 264

adducts, 162
binding, 137, 140
damage, 11, 29, 76–79, 81, 82, 84, 85, 91, 137, 140, 144, 251, 262
double strand breakage, 140
repair, 77–79, 82, 85, 141, 264
Dose–response relation, 210, 256
Downstream legislation, 225
Dust, 8, 9, 14, 15, 40, 42, 43, 45, 46, 49–51, 53–58, 61, 69–71, 85, 103–109, 111–114, 117–119, 121–133, 136, 137, 139, 142, 151, 152, 170, 198, 238, 242
clearance, 118, 119, 121–123, 125, 127, 129, 131
overload, 14, 15, 104–107, 113, 114, 118, 119, 122, 124–130, 132, 133
Dysplasia, 12, 13, 87

Electron microscopy, 18, 51, 58, 60, 205, 247
Employment durations, 194
Endpoints, 252
Environmental Health Criteria (EHC), 214, 215, 217
Enzyme polymorphisms, 12, 86
Epidemiology, vii, 2–6, 12, 15, 16, 39, 49, 50, 54, 56, 60, 62–64, 67, 68, 71, 83, 87, 117, 129, 133, 134, 136, 141, 146, 147, 149, 150–156, 159–163, 165, 166, 168, 170, 171, 173–176, 178, 179, 181, 183–186, 193, 206, 207, 211–215, 221, 235, 239, 241, 254–258, 261–264
Epidermal growth factor receptor (EGRF), 88, 92, 94, 97, 100
Epidermoid carcinomas, 138
Erionite, 41, 43, 48–50, 53, 55, 56, 84
Erythrocyte hemolysis, 41
Essential elements, 7, 21, 23
Essential micro-nutrients, 231
Ethmoid region, 182, 206, 208
European Chemicals Bureau, 225, 226
European Commission, 222, 224, 225, 227, 236, 243–245

Directorate General V, vii, 222, 243
Directorate General XI, 224, 243
European Union legislation, 224, 227, 236
Experimental animal data, 66, 221
Exposure, vii, 1–6, 8–18, 39, 42, 43, 46, 47, 49–51, 53–58, 61, 62, 64–72, 74–76, 80, 83, 86, 87, 92, 103, 104, 106–108, 110–119, 121–123, 125–132, 134, 136, 141–143, 146–158, 160–172, 174–176, 178–182, 189, 192–196, 198, 200, 203, 204, 207, 210–218, 220, 222, 224, 225, 229–232, 234, 235, 237, 239–242, 244–246, 248, 250–252, 254–258, 260–263
assessment, 163, 169, 174, 176, 214, 248, 254, 257, 263
monitoring, 5, 210, 221, 254, 255, 256
standards, 9, 214, 215, 218, 221, 235, 241, 242
Extrapolation, 16, 83, 164, 173, 174, 175, 215, 242, 251, 256, 262

Familial adenomatous polyposis coli, 87
Fanconi's anemia, 87
Feldspars, 59
Fenton reaction, 140
Ferrates, 35
Ferric oxide, 71, 143
Ferric saccharate, 52, 57
Ferrochromium, 150, 166
Fibre length, 66, 81, 85, 111
Fibres, 8–11, 15, 40–56, 58, 59, 61–67, 69–74, 77–85, 110, 111, 115, 116, 131, 132, 158, 251
Fibrogenesis, 137, 139, 142, 144
Fibrosis, 14, 50, 64, 66, 73, 74, 104, 130, 136–138, 140, 144, 151
Field cancerization theory, 87, 94, 96
Fly ash, 104
Free radicals, 35, 137, 140, 141, 144, 252

Furnace, 187, 197, 199, 201–204
Furnacing, 187, 201, 208

Gallium, 29, 34
Gene
  amplification, 88, 89, 93
  expression, 13, 88, 89, 100, 138, 144, 165, 252
Genotoxic carcinogens, 79, 216, 240
Genotoxicity, 11, 75–77, 79, 83, 143, 163, 173, 262
Germanium, 231
Glass fibre, 46, 55, 58
Glutathione S-transferase I, 86
Gold, 17, 20, 74, 155, 159, 211, 212, 234
  miners, 211
Good epidemiological practice, 16, 174, 255, 258, 262, 264
Group 12 elements, 27
Group 2 elements, 23, 26
Groups 3, 4 and 13 elements, 29
Groups 14–17 non-metals, 32
Growth factor, 13, 88, 92–94, 96, 97, 100, 101, 125, 137, 139, 144, 206, 252
GTPase activity, 88
Guidance value, 214, 215, 218
Gypsum, 201

Haber Weiss reaction, 140
Hafnium, 29
Hamsters, 50, 56, 71, 126, 136, 137, 142–144
Hard metal
  pneumoconiosis, 151
  powder, 238
Hat workers, 153
Hazard, 2–4, 17, 30, 149, 161, 193, 196, 206, 210, 214, 220, 223, 225, 229, 230, 232–240, 242, 245, 246, 249, 251, 255, 263, 264
  classification, 232, 237, 240
  identification, 210, 214, 223, 229, 235, 237, 242, 245, 263, 264
Health standards, 220, 221

Hematite, 2, 136, 139, 152, 158
Hepatocytes, 78, 84, 101
Hepatomas, 93
Hereditary nonpolyposis colorectal carcinoma, 94
Hexavalent chromium, 2, 149, 150, 153, 156, 165, 169, 170, 179, 221
Homeostasis, 6, 7, 13, 19, 20, 21, 23, 25–29, 31–33, 35, 37, 77
Hormones, 24–28, 31, 35
Human
  autopsy lungs, 10, 60
  epithelial cells, 92
Human–hamster hybrid cell line, 77
Hydroxyl radical, 53, 140, 141, 144
Hyperplasia, 12, 50, 87, 137, 144

IARC, 1–3, 15, 70, 72, 73, 79, 83, 95, 115, 135, 137, 143, 147, 148, 151, 153, 155–160, 163–165, 167–169, 171, 173, 176, 178, 179, 211, 212, 214, 215, 221, 238, 243, 244
In vitro/vivo studies, 11, 75, 76, 79, 81, 83, 160, 251
Indium, 29
Inert particles, 127
Inflammation, 14, 77, 106, 112, 114, 118, 124, 130, 139, 252
Inhalable dust, 9, 242
Inhaled particles, 40, 59, 70, 73, 75, 76, 113, 115, 131, 154
Initiation, 82, 92, 251, 262
Injection, 3, 46, 48, 49, 52, 53, 57, 73, 116, 136, 250
Inoculation, 49, 52, 56, 67, 75, 78, 80, 138, 142
Inorganic arsenic, 217, 218, 221
Inorganic mercury compounds, 153, 154, 211
Insoluble dusts, 118, 123
Insulation, 62, 67, 69, 74
Insulators, 64, 66
Intergovernmental Forum on Chemical Safety (IFCS), 232, 245
Interindividual variation, 86

Subject Index

Interleukin-1 (IL-1), 137
Interleukin-6 (IL-6), 137
International Labour Organisation (ILO), 213
International Programme on Chemical Safety (IPCS), 213, 243
Interstitial and lymph nodal particulate uptake, 130
Interstitial cells, 114, 123, 129, 131
Intraperitoneal injection, 3, 46, 49, 57, 73, 116
Intrapleural injection, 52, 136
Intrapleural inoculation, 49, 52, 56, 67, 78, 80, 142
Intrapleural/intraperitoneal injection, 46
Iodine, 31, 32
Iron, 2, 7, 17, 25, 28–33, 35, 36, 52, 60, 70, 113, 140, 152, 158, 165, 166, 177, 178, 187, 189, 197, 198, 200, 201, 204–206
Iron ore miners, 151, 152

Job sequences, 17, 197, 202, 203, 206, 207
Joint Meeting on Pesticide Residues (JMPR), 216

Kaolin, 49, 56, 59
Kidney, 93, 152, 153, 156, 171
  cancer, 152–154, 156
Kinases, 7, 23, 27, 98
Kinetochores, 79
Kristiansand, 208

Lanthanum, 29
Large cell carcinoma, 92
Laryngeal cancer, 166, 178
Laser beam ionization microprobe analysis (LIMA), 18, 205
Law and regulation, 264
Lead, 2, 7–9, 13–16, 19–21, 23, 31, 32, 34, 38, 49, 50, 79, 108, 129, 150, 152–154, 156, 158, 160, 162, 165, 167–171, 174, 179, 200, 205, 211, 212, 231, 234, 236, 240, 252

chromate, 150, 156, 169, 170
mining, 152
ores, 168
Legislation, 2, 135, 224, 225, 227, 236, 240, 242
Lewis acid, 34
Limit values, 222, 237
Lipoproteinosis, 104
Lithium carbonate, 21
Liver angiosarcoma, 151
Loss of heterozygocity, 84, 89, 98, 99
Lung
  biopsy samples, 94
  bronchoscopy, 13, 95
  burdens, 108, 118, 119, 125, 126
  cancers, 3, 9, 10, 12, 13, 15, 17, 50–53, 60, 61, 63, 64, 66–69, 71, 73–75, 86–102, 136, 138, 141–144, 146–153, 155–159, 166–172, 177–179, 182–186, 190, 192, 193, 217, 218, 244
  carcinogenesis, 11, 12, 86, 94, 137, 139
  clearance, 13, 103, 107, 110, 112–118, 132, 133
Lymph nodal uptake, 123
Lymph nodes, 14, 73, 96, 106, 108–112, 114, 117, 123
Lymphocytes, 78, 79, 84, 170

Macrophage, 13–15, 59, 103–114, 116, 118, 123, 124, 127, 129, 131, 132, 137, 252
  activity, 129
  mediated clearance, 13
  mobility, 118, 123, 132
Magnesium, 6, 7, 20, 23–25, 28, 30, 32, 127, 261
  silicate, 127
Magnetite, 45
Malignant histiocytic lymphomas, 136
Malignant hypothermia, 27, 32
Man-made mineral fibres, 158, 221
Man-made vitreous fibres, 41, 49, 70, 85, 110, 115

Manganese, 7, 28, 29, 30, 32, 34–36, 155, 217, 219
Mattes, 200, 201, 204
Maximum tolerated dose (MTD), 110, 136, 231
Mercury, 7, 15, 90, 153, 154, 157–159, 178
  vapour, 153
Mesothelioma, 10, 46–53, 55–57, 62–68, 72–75, 80, 85
Metademography, 208
Metal/metal interation, 165
Metallonthionein, 7, 28, 32, 37
Metaplasia, 12, 87
Metastasis, 31, 89, 96, 165
Methylmercury, 37
Mica, 43, 59
Mice, 4, 80, 99, 107, 114, 116, 126, 137–140, 142–144, 154
Micronucleus, 79
Microsatellite instability, 91, 100
Microtubules, 79
Migration of particles, 107
Millers, 51, 63, 64, 66, 67, 72, 172
Miners, 15, 51, 56, 64, 66, 67, 72, 74, 94, 103, 126, 151, 153, 158, 172, 177, 178, 211
Mining, 2, 50–52, 56, 57, 63, 65, 72, 126, 152, 154, 155, 166, 171, 177
Mist, 155
Mixed exposures, 4, 166, 204, 245, 251, 262
Molecular epidemiology, 16, 95, 162, 175
Mond nickel refining process, 181, 197
Monitoring, 1, 5, 9, 10, 161, 168, 210, 214, 221, 239, 242, 246–248, 254–256, 261
Monocytes, 78, 79, 131
Mucociliary function, 119
Mucronuclei, 138
Mutation, 2, 12, 19, 23, 27, 28, 32–34, 76, 77, 79, 81, 84, 87–92, 94, 97–99, 101, 102, 162
myc Family oncogenes, 89

N-acetyltransferase, 86, 95
Nasal cancers, 3, 15, 17, 150, 154, 156, 157, 165, 169, 170, 171, 177, 181–185, 191–197, 202, 203, 206, 207, 209, 217
Nemalite, 45
Neurofibromatosis, 87
NHBE cells, 93
Nickel, 2, 3, 7, 15–19, 28–30, 36, 146–151, 154–157, 159, 160, 163–165, 168, 171–175, 178–191, 193, 197–202, 204–209, 211, 217, 219, 230, 236, 238, 240, 244, 261
  arsenide, 18, 206, 207
  carbonate, 155
  carbonyl, 188, 189, 198–200
  cadmium, 148
  chloride, 173
  compounds, 2, 3, 148, 154, 155, 171–173, 179, 209, 211, 238, 261
  dioxide, 155
  hydroxide, 155
  metal, 17, 18, 181, 186, 188, 198, 199, 240
  oxide, 3, 17, 18, 59, 154, 155, 165, 172, 181, 186, 188, 198, 201, 206, 217, 240
  pellets, 200
  refineries, 3, 154
  salts, 171, 172, 186, 188
  subsulfide, 154, 156
  sulfate, 3, 18, 181, 187, 198, 199, 201, 204, 238, 261
  sulfide, 3, 18, 155
Nickel(II) ions, 261
No observed adverse effect level (NOAEL), 216
Nongenotoxic carcinogens, 79, 240
Non-small cell lung carcinomas (NSCLCs), 88, 96, 97, 99–101
Norway, 154, 169, 171, 179, 208
Nose, 157, 177, 182, 209
Nuclear gene expression, 13, 88
Nuclear weapon industry, 153
Nutritional status, 86

Subject Index

O-binding, 30
Occupational exposure limits, 9, 241, 242
Occupational histories, 257
Occupational Safety and Health Administration (OSHA), 220
Oesophageal cancer, 155
Oncogene, 11–13, 57, 76, 77, 85, 87–89, 91, 93, 96–102, 143, 162, 176
Orcelite, 18, 206, 208, 209
Organotin compounds, 217
Osteocalcin, 28
Overload, 113, 118
  reversibility, 7, 11, 14, 15, 104–107, 113–115, 117–119, 122–133, 238, 264
Oxidants, 20, 35
Oxidation/reduction, 7, 27, 29, 30, 32
Oxidative states, 251
Oxidic sinter, 186
Oxygen consumption, 81
Oxyradicals, 79
Ozone, 249

p53 gene, 82, 85, 87, 89, 90, 91, 94, 98, 99, 100, 102
Painting, 157, 164
Palladium, 17, 198, 200
Parenchyma, 59–61, 67, 69, 73, 145
Parenchymal tissue, 59, 60
Particle
  clearance, 14, 59, 70, 105, 107, 108, 113, 119, 130, 132
  dimension, vii, 39, 54, 73
  morphology, 40
  overload, 11, 14, 113, 114, 117, 132
  size, 8, 9, 39–43, 53, 59, 116, 121, 122, 126, 133, 134, 137, 198, 205, 251, 256
  size distribution, 8, 39–43, 53, 121, 122, 251
Particulates, 8–11, 13–15, 18, 39–46, 48–51, 53, 58–61, 70, 71, 73, 75, 76, 78, 103–119, 123–134, 136, 139–141, 145, 154, 205, 206, 251

Permanganates, 35
Permissible exposure limit, 221
Peroxidase, 31
Peroxide, 30, 53, 140
Petroleum coke, 104
Phagocytosis, 59, 78, 81, 85, 112, 123, 124
Phosphate, 23–25, 27, 28, 29, 31, 32, 33, 34, 37, 141, 166
Phosphorus, 32, 34
  acids, 200
Phosphorylation, 7, 23, 25, 28, 31, 89, 206
Plaque, 66, 68
Plastic microspheres, 125
Platers, 150, 179
Platinum, 17, 155, 198, 200, 217, 219
Pleural plaques, 11, 64, 66, 68, 73
PM_{10}, 10, 60
Pneumoconiosis, 57, 118, 151
Point mutations, 76, 77, 79, 88, 89, 98, 162
Polonium, 61
Poly(ADP) riboslyation, 82
Poly(ADP) ribose polymerase, 78
Polycyclic hydrocarbons, 133, 134, 146, 150, 153, 155, 166, 179, 246
Polyploidy, 79
Polystyrene aerosol, 106
Polyvinyl chloride particles (PVC), 1, 105, 127
Porphyrins, 30
Potassium, 20, 21, 24, 30
Potency, 4, 11, 41, 48, 49, 53, 75, 76, 78–80, 110, 150, 164, 170, 213, 215, 221, 223, 230, 234, 235, 238, 240, 242, 243, 264
Precious metals, 18, 155, 188, 198, 199, 201, 206
Preservation of biological materials, 252
Priority setting, 231, 233
Producer gas, 188
Prostaglandin, 35
Prostatic cancers, 148

Pronto-oncogene, 13, 85–89, 91, 93, 96, 97, 100, 101
Pulmonary fibrosis, 14, 73, 74, 130, 144, 151
Pulmonary interstitium, 112

Quality assessment, 255. 258
Quartz dust, 41–43, 54, 111, 112, 136–144
Québec, 51, 54, 63–65, 67, 72

Radiolabelled polystyrene (*PS) microspheres, 125
Radon, 71, 147, 151–153, 155, 158, 166
Rat, 3, 4, 11, 13–15, 47–49, 52, 55–57, 67, 70, 73, 78, 80, 84, 85, 103, 105–108, 111–116, 118, 119, 123–133, 136, 137, 139, 141–144
  pleural mesothelial cells (RPMC), 78, 80, 84, 85
Rats,
  BALB 3T3, 80
  F-344, 105, 123, 124, 126–128
  Fischer, 111
  Spraque–Dawley, 139
  Wistar, 57, 105, 112, 116, 126
Refining, 2–4, 18, 147, 152, 154, 164, 179, 180–182, 186–189, 193, 196, 197, 199–202, 204, 207, 208, 238
Refractory ceramic fibres (RCF), 49, 50, 53, 56, 116
Registration laws, 255
Registries, 254, 255
Regulatory, 1, 2, 5, 7, 11, 13, 26, 28, 31, 43, 67, 76, 82, 89, 91, 92, 99, 113, 115, 132, 165, 176, 210, 214, 221, 225, 231–233, 237, 240, 243–245, 248, 255, 262, 264
  approach, 232
Respirable dust, 9
Respiratory cancers, 3, 18, 150, 181–183, 186, 190, 193, 202, 207, 209
Retention half-life, 119, 121, 126
Retention kinetics, 108
Retention measurements, 104
Risk, 1–4, 8, 9, 12–14, 16, 19–21, 23, 28, 30–32, 34, 35, 37, 38, 42, 49–53, 55, 65, 66, 68, 71–73, 75, 76, 83, 86, 87, 94–96, 106, 132, 133, 135, 136, 141–143, 147–158, 160–169, 171–181, 185, 189, 193, 196, 200, 202, 210, 211, 213–218, 220–225, 227, 229–243, 245, 246, 251, 255, 256, 260, 262–264
  assessment, 2–4, 14, 19, 95, 106, 132, 133, 164, 175, 176, 210, 213–216, 218, 220, 221, 224, 225, 229, 231, 232, 234–236, 238–243, 245, 251, 260, 262, 263, 264
  characterization, 263, 264
  classification, 233
  communication, 230, 231, 232, 233, 234
  management, 225, 232, 233, 235, 237, 239, 241, 245
RNA, 21, 27, 261
Rubidium, 21
Rodent bio-assay, 2, 4
Rutile, 41, 139

S/N binding, 30
Safety, 1, 5, 72, 83, 213, 216, 220, 222, 223, 225, 227, 230, 232, 236, 237, 241, 243, 245
Sand, 142, 201
Scandium, 29
SCLC cell lines, 91, 93
Selenium, 20, 32, 35, 36, 165, 217, 219
Shipyard workers, 64, 73, 74, 177
Signal transducers, 13, 88
Silica, 32, 41, 42, 56, 59, 60, 111, 116, 134, 136–145, 147, 151, 153, 155, 166
Siliceous slag, 197, 201

Subject Index

Silicon, 32, 36
Silicosis, 136, 140–144, 158
Silicotic granulomas, 136, 139
Silver, 17, 155, 198, 200
Sinters, 187
Small and medium enterprises, 242
Smelter workers, 172
Smelting, 4, 147, 152, 155, 168, 171, 172, 182, 207
Smoking, 10–12, 17, 59, 60, 68–71, 74, 87, 94, 96, 126, 147–149, 152, 153, 159, 166–168, 171, 172, 190, 192, 193, 232, 239, 255, 256, 258
Sodium, 21, 24, 30
Solid particles, 11, 13, 103–105, 108, 111, 113, 114
Solid tumours, 93, 101, 138
Solid ultrafine particles, 122
Soluble nickel, 171, 173, 217, 238
Speciation, 4, 15, 19, 149, 230, 231, 234, 235, 237, 238, 242, 244, 246, 247, 250, 251, 261, 264
Sputum samples, 13, 94, 95
Squamous carcinoma, 92, 94, 100, 182, 183
Stainless steel, 166, 172, 177, 212, 239, 241
Standardization, 250
Stanton hypothesis, 46, 47, 54
Statistical techniques, 162, 190, 193, 202, 254, 255, 257
Steady-state, 20, 121, 122, 126
Sterols, 28, 31
Stokes's diameter, 41
Stomach cancer, 152
Strong acid mists, 166
Strontium, 25
^{85}Strontium, 104, 105, 107, 111
Strontium chromate, 150, 169, 170
Sulfur, 18, 32, 151, 168, 186, 197, 198, 201, 204–206
 dioxide, 151, 197, 198, 201
Sulfuric acid, 17, 146, 166, 178, 187, 198–201, 236
Superoxide, 30, 37, 140

dismutase (SOD), 37, 140
 radical, 140
Surface, 8, 10, 13, 18, 39, 40, 43–46, 48, 53, 66, 111, 124, 137, 141, 144, 149, 182, 200, 206, 208, 246, 251, 252, 261, 262
 area, 39, 40, 43–46, 66, 111, 137, 251
 properties, 8, 40, 46, 53, 149, 246, 251, 261, 262
Surgical implantation, 46
Syrian hamster embryo (SHE) cells, 80, 136, 138, 143

Talc, 47, 48, 59, 78, 127
Technical Progress Committee, 226, 227
Test
 material, 104, 105, 110, 112, 138, 250
 methods, 260
 systems, 78, 81, 250, 252
Thallium, 21, 29, 34, 37
The International Agency for Research on Cancer (IARC), 135, 176, 211, 214, 243
Thetford mines, 64, 65
Threshold, 4, 80, 108, 164, 216, 218, 222, 230, 256, 261
Thymine glycol, 144
Thyroxine, 31, 32
Tin, 155, 177, 189, 217
Tissue analysis, 72, 248
Titanium, 29, 34, 41, 59, 104, 107, 108, 112, 116, 123, 124, 127, 128, 131, 155, 159, 217
 dioxide, 41, 104, 107, 108, 112, 123, 124, 127, 128, 131, 155
 oxides, 59
 tetrachloride, 155, 159
Tolerable intake, 215
Toner, 105–107, 109–111, 114, 115, 124, 126, 127, 132, 143
Total airborne dust, 242
Total dust, 9
Toxicokinetics, 11, 13, 103, 110, 111, 113, 251

Transcription factor, 7, 23, 25, 28, 31, 32, 82, 89
Transcriptional regulation, 89, 91
Transformation, 11, 76, 77, 80–82, 85, 94, 136–139, 141–144, 252
Transforming growth factor, 92
Transition metals, 7, 29, 33
Translocations, 143
Tremolite, 10, 47, 48, 51, 54, 61–67, 72
Trisomy 7, 88
Trivalent chromium, 150, 165
Tumour
　growth factor, 92–94, 99, 137, 139, 140
　necrosis factor, 137, 139, 144
　progression, 90, 91, 99
　suppressor gene, 11–13, 76, 77, 87–90, 94, 95
　susceptibility genes, 12, 86, 87
Tungsten carbide, 151
Tyrosine kinase, 89, 93, 96, 97, 100, 101
　growth factor receptors, 88

United Nations Environment Programme (UNEP), 213, 243
Unscheduled DNA synthesis (UDS), 78

Uranium miners, 94, 178
Urban dust, 129

Vanadium, 22, 29, 217, 219
Vineyard sprayers, 151
Vitamin B, 30
Volcanic ash, 15, 104, 113, 127, 131

Water, 30, 149, 156, 158, 170, 188, 198, 215, 217
Welders, 150, 166, 172, 177, 180, 212
Wilms's tumour, 87
Work records, 189, 191
World Health Organization (WHO), 135, 213, 218, 219, 243

Xeroderma pigmentosum, 87

Ytterbium, 29

Zeolite, 48, 56
ZESTE genotoxic test systems, 79
zinc, 7, 15, 25–34, 36, 37, 90, 147, 150, 155, 156, 165, 168–170, 189
　chromate, 150, 156, 170
　finger transcription factor, 28, 32
　fingers, 7, 28, 32
　protease, 31